INSTRUCTOR'S MANUA

MODERN METALWORKING

by

John R. Walker

Publisher
The Goodheart-Willcox Company, Inc.
Tinley Park, Illinois

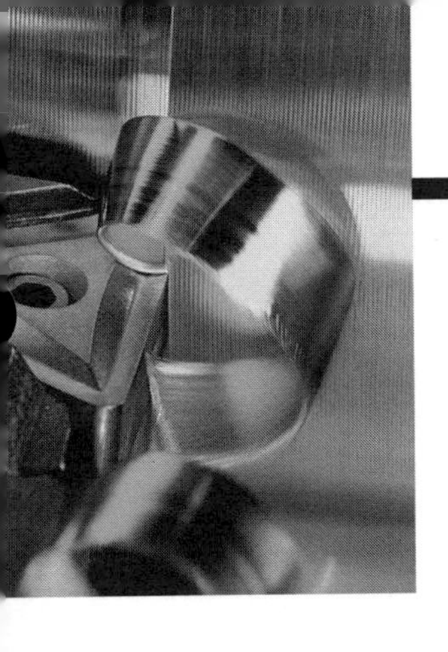

Table of Contents

	Instructor's Manual Page	Textbook Page
Chapter 1		
Technology and Careers	21	9
Chapter 2		
Classifying Metals	33	27
Chapter 3		
Understanding Drawings	40	47
Chapter 4		
Safety Practices	52	63
Chapter 5		
Measurement	55	69
Chapter 6		
Layout Work	79	97
Chapter 7		
Hand Tools	85	109
Chapter 8		
Hand Tools That Cut	89	123
Chapter 9		
Hand Threading	95	141
Chapter 10		
Fasteners	103	151
Chapter 11		
Art Metal	109	165

	Instructor's Manual Page	Textbook Page
Chapter 12		
Sheet Metal	113	179
Chapter 13		
Soldering and Brazing	119	201
Chapter 14		
Sand Casting	123	211
Chapter 15		
Metal Casting Techniques	129	229
Chapter 16		
Wrought Metal	139	249
Chapter 17		
Forging	145	261
Chapter 18		
Heat Treatment of Metals	157	275
Chapter 19		
Gas Welding	165	289
Chapter 20		
Shielded Metal Arc Welding	171	303
Chapter 21		
Other Welding Processes	179	315
Chapter 22		
Metal Finishes	193	329
Chapter 23		
Grinding	197	343
Chapter 24		
Drills and Drilling Machines	207	363
Chapter 25		
Sawing and Cutoff Machines	217	393

	Instructor's Manual Page	Textbook Page
Chapter 26		
Metal Lathe	225	403
Chapter 27		
Cutting Tapers and Screw Threads on a Lathe	237	435
Chapter 28		
Other Lathe Operations	249	449
Chapter 29		
Broaching Operations	257	463
Chapter 30		
Milling Machines	261	469
Chapter 31		
Metal Spinning	275	507
Chapter 32		
Cold Forming Metal Sheet	281	521
Chapter 33		
Extrusion Processes	297	533
Chapter 34		
Powder Metallurgy	303	539
Chapter 35		
Nontraditional Machining Techniques	307	545
Chapter 36		
Quality Control	321	563
Chapter 37		
Numerical Control and Automation	325	575
Product Suggestions	335	593

1.0 INTRODUCTION

Metal has been used in some way during the production of almost everything we eat, see, feel, hear, smell, and touch. With so much of our daily living depending upon metals, it is essential that students learn something about them, how they are worked, and the industry that uses them.

Modern Metalworking emphasizes the important place metals occupy in our everyday lives; it explores the numerous metalworking career opportunities. It is designed to provide a broad experience in metalworking through the use of tools, machines, and materials that are basic to this important area of industry.

Modern Metalworking supplies basic information on tools, materials, and procedures. It covers hand and machine operations, and supplies background knowledge in industrial equipment and processes.

1.1 INSTRUCTIONAL MATERIALS

The following materials have been developed to aid in presenting a dynamic metals technology program.

1.1.1 Textbook

The textbook is a very important part of an instructional program. *Modern Metalworking* provides an introduction to the tools, materials, and procedures used in this important segment of manufacturing technology. It covers safety, hand and machine tool operations, and supplies background knowledge on industrial equipment and state-of-the-art metalworking processes.

Modern Metalworking is written in an easy-to-understand language. There are many color photographs and line drawings to help students clearly visualize metalworking techniques, operations, and procedures.

Color is used throughout *Modern Metalworking* to indicate various materials or equipment features. The color key is on page 4 of the text. The following list identifies colors and materials by name.

Metals (surfaces)	Dark Gray
Metals (in section)	Light Gray
Machines/machine parts	Yellow
Tools	Dark Blue
Cutting Edges	Light Purple
Work-holding and tool-holding devices	Light Green
Rulers and measuring devices	Blue-Gray
Direction or force arrows, dimensional information	Red
Fasteners	Blue-Green
Abrasives	Dark Purple
Fluids	Light Blue
Miscellaneous	Tan

Also included in the text are the following teaching aids:

Learning Objectives

Every chapter opens with a list of objectives. They make the student aware of what he/she will be able to accomplish after studying the chapter.

Because of many factors, it may not be possible to achieve every objective with each class. However, it is better to cover less material than to cover all material poorly.

Technical Terms

The technical terms listed at the beginning of each chapter are drawn directly from the chapter. These terms, along with many others, are printed in bold italic type throughout the text and are included in the glossary. Students should become familiar with these terms and know their meanings.

Summary

The summary is a brief restatement of the main facts or points of the chapter.

Test Your Knowledge

Each chapter ends with a set of questions. They can be used to check student comprehension of the text material. The questions can be used as a quiz, homework, or an out-of-class assignment.

Research and Development

These activities offer an opportunity to bring many of the latest techniques and materials of industry into the classroom. It is recommended that students volunteer for activities.

These activities also provide a unique opportunity for developing student originality and ingenuity. You will be surprised at the accomplishments of many of your students. The problems are broad enough to provide for differing student abilities.

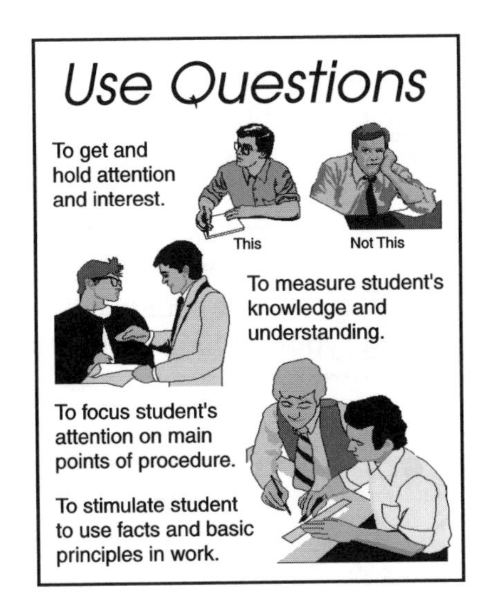

Product Suggestions

Several product suggestions are included in the text. They can be used as class projects if your lab/shop allows or students may choose to make these items on their own.

Reference Section

A reference section of tables and other useful information is provided on pages 607–627. This section contains information on physical properties of metals, their standard shapes and sizes, fastener details, tap drill sizes, drill sizes, cutting speeds for some metals, conventional and metric conversion tables, and other information that will prove useful in class.

Glossary

The glossary provides a quick reference to the definitions of technical terms used in metalworking. Students should refer to this section when they come across unfamiliar terms. All bold italicized terms are included in the glossary.

Index

A comprehensive index completes the text. It allows the student to quickly locate topics in the text. Be sure your students know how to use the index.

1.1.2 Instructor's Manual

The manual has been devised to assist the instructor in improving their metals technology program. For each chapter, it includes *Learning Objectives*, a *Guide for Lesson Planning*, *Chapter Quizzes*, *Research and Development Ideas*, *Answer Keys*, and *Reproducible Masters*.

Learning Objectives

For ease in referencing, text objectives are listed for each chapter. The goals presented involve basic concepts, skills, and understandings that should be stressed while teaching the chapter. The objectives can also be used as an outline when preparing lesson plans.

Chapter Quizzes

Quizzes are included in all chapters. They provide for testing student comprehension of material covered in class. They, along with Test Your Knowledge and Workbook questions, can be used to develop customized tests.

Answer Keys

Answers to *Test Your Knowledge*, *Chapter Quizzes*, and *Workbook* questions are provided within each chapter.

Reproducible Masters

Reproducible masters are designed to be duplicated and distributed to the class as handouts or made into overhead transparencies. Color can be added to the transparencies with felt tip pens. Your teaching preference will determine whether the reproducible masters are used as transparencies, handouts, or both.

The reproducible masters are identified by two numbers. The first number refers to the chapter number and the second number is its number within the chapter. The extent to which you use these materials will vary, based on the extent of the course and the material to be emphasized.

Transparencies should be fitted with cardboard frames for the following purposes:

- Protecting the transparency in use and when storing them.
- Organizing the transparencies by writing the file number/letter on the frame.
- Making the transparency appear more professional.

1.1.3 Workbook

The workbook is an aid for measuring student achievement and comprehension. It employs a variety of questions, problems, and assignments. The chapters are presented in the same order as corresponding material in the textbook.

As a study guide, students should first read and study the material assigned in the textbook, giving careful consideration to the illustrations. Then, without the aid of the textbook, answer the workbook questions. As recommended for several chapters of the text, the workbook assignments can also be divided into parts as seen fit by the instructor.

Answers to the questions and problems consist of letters, numbers, short sentences, and simple sketches. Instruct students that words should be spelled correctly and letters and numbers should be carefully formed. It is highly recommended that the letters and words be printed. Stress to students that most tradespeople follow the same practice since the information will be easier to read and the possibility of errors greatly reduced. Sketches should be carefully drawn in the space provided. When required, mathematical calculations should be made in a neat, organized manner. This makes it easier to check the procedure used in solving the problem.

1.2 TO THE INSTRUCTOR

It is not possible for an author or publisher to provide a detailed metals technology program that will be suitable for every teaching situation. Some school systems require a specific program to be followed. Also, course length, depending upon local and state curriculum guides will vary.

As the instructor, only you can determine what material will best serve your students, and how the material should be presented. Only you know the abilities of your students and the facilities, materials, equipment, and time available. With advanced students in a maximum time program (36 weeks and meeting at least five times a week), you will be able to cover many chapters in depth. In a course that lasts only a few weeks, limited coverage of a few chapters will be a major accomplishment.

Modern Metalworking provides an excellent base for more advanced metalworking technology programs. Student ability, tools and equipment, and time available will determine the number of chapters your class will be able to cover. Start with the chapters that will best prepare your students for the next level metals technology program or other industrial technology classes.

Before the first class session, familiarize yourself with the text. Outline the chapters you plan to teach. After determining what is to be taught, prepare detailed lesson plans and gather or make the teaching aids you will need. Check the tools and machines that will be used. Be sure they are in first class operating condition with all safety features in place. Also, have ample supplies on hand.

This may be the first metals technology class to which your students have been exposed. What is so obvious to you may be completely foreign to the majority of your students. Prepare your lesson plans accordingly.

1.3 SHOP/LAB MANAGEMENT

There are several areas in shop/lab management that must be considered. If properly developed, they can save time that can better be devoted to one-on-one teaching.

1.3.1 Control of Tools and Consumable Materials

Many expensive tools are required in metalworking technology. A method for controlling their disbursement and return to inventory must be devised and continually monitored to reduce the number of damaged tools and to prevent pilferage. The same care must be exercised for issuing stock and other consumable supplies.

You can examine the systems local industry and other instructors have established for control of tools and supplies, and implement one of them, or you can devise one of your own.

1.3.2 Scheduling

Because of tool and equipment limitations, you will not be able to have all students working on the same assignment at the same time. Assignments must be organized and scheduled so equipment will be used as much as practical.

1.3.3 Shop/Lab Management System

Good housekeeping and cleanliness are important in all shop/lab situations. Insist that students clean workstations after use. To aid in overall shop/lab cleanliness, each student should be assigned, on a rotating basis, a specific cleanup task each week. Each student will be able to experience each position at least once during the term. Praise them when the job is well done.

At least once every two weeks, a more thorough cleanup should be done. This includes cleaning lockers and getting shop coats and aprons washed.

1.4 IMPROVING INSTRUCTION

Whether teaching metals technology, or any other subject, you make use of certain universal instructional tools. All good teachers apply these concepts, either consciously or unconsciously. When making your lesson plans, try to implement the following ideas:

Reinforce. The more ways a student is exposed to a given concept, the greater the understanding and retention of the material. A variety of learning experiences are designed to meet the reinforcement needs of students.

Extend. The teaching suggestions in this Instructor's Manual are directed at students with a variety of ability levels. You may choose some of the assignments to encourage highly-motivated students to extend their learning experience outside the classroom. These types of activities allow students to relate text information to other experiences, especially life skills.

Enrich. Enrichment activities are designed to help students learn more about topics introduced in the text. These types of activities, such as research and survey activities, give students the opportunity to enrich their learning through more in-depth study.

Reteach. Studies have shown that students respond differently to different teaching methods and techniques. Therefore, these materials provide suggestions for several strategies that can be used to teach the concepts in the text. This allows you to choose a different strategy to reteach students who respond poorly to a previously used strategy.

1.5 PLANNING AN INSTRUCTIONAL PROGRAM

No matter what approach you take in teaching a basic metals technology program, the importance of careful planning and organizing cannot be overemphasized. Planning to achieve program goals will be easier if you know *why*, *when*, and *how* to plan.

Knowing Why to Plan

Planning is the process of carefully selecting and developing the best course of action to achieve program goals and objectives. This action cannot be a hit-or-miss situation. How well you meet your teaching responsibility depends

largely upon your consistency. Plan and prepare carefully for each class.

Planning is done so action will take place at the right time. Planning helps anticipate problem areas of learning, and makes adapting to and handling emergencies easier. Planning saves time and ensures a higher quality product—your student.

Planning helps ensure quality results, saves valuable class time, helps reduce discipline problems, and makes it easier to adapt to changes. Planning is also good teaching—knowing what has to be taught, what has been taught, and what needs to be taught to reach program goals and objectives.

Knowing When to Plan

In lesson planning, it is best to be flexible. That is, planning should be continuous. Otherwise, you may be unable to adjust your lessons for differences in classes or for classes missed for assemblies, bad weather, and other unplanned interruptions.

Rigid planning means that a specific lesson will be taught on a specific day. If for some reason a class is missed, there is no way to make it up. Although preferable to not planning at all, rigid planning is less effective than flexible planning.

When developing plans, you must consider the time available. Once you know this, you can structure this time by what must be done, what should be done, and what need not be done.

After a course of action has been determined, specific planning can begin by:

- Establishing goals for each class. Lessons must start with what students know and progress from there. Some classes may require more time to master the same skills and information than other classes.

- Taking inventory of supplies and equipment. Be sure supplies are adequate.
- Considering different teaching techniques. What other methods are available that will be equally effective?
- Providing specific learning experiences in an interesting manner.
- Setting plans into action. Once a plan of action has been developed, it should be followed.

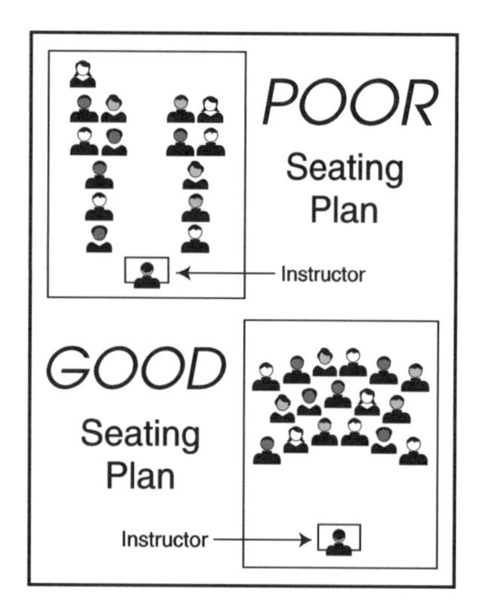

Knowing How to Plan

A lesson plan is an outline for teaching. It keeps the essential points of the lesson in front of you and ensures an orderly presentation of material. Such a plan does much to prevent important aspects of a lesson from being omitted. It also prevents you from straying from the lesson and introducing irrelevant material.

Your lesson plan should include the material to be taught, the methods and techniques best suited to teach this material, the supplies or equipment needed, and teaching aids to be used.

Review your material before class time. Be familiar with all the material related to the subject. You should be prepared to teach more material than you may have time to cover. Check all the supplies and equipment to be sure they are ready to use, in safe working condition, and arranged so they will be handy for the lesson. Seating and viewing arrangements should be such that every student can see and hear the lesson.

Finally, keep your plans up to date. This can be done by observing student progress and achievement. Make changes when you think they are necessary for curriculum improvement.

In general, *always* prepare a lesson plan and *follow* it.

Modify as necessary. Start with what is familiar to the student, then move into new material in short easily understood steps. Avoid long boring lessons and lessons that do not allow student participation. A sample lesson plan sheet is included in this manual (Reproducible Master 1-3). Use it as it is or as a guide in developing your own planning sheet.

1.6 TEACHING METHODS FOR METALS TECHNOLOGY

No learning takes place until a student wants to learn. Getting them interested is the most difficult part of teaching. Motivation, therefore, is the first step in good teaching. You want to stimulate students so they will want to learn.

One of the quickest ways to lose students' interest is to be unprepared and present a meandering lesson. For effective teaching, you must carefully prepare the material to be taught, the situation in which it will be taught, and the student who will receive the new information.

Start by learning and mastering the material yourself. Have a clear picture of what you want to teach. Be sure all necessary teaching materials and teaching aids are readily available and in good working condition.

Decide how you want to motivate your class. You can use curiosity. Students want to see, hear, and know about new and different things. Competition is another way to create interest. Some students may want to take on a challenge to surpass another person or group.

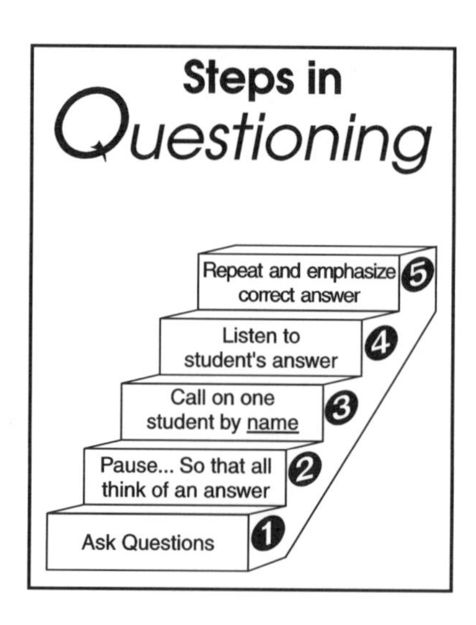

People do not want to do anything the hard way. They are motivated by the promise that they will learn an easier way of doing something. A word or two of praise can often change a person or class into success seekers. Emphasize a reward or personal gain. Quite often, the opportunity to secure a better job with greater earnings can act as a motivating force.

Above all, *you* must be interested in what you are doing. If you are not interested in the subject matter, it is unlikely that your students will be interested in it. Also, your being interested will not guarantee the students' interest.

Although careful preparation will greatly improve your teaching effectiveness, it still does not guarantee success. You must be prepared to adapt every lesson to student behavior. The amount of effort required to hold their interest will vary from class period to class period. You may have many habits that distract your students. Do you:

- Use and pronounce words correctly? Use a dictionary if necessary.
- Avoid the use of localisms, slang, the monotonous connectives such as "you know" and "now we'll do this"? These momentary pauses in the flow of information give the students' minds a moment to begin wandering.
- Look directly at and speak directly to the student? Maintaining good eye contact will help to keep students interested and will also help you notice when the students are losing interest (wandering eyes).
- Use a vocabulary that is on the students' level? If they cannot understand the words you use, students cannot learn the concepts you are trying to explain.
- Prepare yourself? If you are prepared, you do not have to make excuses.
- Gain emphasis by forceful presentation, repetition, gestures, pauses, and variations in the rate, pitch, and intensity in your voice?
- Give credit when credit is due? If students do well, praise them.
- Stimulate thinking when you ask a question by phrasing it to bring out the *why* and *how?*
- Know all students by name? Always acknowledge each student as an individual.
- Summarize frequently, as each major point of the lesson is made?
- Direct questions at inattentive students? Questioning students frequently will keep the class alert and aid you in gauging their understanding.

- Delay the entire class when one student is causing a problem? It is better to deal with the student individually.
- Continually check class reaction?
- Use teaching aids whenever possible?
- Use teaching aids that contain thought-provoking elements? They should hold attention, be timely, appropriate, and direct.
- Talk to the class, not the teaching aids or chalkboard?
- Check training aids like projectors, screens, and tapes before class starts? The students will be distracted if you waste time checking these things during class.
- Preview and select audiovisual material for specific instructional material?
- Prepare students for the film or tape? Tell them what it is about and what they should look for while watching.

The list could go on and on. However, remember the following—*be prepared.* Know what you are going to do and how you are going to do it. You will not develop student interest in what you are teaching if you, as the teacher, are not interested in your students and what they are doing.

Some tips for teaching effectively include the following:

- Put the student at ease. Humor is an effective tool to help your students relax. Remember, a tense atmosphere will inhibit learning.
- Before placing responsibility of failure on a student, you must review and evaluate the lesson taught to determine whether the goals

of the lesson were reached. If a large group of students are doing poorly in the course, the blame more than likely falls squarely on the shoulders of the instructor.

- Some teachers avoid giving reading assignments. There is no reason for this if you help students while they read. When reading assignments are given, you may want to make them *directed reading assignments.* That is, you ask questions that will guide them and help them better understand the material you want them to learn.
- Students have five senses. The more senses you can involve in the teaching process, the more likely that the student will remember what is taught.
- You should direct questions to your class for a variety of reasons:
  To attract and hold attention and interest.
  To measure knowledge and understanding.
  To focus attention on the main concepts.
- Students are people, with different personalities and attitudes. You will be more successful if you get to know your students individually. This will allow you to see the areas in which each student needs help and encouragement.
- Maintain two-way contact between you and the students. Do not allow your class to become a monologue in which you spend the entire period lecturing. You must listen to your students and encourage them to ask questions and discuss topics. Telling isn't teaching, and listening isn't learning.
- Remember topics that appear simple to you may be the most complex for the students. Avoid covering a topic too quickly if the students are not understanding it.
- When using a chalkboard, stand to the side of the material. Also, talk to the students not the chalkboard.
- You must speak clearly enough, loudly enough, and slowly enough for your students to follow your lesson. Use a tape recorder and review your own verbal presentation.

1.7 MAINTAINING DISCIPLINE

Discipline is one of the primary responsibilities of the teacher. Good class discipline can make teaching a pleasure while poor class discipline can make teaching agony.

It would be foolish to assume that there is a set of rules, magic formulas, or chants that will guarantee a teacher good discipline and respect. There are, however, ideas that have been successfully used by many teachers. Remember, they are only recommendations. It is necessary to work at maintaining good discipline every minute you are in the classroom. There is no one way to do it. What works today may not work tomorrow.

- The purpose of discipline is correction. Discipline is *not* chastisement. Discipline is a systematic training for improvement of the student's attitude and action.
- Make reprimands with justice and tact. When you are angry or not feeling well, it is wise to refrain from any drastic action until you have had the opportunity to review the situation in a better state of mind.
- Be consistent in disciplinary actions.
- Consider the student's mental and physical condition. Seek the actual cause of the student's poor work or attitude.

- Never be influenced by a student's reputation.
- Control order through an interest in your work. Students sense when you do not like your job.
- Provide sufficient equipment and working materials.
- Keep machines and tools in good working condition.
- Be sure all students can see and hear when you present a lesson.
- Handle all disciplinary cases yourself whenever possible. A good teacher is one who seldom has to call upon higher authority to maintain discipline in his/her class. Report only major infractions to higher authority. Be able to document problems with your records.
- Stop problems at their origin.
- Make only necessary rules and enforce them every day.
- Avoid assigning schoolwork as punishment.
- Do not discipline the entire class for the acts of an individual.
- Make disciplinary actions fit the deed.
- Do not become excited over misdeeds.
- Never use abusive language or profanity.
- Plan your work. Know what you are going to teach. Grade papers and tests so they can be returned the next class period.
- Have work prepared for the substitute when you must miss a class. The time should not be wasted, and you will know where to start when you return to class.

Discipline does require more than the suggestions listed above. Talk with your principal and supervisor on what their ideas are on discipline and what you can expect from them should a problem arise that you are not able to handle.

1.8 PREPARING FOR THE SUBSTITUTE TEACHER

Substitute teachers have one of the most difficult jobs in education. Some of us expect a substitute teacher to do things we cannot do ourselves. For example, they are to keep a class under control with a lesson plan as brief as, "Read Chapters 1, 2, and 3 and answer the questions at the end of the chapter."

The instructor who maintains a well organized metals technology shop/lab will provide carefully planned material for the substitute teacher. Advanced preparation will permit the program to continue with minimum interruption.

Materials and information that will be required for the substitute teacher include:

- A carefully planned lesson outline of the material to be covered.
- An attendance book and a seating chart for each class.
- Location of keys.
- Brief description of the daily class routine and the student organization.
- Procedure for issuing equipment and materials (textbooks, workbooks, paper, etc.).
- Procedure for cleanup and dismissal.
- The operation of equipment should *not* be allowed.
- Any other information you deem necessary for the substitute teacher.

The best plans are of little value if the substitute cannot locate them. To avoid such a situation, a specific place for keeping the plans should be maintained. It is recommended that several instructors in the shop/lab area be informed where the plans are stored.

Request that the substitute leave a report on accomplishments and problems for each class. The substitute's report should include the material covered, work completed, and class conduct. They should also include the strengths and weaknesses of material prepared for his/her use as well as any suggestions for improvement.

1.9 VARYING STUDENT ABILITIES

Students have a wide range of abilities in comprehension and performance The challenge for a teacher is to use the many learning techniques and devices available to achieve the maximum learning for each student in the class.

Achievement in metals technology depends not only on a person's communication skills (reading, listening, questioning) and dexterity, but also on their ability to visualize solutions to problems encountered when working metals. You must recognize the degree to which your students possess these skills. Different teaching methods are required for students with different ability levels. Following are some suggestions that may be helpful in serving all students, especially special needs students.

1.9.1 Identifying Students with Special Needs

Ideally, special needs students will be identified by the school psychologist. The school nurse should also notify the teacher of students who, because of the medication they are taking, or because of other physical disadvantages, should not be permitted to operate machinery. Actions of some of these students may endanger themselves or other members in the class. Work out a plan with your principal and supervisor well in advance on how this type of situation should be handled. *If you have not had such a prior understanding, you may be held liable if such students are injured, or through their actions, they injure another student while operating machinery or other equipment in the shop/lab.*

Characteristics of special needs students include the following list. An infrequent occurrence does not connote that someone is a special needs student.

- Impaired speech, hearing, or vision.
- Limited skills in reading, writing, and communication.
- Lack of interest in subject presented.
- Lack of self-confidence.
- Poor math skills.
- Frequent disruptive behavior.
- Frequently absent from class.
- General appearance of poor health.
- Poor achievement in early class work.
- Limited understanding of the English language.

When working with a hearing impaired student, talk in a normal voice. Face the class, speak distinctly, and pronounce words clearly. With this approach, a student with lip reading skill will find it easier to understand the lesson. At the same time, you will be heard more clearly by the rest of the class.

There is also the possibility that there will be students who may have limited ability to communicate effectively using the English language. Speak slowly and clearly to them. Encourage them to communicate as best they can in English. Help such students to develop their English speaking and writing skills so they will feel more comfortable with the English language.

Most schools attempt to provide for students of varying abilities through the less restrictive educational environment—the regular classroom. This is especially true for such students enrolling in industrial-technical classes, where separate classes are expensive and difficult to provide.

You may be able to help some students with special needs in your class using the following techniques:

- In seating students, be aware of any special needs of students with physical, hearing, or vision impairments.
- Motivate student interest in the subject by clearly stating its importance, such as its use in industry or the community. An example is the importance of accurate measuring in the manufacture of a part versus the cost of rework or scrap parts.
- Involve as many students in class discussion or activities as time permits. Place the responsibility for learning on each student through the use of questions, discussion, and hands-on demonstrations.
- Adjust the length of the lesson presentation to the attention span of the students.
- Use two or more instructional techniques in presenting a subject, such as lecture and demonstration or video and discussion.
- Permit the students to use the subject presented soon after the presentation by discussion, application, or performance.
- Arrange for evaluation of students in a reasonable time and adjust learning activities appropriately.
- Reinforce learning activity in successive presentations by relating a previous topic to the present topic.

Note: You must decide, with the help of the school nurse, principal, and supervisor, whether some students in this category would be in danger of injuring themselves or other students in a shop/lab situation.

1.9.2 Identifying the Gifted Student

The gifted student may be more difficult to recognize than the special needs student. You should be familiar with gifted student characteristics and modify teaching methods accordingly. Common characteristics include:

- When properly motivated, the gifted student displays more interest and a longer attention span than others.
- More inclined to question or comment on the topic.

- At times, the student may seem disinterested or restless.
- Disruptive behavior may occur when learning activity lags.
- Usually completes assignments ahead of others and desires additional work.
- Scores higher on technical and performance evaluation activities.

Gifted students can be motivated and challenged by being assigned work with greater complexity, asked to assist slower students, pressed into service as assistants in the shop/lab, given problem-solving activities that involve creative thinking, and given special topics to research and report or demonstrate to the class.

1.10 ADDITIONAL RESOURCES

A variety of supplemental materials and experiences may be used to allow students to further develop their interest in metals technology. Not all learning occurs in the shop/lab. When possible, outside activities should be followed up by discussion and evaluation.

1.10.1 Audiovisual Aids

Many types of teaching aids are available. Use only those that fit into your program. Do not overlook the advantages of preparing your own. Let your students help.

Investigate the possibility of developing a teaching kit for each chapter. This would include a collection of aids and materials (transparencies, slides, tapes, industrial materials, etc.). The "kit" should be stored in specially designed boxes (all the same size for easy storage) and be readily available to the teacher and students.

1.10.2 Field Trips

Field trips are an excellent way to motivate students. Almost every community offers opportunities in the metalworking field. Field trips should be carefully planned and arrangements made ahead of the visit. Invite the principal or supervisor to accompany the class.

One such trip should be a visit to the Vocational-Technical school/center serving the school system, or local community college technical programs. This will make students aware of available advanced technical programs.

Brief the students before any field trip so they will know what to look for and can better understand what they see. A follow-up evaluation and discussion is a must after a field trip. Often having a student write a short report on what they have observed can be helpful. Such a report will also improve their observation and writing skills.

1.10.3 Mass-Production Problems

Many projects in metals technology are adaptable to a mass-production problem. Such a problem can be accomplished by a group of students, or the entire class can participate in the project. Properly planned, mass-production problems provide interesting and challenging experiences. They also can be used to develop leadership and entrepreneurial skills.

Before starting a production problem, construct several prototypes to be certain the project will operate according to specifications.

1.10.4 Industrial Technology Fairs

Many industrial technology programs use fairs to motivate students. Students should be involved when planning such an exhibit. The displays may be in the shops/labs, local mall, or some other location.

Students may want to display completed assignments and projects and give demonstrations.

1.11 SHOP/LAB MANAGEMENT

Shop/lab management starts several days *before* the first class session. It should include conservation (time, effort, and energy), plus student safety, so the major endeavor of the teacher can be directed toward improving instruction, which in turn, enhances learning.

Shop/lab management also includes some paperwork. The better you prepare for the first day of school, the more time you will have to take care of the many minor problems that crop up when the students arrive. Get all the forms you plan to use made up and reproduced in advance. Be sure they are readable and prepare enough for all of your students.

1.11.1 Student Information Form

Know your students. Prepare a student information sheet similar to a job application. (A sample form is included as a reproducible master with Chapter 1.) It will not only provide student experience in filling out a job application, but will also make them aware that they will need references and a social security number when applying for a job.

At the first class session, after all the administration paperwork is completed, have students complete an information sheet designed similar to a job application. Besides giving students experience in filling out a simulated job application, it will provide you with information that can help you better aid them.

It will be more interesting if samples of actual job applications are on hand for students to examine. Discuss the importance of completing a job application accurately. A sample job application is included in this Instructor's Manual. You may want to design a different one for use in your class.

If permissible, the student information sheet might also have a space where students list their parents' or guardians' occupation(s). This will be of benefit to those students who are not sure exactly what it is their parents or guardians do at work. It is advised that you check with your principal or supervisor to learn the guidelines on this matter.

1.11.2 Seating Chart

A seating chart will aid in taking attendance. Seating charts also help you learn students' names.

A temporary roll should be used until all scheduling changes have been made. Keep the chart up-to-date. If a student does not show up for class, and is not listed as absent, there is the possibility that you may be legally responsible for that student until you notify the proper school personnel.

1.11.3 Progress Chart

A progress chart is helpful in many teaching situations. With it you can show the principal and supervisor student progress. It is also helpful for students to compare their progress with the progress of others in the class.

1.11.4 Accident Report

In many learning situations, it is mandatory that each accident or injury be reported. Fill out a report immediately in those situations. Check with your principal or supervisor for specific instructions on filling out required forms. A sample form is included in Chapter 1 of this manual and in the back of the workbook.

1.11.5 Report of Money Collected

Should it be required that money be collected for supplies from students, be aware of the school's policy on reporting money collection. Give a receipt to student's who pay for supplies. Secure a receipt for any money turned into the school's financial officer.

1.12 INSTRUCTOR RESPONSIBILITY

Most states now have laws relating to safety requirements in various areas of the school curriculum. Contact the necessary authorities to check your state's laws and local requirements. In metals technology they usually relate to mandatory eye protection for special areas of the shop/lab. These include:

- Turning, shaping, cutting, or stamping solid materials with hand tools and power driven tools, machines, or equipment.
- Casting, welding, and heat treating.
- Handling caustic materials.

While not usually mandated by state law, some school systems have established policies on keeping certain problems under control or eliminating them as a possible source of student injury. They include:

- Appropriate footwear must be worn in the metalworking areas.
- Long hair must be contained while operating power equipment.
- The use of drugs (medication and illegal) that impair muscular coordination, time perception, etc.

You will find it to your advantage to know and understand state laws and school policy affecting your program.

Safety of students is paramount. Dangers from toxic gases, flying particles, sharp metal edges, hot metal, electrical shock, and falling objects are ever present. A strong, firm safety policy is mandatory for both students and teacher. Safety warnings, and cautions (printed in red) are located throughout the *Modern Metalworking* textbook. These warnings are *not* exhaustive.

The office of Occupational Safety and Health Administration (OSHA) requires assurance that working (learning) conditions are kept healthy as far as vapors, fumes, temperature, dust, air contaminants, and noises are concerned.

Safety should be a top priority for everyone. It is the best and most efficient way to do any task. Common sense safety rules include the following:

- Keep workstations clean and orderly.
- Wear correct clothing for the job. Students should wear gloves when handling corrosive substances, materials with sharp edges, and materials heated from welding, forging, and casting operations. Protect clothing by wearing an apron. Loose clothing can be caught, and a careless movement can mangle fingers. Advise students that they must wear appropriate safety gear while in the shop/lab.
- Respect fire.
- Use proper lighting.
- Keep aisles and exits clear.
- Be sure the electrical circuit is locked open before attempting to repair equipment. Avoid standing in damp or wet places when using electrically powered equipment. Do not allow students to repair electrical equipment.
- Report all accidents and injuries.
- Wear safety glasses, goggles, or a face shield when appropriate. Proper eye protection should be worn at all times in the shop/lab. Goggles and safety glasses should meet USASI Safety Code Z17.1-1968 (United States of America Standards Institute). When not being used, goggles and safety glasses should be stored in a cabinet and exposed to a germicidal lamp to destroy bacteria.
- Use adequate respiratory protection. Proper ventilation is important in any shop/lab environment. Toxic vapors and fumes must be minimized. Exhaust blowers, along with makeup air units, should be used.
- Maintain and store equipment, tools, and supplies correctly. Never operate machinery without *all* safety guards in place and functioning properly.
- Disconnect all power when tools or equipment are not in use.

The Occupational Safety and Health Act (OSHA) establishes standards affecting the safety of workers in any occupation. OSHA is a federal law that mandates most safety recommended by safety organizations. Graduates who find employment in industry should know and practice these regulations. The act is enforced by the U.S. Department of Labor.

A copy of the act should be available to all instructors. Copies of the Federal Act are available from the Superintendent of Documents, U.S. Government Printing Office, Washington, DC 20402. Copies of the state Act are available from state Departments of Safety.

Job Safety and Health, published by the U.S. Department of Labor, Occupational Safety and Health Administration, may be used in the classroom for discussing safe working conditions. It is available from the U.S. Government Printing Office.

The National Safety Council, 1121 Spring Lake Drive, Itasca, IL 60143-3201, publishes safety education data sheets and posters that cover safety on the job. A list of the available safety education data can be obtained by writing to the preceding address.

1.12.1 Other Areas of Responsibility

We live in a litigious society. Teachers of industrial-technical education must be acquainted with situations that might cause legal actions to be brought against them and their school system.

Often times, an act of negligence is caused by ignoring basic safety rules or not using common sense. The following list is provided to enlighten your sense of responsibility. It is not all-inclusive. If you have serious concerns concerning legal matters and liabilities, contact the appropriate departments of your school system, and state and national teachers associations.

Acts Which May be Considered Negligence

- The act is not properly performed; the teacher does not employ proper care.
- The circumstances under which the act is performed create risks although the act is performed with due care and precaution.
- The teacher performs acts that involve unreasonable risks of direct and immediate harm to pupils.
- The teacher sets into motion a force, the continuous operation of which may be unreasonably hazardous to pupils.
- The teacher creates a situation that is unreasonably dangerous to pupils because of the

likelihood of the action of third persons or of inanimate forces.

- The teacher entrusts dangerous devices or tools and equipment to pupils who by reason of incapacity or abnormality he/she knows to be likely to inflict intended harm upon others.
- The teacher neglects a duty of control over pupils who by reason of some incapacity or abnormality he/she knows to be likely to inflict intended harm upon others.
- The teacher fails to employ due care to give adequate warning.
- The teacher fails to exercise the proper care in looking out for pupils whom he/she has reason to believe may be in a danger zone.
- The teacher fails to use appropriate skill to perform acts undertaken.
- The teacher fails to give adequate preparation to avoid harm to pupils before entering certain conduct where such preparation is reasonably necessary.
- The teacher fails to inspect and repair instruments or mechanical devices used by pupils.

1.13 KEEPING UP-TO-DATE

As with all technical areas, there are constant changes and improvements in metalworking technology. The following material may help keep your program in tune with the changes taking place in the metalworking field.

1.13.1 Means for Keeping Up-to-Date

To keep pace with the changes in metalworking technology, teachers should subscribe and read monthly trade magazines and journals; visit modern metalworking facilities and talk with management and personnel; and attend trade shows and state and national professional technology organization conventions

Summer employment in a modern metalworking shop is also helpful. Summer sessions at the college level can be beneficial in improving teaching techniques and developing a better understanding of the overall educational picture.

Keeping abreast with modern metalworking technology is mandatory if you want to offer a course that will equip your students with the skills, knowledge, and aptitudes needed today.

1.14 RESOURCE MATERIAL

There is much resource material available on metalworking technology. Maintain a reference library in the shop/lab or request that the library carry a selection of reference books and publications on the field. The following books, catalogs, manuals, and periodicals are only a brief listing of available material.

1.14.1 Reference Books

Applied Mathematics, Phagan, R. Jesse, Goodheart-Willcox Publisher

Arc Welding, Walker, John R., Goodheart-Willcox Publisher

Basic Mathematics, Brown, Walter C., Goodheart-Willcox Publisher

The Book of Trades (Historical), Amman, Jost: Dover Publications, (Original publication date 1658)

CIM Technology, Biekert, Russell, Goodheart-Willcox Publisher

Contemporary Manufacturing Processes, DuVall, J. Barry, Goodheart-Willcox Publisher

Design Dimensioning and Tolerancing, Wilson, Bruce A., Goodheart-Willcox Publisher

A Dideroit Pictorial Encyclopedia of Trades and Industry, Vol. 1 and 2 (Historical), Dideroit, Denis, Dover Publications (Original publication date 1752)

Exploring Technology, Wanat, John A.; Pfieffer, E. Weston; and Van Gulik, Richard M., Goodheart-Willcox Publisher

Exploring Technology, Wright, R. Thomas and Hanak, Richard M. Goodheart-Willcox Publisher

From School to Work, Littrell, J. J., Smith, Harry; Morenz, James Goodheart-Willcox Publisher

Geometric Dimensioning and Tolerancing, Madsen, David A., Goodheart-Willcox Publisher

Machine Trades Print Reading, Barsamian, Michael, and Gizelbach, Richard A., Goodheart-Willcox Publisher

Machinery's Handbook, Green, Robert E., Editor, Industrial Press Inc.

Machining Fundamentals, Walker John R., Goodheart-Willcox Publisher

Metallurgy Fundamentals, Brandt, Daniel A., Goodheart-Willcox Publisher

Oxyacetylene Welding, Baird, Ronald J., Goodheart-Willcox Publisher

Print Reading for Industry, Brown, Walter C., Goodheart-Willcox Publisher

Processes of Manufacturing, Wright, R. Thomas, Goodheart-Willcox Publisher

Robotics Technology, Masterson, James; Towers, Robert; and Fardo, Stephen, Goodheart-Willcox Publisher

Technology Shaping Our World, Gradwell, John; Welch, Malcom; and Martin, Eugene, Goodheart-Willcox Publisher

Understanding Technology, Wright, R. Thomas and Smith, Howard, Goodheart-Willcox Publisher

Welding Print Reading, Walker, John R., Goodheart-Willcox Publisher

Welding Technology Fundamentals, Bowditch, William A., and Bowditch, Kevin E., Goodheart-Willcox Publisher

1.14.2 Catalogs and Manuals

American National Standards Institute (ANSI)
1430 Broadway
New York, NY 10013
(Catalog of Standards and price list)

The Association of Manufacturing Technology
7901 Westpark Drive
McLean, VA 22102-4206

Dover Publications, Inc.
31 East 2nd Street, Mineola, NY 11501
(Catalog on technology history)

Hanser Gardner Publications
6915 Valley Avenue, Cincinnati, OH 45244-3029
(Catalog of technical books and audio visual material)

The Industrial Press
93 Worth Street, New York, NY 10013
(Machinery's Handbook)

The M.I.T. Press
Massachusetts Institute of Technology,
Cambridge, MA 02142
(Books on technology)

National Tool & Machining Association
9300 Livingston Road
Fort Washington, MD 20744
(Catalog of books and audio visual material)

U.S. Bureau of Labor Statistics
Government Printing Office, 200 Constitution Ave.
NW, Washington, DC 20210
(Occupational Outlook Handbook)

1.14.3 Periodicals

American Machinist
Penton Publishing
1100 Superior Avenue
Cleveland, OH 44114

Automation News
155 E. 23rd Street
New York, NY 10010

CAD/CAM & Robotics
Kerrwil Publications Ltd.
501 Oakdale Road, Downsview
ON, Canada M3N 1W7

The Home Shop Machinist
The Village Press
2779 Aero Park Drive
Traverse City, MI 49686

Industrial Education
Cummins Publishing Company
26011 Evergreen Road
Southfield, MI 48076

Industrial Machinery Digest
One Chase Drive #300
Hoover, AL 35244

Machine Design
1100 Superior Avenue
Cleveland, OH 44144

Metalfax
29100 Aurora Road
Solon, OH 44139

Metalworking Digest
1350 East Touhy Avenue
Des Plaines, IL 60017

Modern Machine Shop
6600 Clough Pike
Cincinnati, OH 45244-4090

1.14.4 Agencies and Associations

American National Standards Institute (ANSI)
1430 Broadway
New York, NY 10018

American Vocational Association
1410 King Street
Alexandria, VA 22314

International Technical Education Association (ITEA)
1914 Association Drive
Reston, VA 22901

National Association of Industrial Technology (NAIT)
3300 Washenaw Avenue, Suite 220
Ann Arbor, MI 48104-4200

National Institute of Metalworking Skills
2209 Hunter Mill Road
Vienna, VA 22181

SkillsUSA—Vocational Industrial Clubs of America (SkillsUSA—VICA)
Box 30, Leesburg, VA 22075

1.14.5 Audiovisual Materials

A variety of audiovisual materials is available for use in metalworking technology. Contact the following companies and associations for listings of available materials.

DCA Educational Products, Inc.
1814 Keller Church Road
Bedminster, PA 18910

Hanser Gardner Publications
6915 Valley Avenue
Cincinnati, OH 45244-3029

Minnesota Mining and Mfg. Co.
3M Center, Visual Systems Division
Austin, TX 78769

National Tooling & Machining Association
9300 Livingston Road
Fort Washington, MD 20744

L.S. Starrett Company
121 Crescent Drive
Athol, MA 01331

Sterling Educational Films
241 E. 34th Street
New York, NY 10016

1.14.6 On-Line Resources

Many resources for metalworking technology and education are available over the information superhighway. They are sponsored by corporations, private organizations, and individuals.

On-line addresses for information on all areas of metalworking technology can be found in trade magazines and journals. Using the Internet can keep you current on new metalworking developments and manufacturing techniques.

Most of the sites provide information free of charge or for a minimal fee. Since the information superhighway is constantly expanding, the addresses for some of the following sites may have changed since the publication of this manual. The following is only a sampling of companies with on-line sites.

Bridgeport Machines, Inc.
www.bpt.com

Cincinnati Machine (CNC machine tools)
www.cinmach.com

Mastercam (software)
www.mastercam.com

Modern Machine Shop Magazine
www.gardnerweb.com/mms

Nikon (optical comparators and CNC measuring systems)
www.nikonusa.com

Sharnoa Corp. (CNC machine tools)
www.sharnoa.com

South Bend Lathe Corp. (machine tools)
www.southbendlathecorp.com

GOODHEART-WILLCOX

If you have comments, corrections, or suggestions regarding the textbook or its supplements, please send them to:

Managing Editor, Goodheart-Willcox Publisher, 18604 West Creek Drive, Tinley Park, IL 60477-6243

1 Technology and Careers

After studying this chapter, students should be able to:

- Define the term *technology.*
- Explain how various types of technology contribute to advances in industry.
- List the types of careers available in the metalworking industry.
- Describe the importance of leadership skills and how they can be developed.
- Explain how to enter the workforce and get a job in the metalworking industry.
- Identify what industry expects of an employee.
- Recognize what an employee should expect from industry.

Chapter Resources

Text, pages 9–26
- Test Your Knowledge, page 25
- Research and Development, page 26

Workbook, pages 7–14

Instructor's Manual
- Answer keys for:
  - Test Your Knowledge Questions
  - Workbook
  - Chapter Quiz
- Reproducible Masters:
  - 1-1 Employment Application
  - 1-2 The History of Modern Technology
  - 1-3 Lesson Plan
  - 1-4 Accident Report
  - 1-5 Shop Cleanup Assignments
  - 1-6 Chapter 1 Quiz

Guide for Lesson Planning

Introduce this chapter by using one of the following approaches:

- Prepare a large drawing or use an overhead transparency of a clock face. Let it represent the development of technology since humans first used tools. To show how brief of time modern technology has existed, question students on what part of the twelfth hour the automobile and airplane appeared in history. They are both slightly more than a hundred years old. On the Clock Face of History, both were invented only one-thousandth (that is 1/1000) of a second before 12 o'clock. Use Reproducible Master 1-2.
- A calendar may also be used to show how *young* modern technology is. Tape each month of a calendar on the chalkboard or wall. Let the months represent the history of technology. Ask the same questions as asked above. The automobile and airplane were invented only 10 seconds before the end of the year.

Class Discussion

Have students read and study all or part of Chapter 1, paying attention to the illustrations. Review the assignment and discuss the following:

- Student definition of *technology.*
- What do students think are the most important contributions of technology? (Many experts believe it was fire and the wheel.)
- With such great advances in modern technology, what type of workers will be required in the future? What will happen to people with minimum education?
- What will they have to do to secure a good job when they graduate?
- What can they expect when they enter the world of work?
- How should they go about getting that first *big job?*

The importance of accurately completing a job application has been covered in the text. No question on an application should be left blank. If a question does not apply, have the students write NA (not applicable). However, filling out a job application is only the first step in securing a job. The second step is usually a job interview. For this reason, you want to discuss how to prepare for an interview and what can leave a bad impression on the interviewing person.

Using the chalkboard, have students list what actions they think will fall into each of the two categories. Some instructors use a mock interview to emphasize the differences between what constitutes a good interview and a poor interview.

Assignments

1. Assign Test Your Knowledge questions at the end of the chapter.
2. Assign Chapter 1 in the *Modern Metalworking Workbook.*
3. Assign the chapter quiz. Copy and distribute Reproducible Master 1-6.
4. Permit students to volunteer for one or more of the Research and Development activities at the end of the chapter.

Test Your Knowledge

1. Applying knowledge through the use of tools, machines, and processes to change existing conditions and control the environment.
2. stone
3. Farming
4. Any order: water wheel, steam engine.
5. flexible manufacturing system
6. Evaluate responses individually.
7. Industrial
8. profit
9. Any order: semiskilled, skilled, technical, professional, management.
10. Instructional programs that consist of on-the-job training under the supervision of a skilled worker.
11. Technicians
12. Any order: teaching, engineering.
13. (d) Civil
14. Management
15. Answer may include any four of the following: vision, strength in communication, persistence, ability to organize, responsibility, ability to delegate authority.
16. In the process of applying advanced technology, there is a constant development of new ideas, materials, and manufacturing techniques that require the ability to keep skills up-to-date.
17. Answer may include any three of the following: school guidance office, technology education department, career information center, library, state employment service.
18. Evaluate responses individually.
19. honesty
20. Evaluate responses individually.

Workbook

1. Technology
2. Human survival
3. Industrial Revolution
4. Evaluate individually.
5. Student answers will vary. Evaluate individually.
6. (b) By innovative technology.
7. Any order: agricultural, medical, industrial, metalworking.
8. Evaluate individually.
9. Profit
10. It is necessary if people are to risk their time, ability, and money in a business venture.
11. semiskilled
12. skilled
13. An instructional program consisting of on-the-job training under the supervision of a skilled worker. It requires four or more years to complete.
14. technician
15. design, development
16. (d) All of the above.
17. (b) methods and means needed to manufacture and assemble a product
18. aerospace
19. Metallurgical
20. Management
21. To be competitive with other technologically-oriented countries.
22. success, failure
23. Evaluate individually.
24. Evaluate individually. Refer to Section 1.4.1 in the text.
25. Student responses will vary. Evaluate individually.
26. Evaluate individually.
27. Evaluate individually. Refer to Section 1.5.1.

28. Evaluate individually. Refer to Section 1.5.2.
29. Evaluate individually.
30. Evaluate accuracy and completion of job applications individually.
31. Evaluate individually.

Chapter Quiz

1. (f) Skilled worker
2. (a) Industrial Revolution
3. (j) Leader
4. (b) Technology
5. (k) Leadership
6. (e) Semiskilled
7. (d) Apprentice
8. (c) Profit
9. (i) Management
10. (g) Technician

Notes

APPLICATION FOR EMPLOYMENT

PERSONAL INFORMATION

Date ______ Social Security Number ______

Name ______
Last First Middle

Present Address ______
Street City State

Permanent Address ______
Street City State

Phone No. ______

If related to anyone in our employ, state name and department ______ Referred by ______

EMPLOYMENT DESIRED

Position ______ Date you can start ______ Salary desired ______

Are you employed now? ______ If so may we inquire of your present employer? ______

Ever applied to this company before? ______ Where ______ When ______

EDUCATION

	Name and Location of School	Years Completed	Subjects Studied
Grammar School			
High School			
College/University			
Trade, Business or Correspondence School			

Subject of special study or research work ______

Activities other than religious (Exclude organizations the name or character of which indicates the race, creed, color or national origin of its members.)

What foreign languages do you speak fluently? Read fluently? Write fluently?

U.S. Military service Rank Present membership in National Guard or Reserves

FORMER EMPLOYERS List below last three employers starting with last one first

Date Month and Year	Name and Address of Employer	Salary	Position	Reason for Leaving
From				
To				
From				
To				
From				
To				

REFERENCES Give below the names of two persons not related to you whom you have known for at least one year

	Name	Address	Job Title	Years Acquainted
1				
2				

PHYSICAL RECORD

Have you any disabilities that might affect your ability to perform this job? ______

In case of emergency notify ______

Name Address Phone No.

Date ______ Signature ______

The History of Modern Technology

Lesson Plan

Section ______________________ Date ____________

Unit Assignment	Topic	Technique(s)	Tools and Supplies Required
Teaching Aids	**Student Activities**	**Quizzes**	**Demonstration(s)**
Next Assignment	**Evaluation**		

Accident Report

Name: ______________________ **Date:** ______________________

Instructor: ______________________

1. Injured person: ______________________

   Address: ______________________

   Telephone: ______________________

   Homeroom: ______________________

2. Accident witnesses: ______________________

   Name: ______________________

   Address: ______________________

   Telephone: ______________________

   Name: ______________________

   Address: ______________________

   Telephone: ______________________

3. Type of injury: ______________________

   ____

   ____

   ____

4. Treatment: ______________________

   First aid: ______________________ By whom: ______________________

   Physician: ______________________ Address: ______________________

   Hospital: ______________________ Address: ______________________

5. Cause of accident: ______________________

   ____

   ____

6. Tools/machines involved: ______________________

   ____

   ____

7. Action taken to prevent recurrence of accident: ______________________

   ____

   ____

   ____

Shop Cleanup Assignments

Period: ______________________ **Week of:** ______________________

Assignment	Students	Comments

Students must return their tools to tool panel and clean equipment they have been using.

Chapter 1 Quiz
Technology and Careers

Name: ______________________ **Date:**______________ **Period:**______________

Match each word or phrase with the sentence that best describes it.

(a) Industrial Revolution
(b) Technology
(c) Profit
(d) Apprentice
(e) Semiskilled
(f) Skilled worker
(g) Technician
(h) Professional
(i) Management
(j) Leader
(k) Leadership

_________ 1. A worker capable of performing the exacting work and skills of a trade.

_________ 2. Industry was taken out of the home and handwork was replaced by machinery.

_________ 3. A person who is in charge or command.

_________ 4. The know-how linking science and the industrial arts.

_________ 5. The ability to be a leader.

_________ 6. Are told what to do and how to do it.

_________ 7. A person learning a trade. Usually requires four years of study and training with experienced craftpersons.

_________ 8. Necessary if a company is to remain in business.

_________ 9. People who plan, direct, and supervise the operation of an industrial organization.

_________ 10. A specialist who repairs and maintains computer controlled machines and robotic equipment.

2 Classifying Metals

Learning Objectives

After studying this chapter, students should be able to:

- Explain how metals are classified.
- Describe the properties and characteristics of many different metals.
- Identify how metals and alloys are developed for specific applications.
- Describe the characteristics of different types of steel and list the methods used to identify steels.
- List the hazards posed by metals and use the safety precautions followed in industry.
- Recognize how metals are measured and purchased for industrial use.

Chapter Resources

Text, pages 27–46
  Test Your Knowledge, page 45
  Research and Development, pages 45–46
Workbook, pages 15–18
Instructor's Manual
Answer keys for:
  Test Your Knowledge
  Workbook
  Chapter Quiz
Reproducible Masters:
  2-1 Metals used in Manufacture (chart)
  2-2 Chapter Quiz

Guide for Lesson Planning

Before students read and study the chapter, have them define the term *metal* in their own words. Is there a term that will describe all metals? Have samples of as many different metals as possible on hand. They should be identical in size. Allow students to handle the samples, feel their weights (the weights of two types of aluminum will differ slightly), and note how the metals differ in color.

Class Discussion

Have the students read all or part of the chapter paying attention to the illustrations and their captions. Review the assignment and discuss the following:

- The use of metals in history, especially in our modern world. Use the chalkboard (or Reproducible Master 2-1) and prepare two columns. List products that use metal in their manufacture in the first column. List products that do not use metal in their manufacture in the second column. It will be difficult for them to name a modern product that does *not* use metal somewhere in its manufacture.
- How metals are classified.
- Characteristics of some metals.
- Reinforced composite materials. The body of the Chevrolet Corvette and aircraft parts are made from these materials.
- How to safely handle metals in the shop/lab.

Assignments

1. Assign Test Your Knowledge questions at the end of the chapter.
2. Assign Chapter 2 of the *Modern Metalworking Workbook.*
3. Assign the chapter quiz. Copy and distribute Reproducible Master 2-2.
4. Encourage students to volunteer for one or more of the Research and Development activities at the end of the chapter.

Test Your Knowledge

1. Evaluate responses individually.
2. metallurgist
3. Any order: ferrous metals,

nonferrous metals, precious metals, high-temperature metals, rare metals.
4. iron
5. Mixtures of two or more metals.
6. (b) Wrought
7. By their percentage of carbon in points, or hundredths of one percent.
8. (a) Chromium
9. Alloy steels that resist wear and retain strength at high temperatures.
10. Any order: austenitic, martensitic, ferritic.
11. (c) bloom
12. black oxide
13. First, the metal is pickled or passed through a dilute acid to remove the oxide coating. It is then rolled into a final shape and size while it is cold.
14. Any order: AISI/SAE codes, color coding, spark test.
15. nonferrous
16. Answer may include any three of the following: they are lighter than most other metals designed for commercial use; they do not rust or corrode under normal conditions; they can be shaped and formed easily; they are available in a large assortment of shapes, sizes, and alloys.
17. (d) Titanium
18. beryllium copper
19. (c) copper
20. copper, zinc
21. galvanized
22. precious
23. nuclear, aerospace
24. (b) Tungsten
25. Honeycomb structures are bonded sections of thin metal. They are joined with an adhesive, or fused by brazing or resistance welding, to form a panel between two metal or composite sheets.

Workbook

1. Evaluate individually.
2. Nonferrous metal does not contain iron, while ferrous metal contains iron.
3. A single metal with no added elements.
4. A mixture of two or more metals.
5. Evaluate individually. Refer to Section 2.8.
6. Ferrous metals: cast iron, steel, wrought iron, stainless steel, carbon steel, low-carbon steel, high-carbon steel.
   Nonferrous metals: aluminum, copper, tin, chromium, manganese, tungsten, pewter, lead, brass, zinc, bronze, nickel, molybdenum, tungsten carbide, magnesium, beryllium, columbium, tantalum, gold, silver.
7. Base metals: aluminum, copper, tin, chromium, manganese, tungsten, lead, zinc, nickel, molybdenum, magnesium, beryllium, columbium, tantalum, gold, silver.
   Alloys: cast iron, steel, wrought iron, stainless steel, carbon steel, pewter, brass, bronze, tungsten carbide, low-carbon steel, high-carbon steel.
8. Evaluate individually.
9. It is fluid at room temperature.
10. A structure made up of sections of thin metal that are bonded together to form a structure similar in appearance to the wax comb bees make to store honey. To give existing metals greater strength and rigidity while reducing their weight.
11. Strong, lightweight structures composed of metal fibers bonded together in a matrix through the application of heat and pressure.
12. Carbon
13. Ferrous metal
14. Nonferrous metal
15. Alloy
16. Base metal
17. Precious metal
18. (n) Zinc
19. (p) German silver
20. (k) Bronze
21. (j) Brass
22. (m) Tin
23. (a) Steel
24. (e) Stainless steel
25. (b) Manganese
26. (d) Tungsten carbide
27. (h) Magnesium
28. (i) Titanium
29. (f) Hot-rolled steel
30. (g) Spark test
31. (l) Copper
32. (q) Platinum
33. (o) Pewter
34. (c) High-speed steel
35. Evaluate individually.

CHAPTER QUIZ

1. (j) Copper
2. (m) Tin
3. (k) Bronze
4. (r) Brass
5. (t) Galvanized sheet
6. (n) Pewter
7. (i) Beryllium
8. (b) Nonferrous metal
9. (f) Aluminum
10. (a) Ferrous metal
11. (g) Magnesium
12. (h) Titanium
13. (e) Alloy
14. (l) Zinc
15. (q) Platinum
16. (s) Tungsten
17. (p) Gold
18. (o) Sterling silver
19. (c) Base metal
20. (u) Cast iron

Notes

Metals used in Manufacturing

Products Using Metal in Their Manufacture	Products Not Using Metal in Their Manufacture

Chapter 2 Quiz

Classifying Metals

Name: ______________________ **Date:**______________ **Period:**______________

Match each word or phrase with the correct sentence.

_________ 1. Reddish-brown in color. Oldest metal known.

_________ 2. Soft, shiny, silvery metal.

_________ 3. An alloy of tin and copper.

_________ 4. An alloy of copper and zinc.

_________ 5. Steel sheet with a zinc coating.

_________ 6. An alloy of tin, copper, and antimony.

_________ 7. Used in nuclear applications. Extreme care must be observed when machining this metal.

_________ 8. Metals with no iron as a basic ingredient.

_________ 9. A large family of metals, not just a single metal.

_________10. Metals containing iron.

_________11. Lightest of the structural metals.

_________12. As strong as steel but only half as heavy.

_________13. A mixture of two or more metals.

_________14. Resists many forms of corrosion. Used as a protective coating on steel and iron.

_________15. A heavy grayish-white metal that is resistant to most chemicals.

_________16. Melts at a higher temperature than any known metal.

_________17. Used to make jewelry and coinage.

_________18. A very shiny silvery-white metal when polished.

_________19. The principal alloying agent in a metal.

_________20. Contains 3% to 4% carbon.

(a) Ferrous metal
(b) Nonferrous metal
(c) Base metal
(e) Alloy
(f) Aluminum
(g) Magnesium
(h) Titanium
(i) Beryllium
(j) Copper
(k) Bronze
(l) Zinc
(m) Tin
(n) Pewter
(o) Sterling silver
(p) Gold
(q) Platinum
(r) Brass
(s) Tungsten
(t) Galvanized sheet
(u) Cast iron

Understanding Drawings

Learning Objectives

After studying this chapter, students should be able to:

- Read drawings dimensioned in fractional and decimal inches and metric dimensions.
- Identify and understand the different types of information indicated on a typical drawing.
- Describe how detail, assembly, and subassembly drawings differ and identify standard drawing sheet sizes.
- List the types of prints developed for industrial use.
- Indicate how tolerances are indicated on drawings.
- Explain the geometric dimensioning and tolerancing (GD&T) system.
- Complete a project plan sheet.

Chapter Resources

Text, pages 47–62
Test Your Knowledge, page 61
Research and Development, page 61
Workbook, pages 19–24
Instructor's Manual
Answer keys for:
Test Your Knowledge Questions
Workbook
Chapter Quiz
Reproducible Masters:
3-1 Project Plan Sheet
3-2 Alphabet of Lines
3-3 Information on a Typical Drawing
3-4 Typical Assembly Drawing
3-5 Dual-Dimensioned Drawing
3-6 Standard Symbols Used in Dimensioning
3-7 Chapter Quiz

Guide for Lesson Planning

Secure drawings used by industry along with samples of the parts made from the drawings. They will be of great help in teaching this chapter.

Class Discussion

Have students read and study the chapter paying particular attention to the illustrations and their captions. Review the assignment. Asking the appropriate questions, discuss the following:

- The importance of drawings in industry.
- Alphabet of Lines.
- Information found on a typical drawing.
- How to read drawings that are dimensioned in fractional and/or decimal inches and metrics.
- Why dimensional tolerances are needed.
- How detail, subassembly, and assembly drawings differ.
- Geometric dimensioning and tolerancing and why it is used.
- How to complete a project plan sheet and bill of materials.

Assignments

1. Assign the Test Your Knowledge questions at the end of the chapter.
2. Assign Chapter 3 of the *Modern Metalworking Workbook.*
3. Assign the chapter quiz. Copy and distribute Reproducible Master 3-7.
4. Encourage students to volunteer for the Research and Development activities at the end of the chapter.

Test Your Knowledge

1. Drawing standards are needed because of the many different applications and manufacturing processes used in industry, and because parts for various products are frequently assembled in different locations.
2. language
3. dual-dimensioned
4. millimeters
5. Lines that identify the materials making up a part on a drawing.
6. surface finish
7. Allowances that indicate how much the dimensions of a machined part can vary from the original dimensions on a drawing.
8. Bilateral tolerances allow a variance of measurement in either direction from the given dimension and are expressed as "plus" and "minus" tolerances. Unilateral tolerances allow variance of measurement in only one direction from the given dimension and are expressed as a "plus" or "minus" tolerance.
9. Title block
10. (c) Subassembly
11. (b) the drawing might be lost, damaged, or destroyed
12. (d) diazo
13. computer-aided design
14. A standard system of tolerances used to control interpretation of the form, profile, orientation, location, and runout of features on drawings.
15. bill of materials, plan of procedure

Workbook

1. To provide metalworkers and machine operators the information needed to produce a part or assemble a product.
2. The sizes and locations of drawn features that are needed to produce a part. To indicate the measurements used to assemble a product and their proper relation to one another.
3. Dual-dimensioned
4. Allowances indicating how much the dimensions of a machined part can vary from the original dimensioned sizes on a drawing. They indicate how much larger or smaller the part can be machined compared to what the dimensions call for and still work to specifications.
5. bilateral
6. unilateral
7. For convenience in filing and locating drawings.
8. Detail drawing
9. Assembly drawing
10. To control interpretation of the form, profile, orientation, location, and runout of features on drawings. This type of tolerancing provides the necessary precision for the most economical manufacture of parts and interchangeable parts.
11. Evaluate individually. Refer to Section 3.7.
12. An outline, in sequential order, of the operations that should be followed to complete the job or project.
13. Study the prints or drawings for the job.
14. All the items that will be needed to complete the job, along with the costs for each part.
15. (a) Dimension line. Locates points dimensioned.
    (b) Hidden object line. Indicates or shows hidden features of a part.
    (c) Extension line. Indicates point from which dimension is given.
    (d) Centerline. Used to locate centers of symmetrical objects.
    (e) Object line. Indicates visible edges of part.
    (f) Cutting plane line. Indicates where the part has been cut to show interior features.
    (g) Section lines. Used when showing interior features of part exposed by the cutting plane line.
16. Adapter
17. B54-2345
18. Full size
19. Steel bar CR-1020
20. ±1/16″
21. 4″
22. 2″ high, 2″ thick
23. 2
24. 0.50″ DIA
25. 1″
26. 1″
27. 3″
28. 0.37″
29. In order: 4″, 1″, 0.25″
30. Evaluate individually. Refer to the drawing in the workbook.

Chapter Quiz

1. Rotary loader.
2. B3345.
3. Full and 1.5X size.
4. Steel, AISI 1020.
5. ±1/64″
6. ±0.010″
7. 5.00″
8. .50″
9. 2.25″
10. .750″ diameter.
11. R.50″
12. Yes.
13. Sheet 2 of 7.
14. Machining done after welding and heat treat.
15. Remove all sharp edges R.01″ max.

Notes

Project Plan Sheet

Name:________________ **Section:**____________ **Name of Project:**________________

Date Started:________ **Date Completed:**________ **Source of Idea for Project:**________________

Bill of Material

Part Name	No. of Pieces	Material	Size (T×W×L)	Unit Cost	Total
				Total Cost	

Plan of Procedure

List the operations to be performed in the order they are to be done and the tools and equipment required to do the job.

Operation No.	Operation	Tools and Equipment Required

Operation No.	Operation	Tools and Equipment Required

Student Evaluation of Project:

Teacher Evaluation:

Grade: ___________

Alphabet of Lines

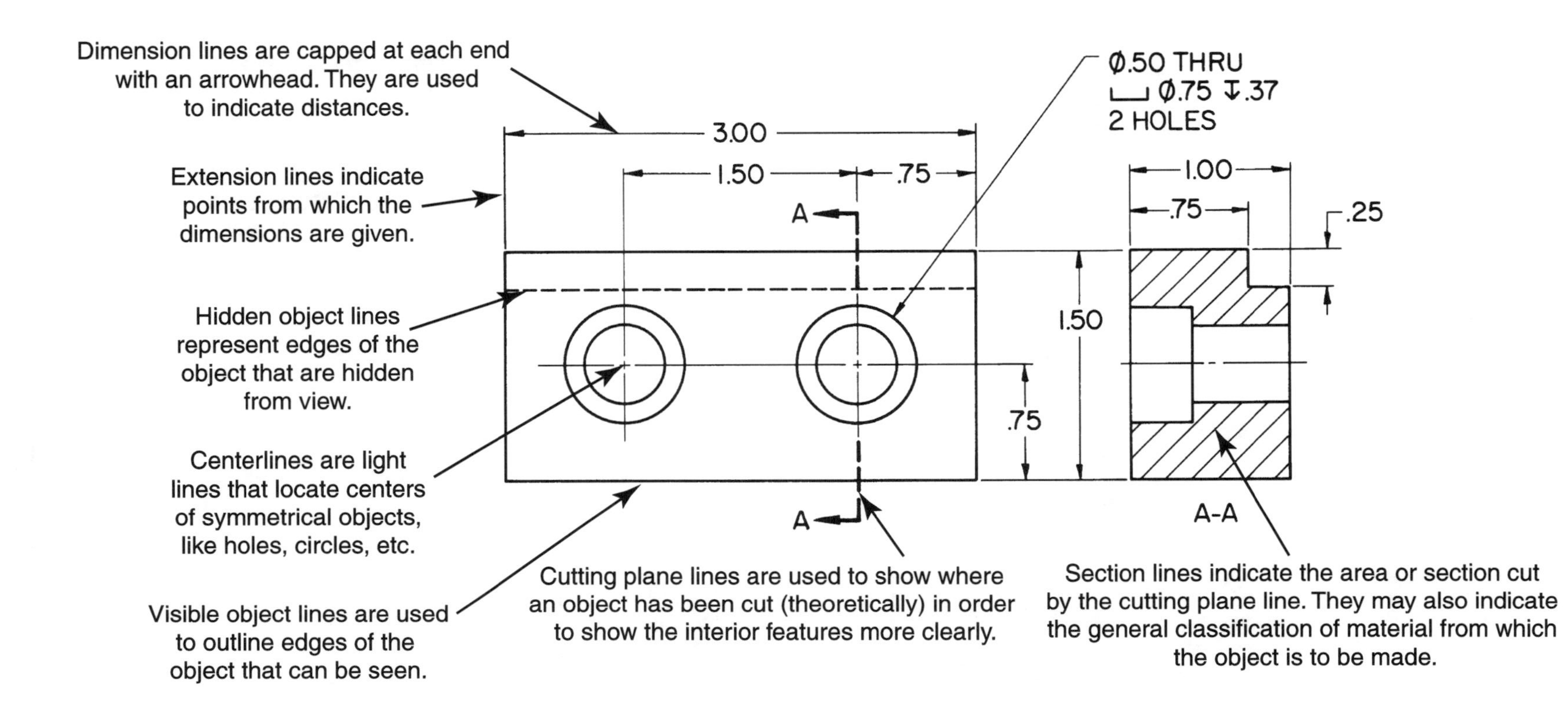

3-2

Information on a Typical Drawing

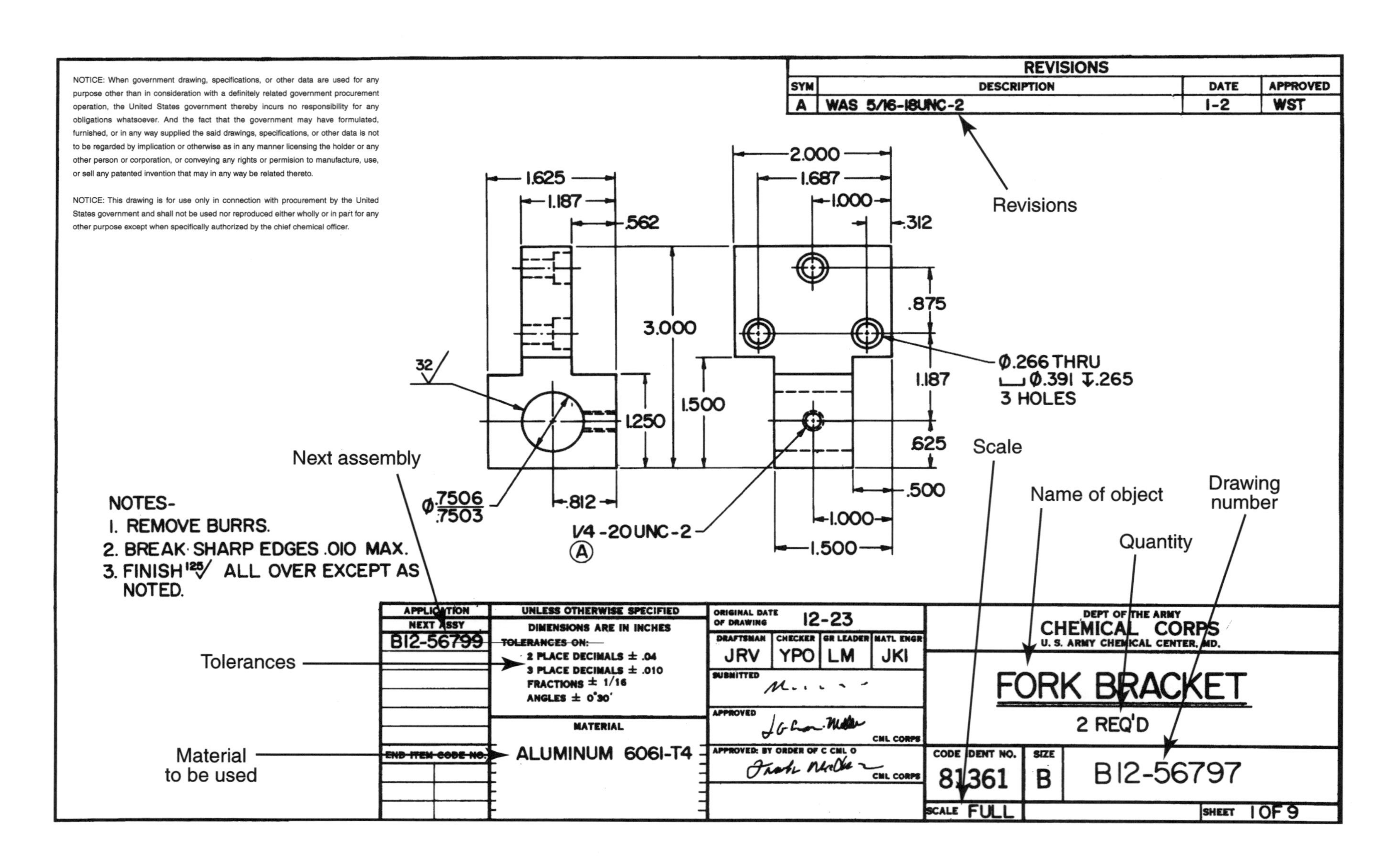

Typical Assembly Drawing

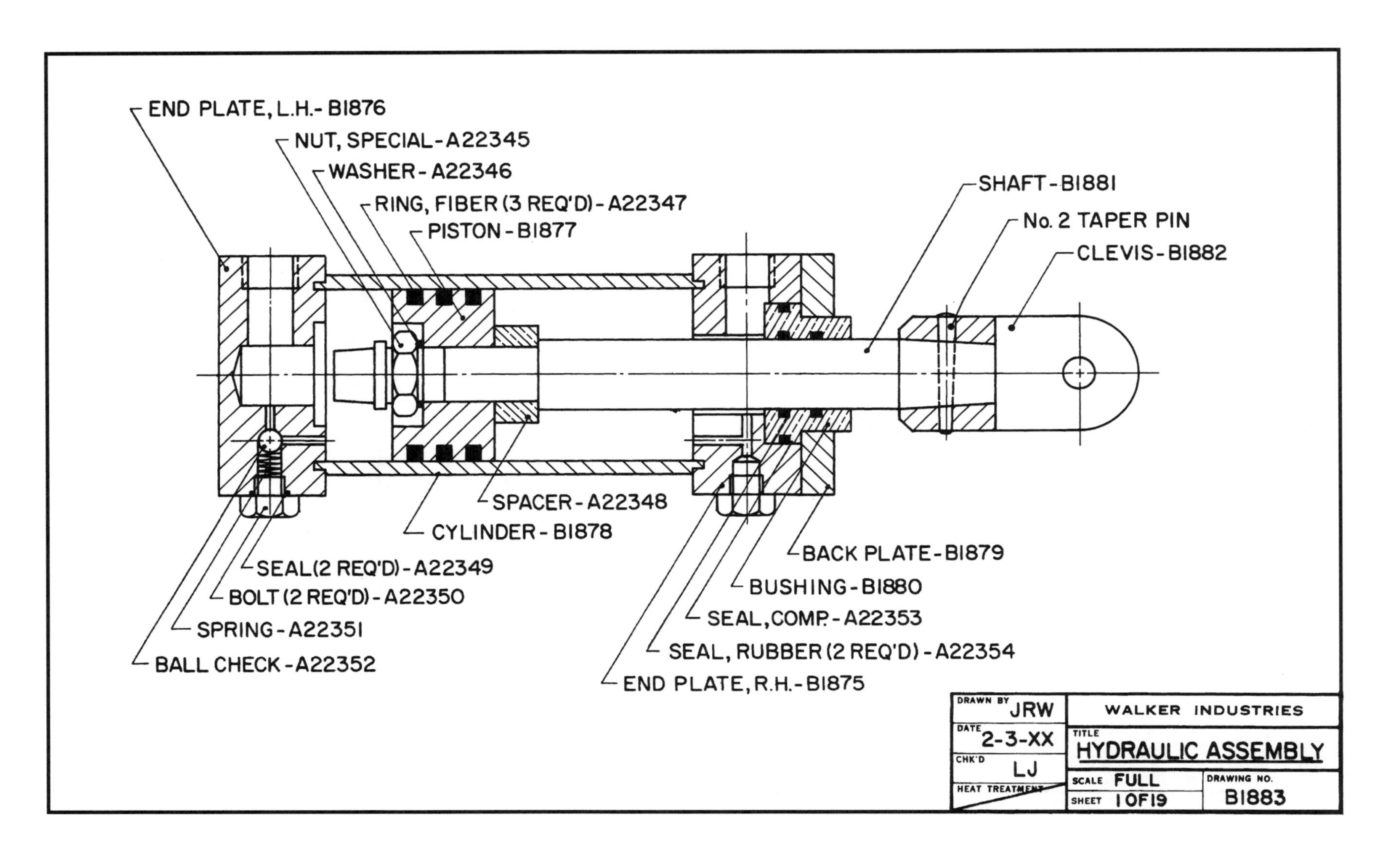

Dual Dimensioned Drawing

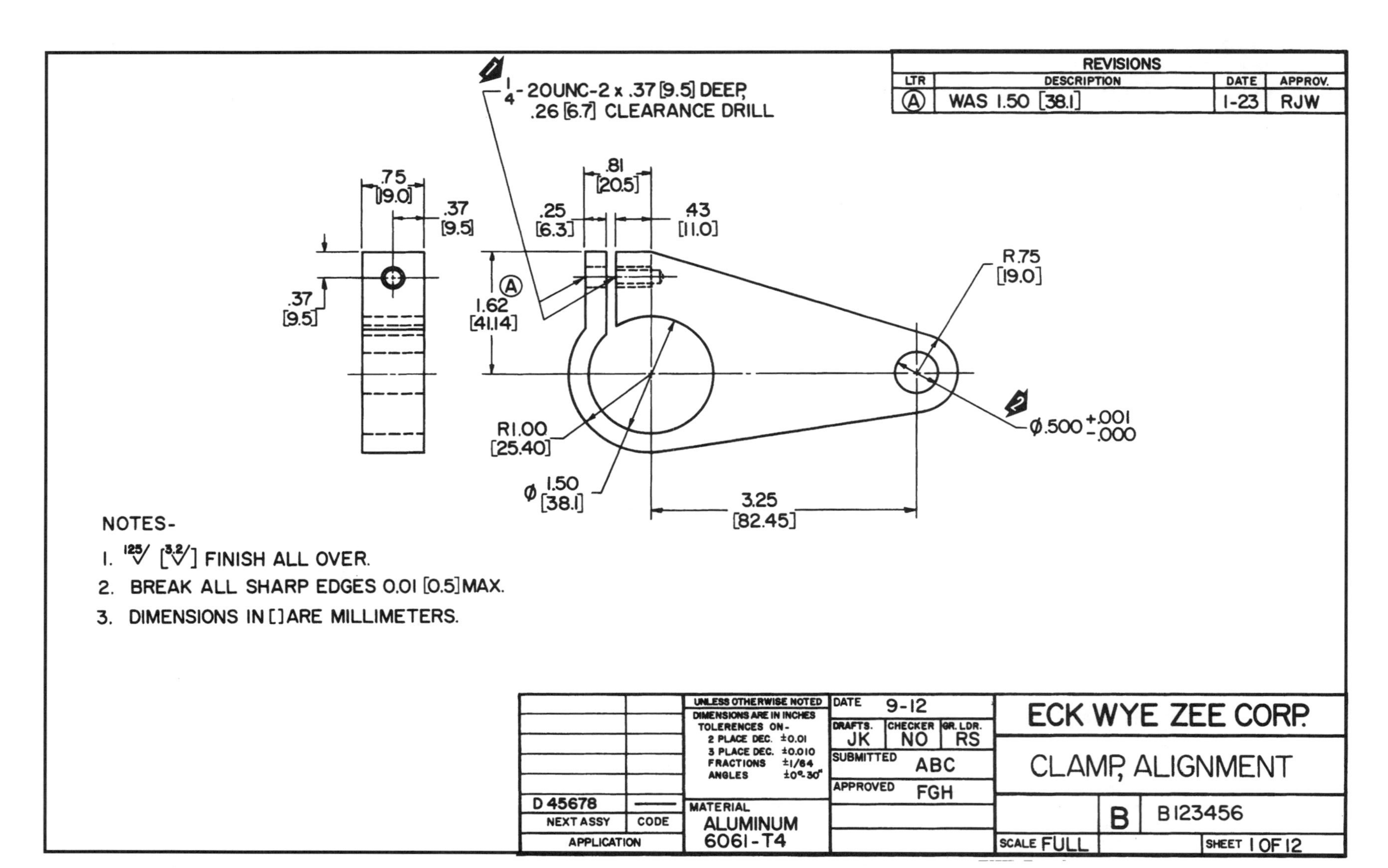

➡ A metric thread size has not been given because there is none equal to this size fractional thread.

➡ There is not metric reamer equal to this size.

Standard Symbols Used in Dimensioning

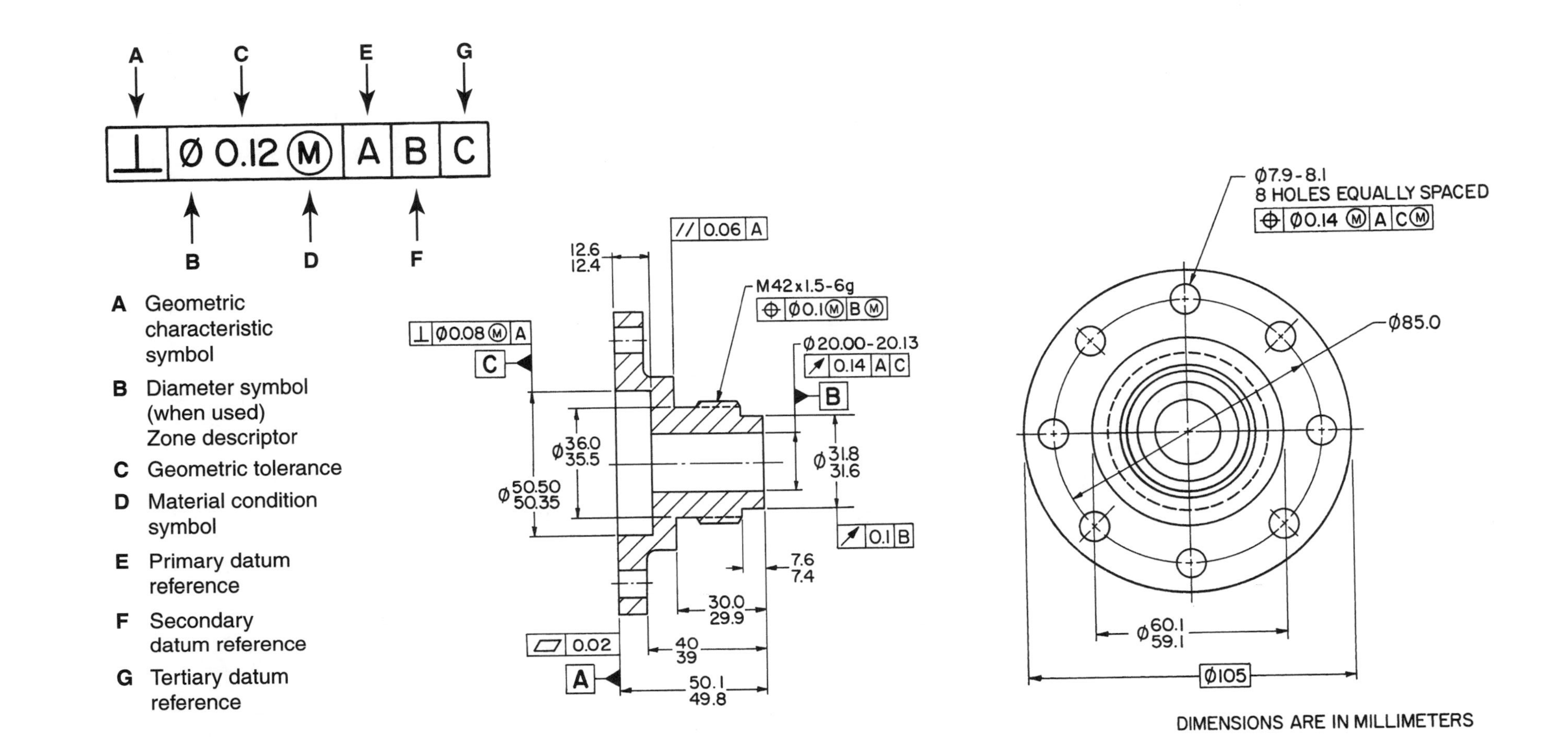

3-6

Chapter 3 Quiz

Understanding Drawings

Name: ______________________________ **Date:**________________ **Period:**________________

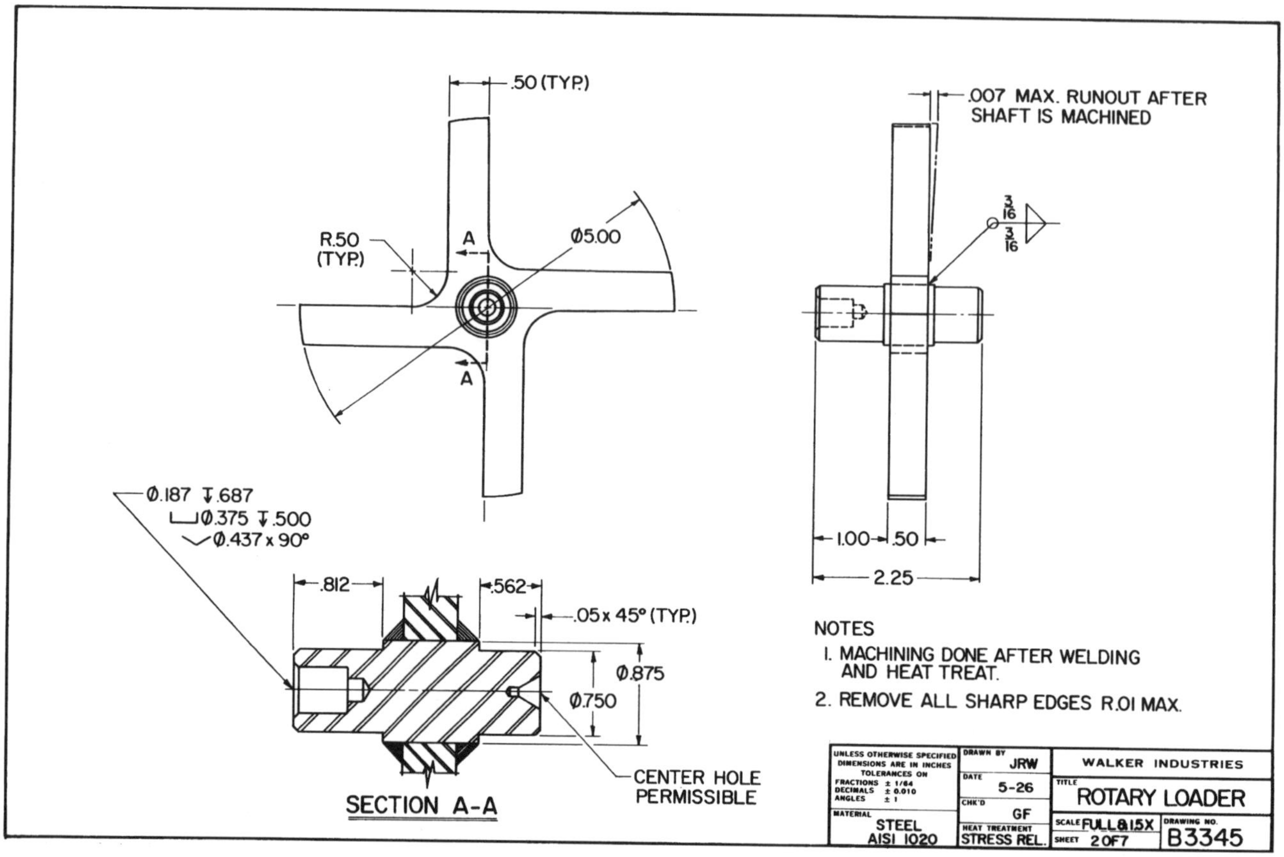

.50 (TYP.)

R.50 (TYP.)

A

Ø5.00

A

.007 MAX. RUNOUT AFTER SHAFT IS MACHINED

3/16

3/16

Ø.187 ↧.687
⌴Ø.375 ↧.500
⌵Ø.437 x 90°

1.00 .50

2.25

.812

.562

.05 x 45° (TYP.)

Ø.875

Ø.750

CENTER HOLE PERMISSIBLE

SECTION A-A

NOTES

1. MACHINING DONE AFTER WELDING AND HEAT TREAT.
2. REMOVE ALL SHARP EDGES R.01 MAX.

UNLESS OTHERWISE SPECIFIED DIMENSIONS ARE IN INCHES TOLERANCES ON FRACTIONS ± 1/64 DECIMALS ± 0.010 ANGLES ± 1	DRAWN BY JRW	WALKER INDUSTRIES	
	DATE 5-26	TITLE ROTARY LOADER	
	CHK'D GF		
MATERIAL STEEL AISI 1020	HEAT TREATMENT STRESS REL.	SCALE FULL & 1.5X / SHEET 2 OF 7	DRAWING NO. B3345

Carefully study the working drawing and answer the following questions.

1. What is the name of the part? ____________________
2. What number has been assigned to the drawing? ____________________
3. The original drawing was drawn to what scale? ____________________
4. The part is made of what material? ____________________
5. What tolerances are allowed for fractional dimensions? ____________________
6. What tolerances are allowed for decimal dimensions? ____________________
7. What is the largest diameter of the part? ____________________
8. How thick are the arms? ____________________
9. The shaft is how long? ____________________
10. How large is the machined diameter of the shaft? ____________________
11. Indicate the typical radius at the base of the arms. ____________________
12. Are center holes permissible in the shaft?____________________
13. The drawing is sheet ______ of ______ needed to make and assemble the part. ____________________

There are two items of information on the drawing that concern how the part is to be finished. List them.

14. ____________________
15. ____________________

4 Safety Practices

Learning Objectives

After studying this chapter, students should be able to:

- Explain the importance of practicing safe work habits.
- Summarize the general safety practices observed in metalworking.
- List common safety equipment and protective clothing used in the metals lab.
- Apply safe work habits when operating machinery.
- Recognize and avoid unsafe work practices.

Chapter Resources

Text, pages 63–68
  Test Your Knowledge, page 67
  Research and Development, pages 67–68
Workbook, pages 25–28
Instructor's Manual
Answer keys for:
  Test Your Knowledge Questions
  Workbook
  Chapter Quiz
Reproducible Master:
  4-1 Chapter Quiz

Guide For Lesson Planning

Introduce the chapter with a display of safety equipment and student-made safety posters. Discuss the duties of a student safety officer.

Class Discussion

Have students read and study the chapter. Review it with them in relation to the general safety rules. Be specific about such things as:

- State and school safety requirements.
- Students must assume responsibility for their safety and the others in the shop/lab.
- Approved eye protection must always be worn while working in the shop/lab.
- No machine is to be operated until instructions have been given in its operation.
- Permission must be received before operating a machine tool.
- Dress must be appropriate for the metals area in which students are working.
- Machines must not be used unless all guards and safety devices are in place and functioning properly.
- Students must avoid operating equipment while their senses are impaired by medication or other substances.
- Safe techniques for handling metal chips and cuttings produced while operating machine tools.
- Importance of washing hands thoroughly after working in the metals shop/lab.
- Safe disposal of rags used to clean machines.
- Procedure to be followed for reporting and taking care of any cut, burn, bruise, scratch, or puncture, no matter how minor it may appear.
- Shop/lab safety *cannot* be overemphasized!

Assignments

1. Assign Test Your Knowledge questions at the end of the chapter.
2. Assign Chapter 4 in the *Modern Metalworking Workbook.*
3. Assign the chapter quiz. Copy and distribute Reproducible Master 4-1.
4. Encourage students to volunteer for the Research and Development activities at the end of the chapter.

Test Your Knowledge

1. safety rules
2. Any order: safety glasses, goggles, face shields.

3. Certain forms of medication can cause drowsiness, making it unsafe to operate machinery.
4. Answer may contain any four of the following: eye protection, hearing protection, plastic gloves, apron, dust mask.
5. Flying metal chips can cause injuries, and oil that has been vaporized by compressed air can ignite and cause serious burns.
6. (b) metal safety container
7. inside
8. All machine guards and safety devices must be in place and the machine must be checked to make sure it is in good working order.
9. measurements, adjustments
10. brush

Workbook

1. Any order: safety glasses, goggles, face shields.
2. The failure to follow simple safety rules.
3. It can get caught in moving machinery or conduct electricity.
4. For jobs where dust and fumes are a hazard. During activities such as grinding, buffing, and foundry work.
5. To prevent workers from tripping or falling on or over scraps and spills.
6. fire, explosion
7. Evaluate individually.
8. Evaluate individually. Refer to Section 4.3.
9. Evaluate individually.
10. Evaluate student posters individually.

Chapter Quiz

1. inside
2. It can get caught in moving machinery or conduct electricity.
3. thinners, solvents
4. Flying metal chips can cause injuries, and oil that has been vaporized by compressed air can ignite and cause serious burns.
5. To prevent workers from tripping or falling on or over scraps and spills.
6. All machine guards and safety devices must be in place and the machine must be checked to make sure it is in good working order.
7. For jobs where dust and fumes are a hazard. During activities such as grinding, buffing, and foundry work.

Chapter 4 Quiz
Safety Practices

Name: ______________________ **Date:**______________ **Period:**______________

1. When securing material in a vise, position the work and vise handle to the ______ of the bench. 1.______________

2. Why should all types of jewelry be removed before working in a machining lab/shop?

____

____

____

____

3. Portable power tools should *not* be operated in areas where ______ and ______ are in use because a serious fire or explosion could result. 3.______________

____

4. What is the danger of using compressed air to remove metal chips and cutting oil from machinery?

____

____

____

____

5. Why is it important to keep the metalworking area clean?

____

____

____

____

6. What are two precautions to observe before operating a machine?

____

____

____

____

7. For what types of metalworking applications should a mask be worn?

____

____

____

____

Measurement

Learning Objectives

After studying this chapter, students should be able to:

- Identify basic measuring tools and gages and explain how they are used.
- Measure to 1/64″ and 0.5 mm with a steel rule.
- Make measurements to 0.001″ and 0.01 mm with a micrometer caliper.
- Use a Vernier micrometer caliper to read measurements to 0.0001″ and 0.002 mm.
- Explain how to make readings with common Vernier measuring tools.
- Recognize different types of gages and demonstrate how they are used to check sizes.

Chapter Resources

Text, pages 69–96
Test Your Knowledge, page 93
Research and Development, page 96
Workbook, pages 29–38
Instructor's Manual
Answer keys for:
Test Your Knowledge Questions
Workbook
Measurement Pretests
Chapter Quiz
Reproducible Masters:
5-1 US Conventional and Metric Rules
5-2 Inch-Based Rule and Metric Rule
5-3 Inch-Based Micrometer
5-4 Inch-Based Vernier Micrometer
5-5 Metric Micrometer
5-6 Metric Based Vernier Micrometer
5-7 25-Division Inch-Based Vernier Caliper
5-8 50-Division Inch-Based Vernier Caliper
5-9 25-Division Metric Vernier Caliper
5-10 50-Division Metric Vernier Caliper
5-11 Universal Bevel Protractor
5-12 Chapter Quiz (Part I)
5-13 Chapter Quiz (Part II)
5-14 Chapter Quiz (Part III)

Guide for Lesson Planning

This chapter is extensive and should be divided into four segments. If program time is limited, teach students how to read and make accurate measurements with rules and micrometers.

Part I—Rules

The ability to make accurate measurements is basic to all types of skilled occupations. Use Reproducible Master 5-1 to show what the graduations are on inch-based and metric rules. Use Reproducible Master 5-2 as a pretest to determine a starting point for teaching measurement.

Class Discussion

Have students read and study Section 5.1. Review the assignment and demonstrate and discuss the following:

- The various types of rules.
- How to read the various types of rules.
- How to make accurate measurements with a rule.
- How to handle and care for rules so they will retain their accuracy.

Assignments

1. Assign Part I of the *Modern Metalworking Workbook.*
2. Administer Part I of the chapter quiz. Copy and distribute Reproducible Master 5-12.

Part II—Micrometers

When students become proficient measuring with rules, introduce the micrometer caliper. Overhead transparencies and handouts can be used to teach students how to use a micrometer. Use Reproducible Masters 5-3, 5-4, 5-5, and 5-6.

Have students read and study Section 5.2. Review the assignment. Discuss and demonstrate the following:

Class Discussion

- The various types of micrometers.
- How to read inch- and metric-based micrometers.
- The proper way to use micrometers.
- The proper way to care for micrometers so they will retain their accuracy.
- Allow students to examine and use micrometers. Have several sizes of material on hand for them to use for practice in reading micrometers.

Assignments

1. Assign Part II of the *Modern Metalworking Workbook.*
2. Administer Part II of the chapter quiz. Copy and distribute Reproducible Master 5-13.

Part III—Vernier Measuring Tools

Have students read and study Section 5.3. Use Reproducible Masters 5-7, 5-8, 5-9, 5-10, and 5-11 to make handouts and overhead transparencies. Review the assignment and discuss and demonstrate the following:

- The various types of Vernier measuring tools.
- How to read inch and metric based Vernier measuring tools.
- The correct way to use Vernier measuring tools.
- The proper way to care for Vernier measuring tools so they will retain their accuracy.

Assignments

1. Assign Part III of the *Modern Metalworking Workbook.*
2. Administer Part III of the chapter quiz. Copy and distribute Reproducible Master 5-14.

Part IV—Other Measuring Tools

Have as many of the tools described in this section available for student examination.

Students should read and study Sections 5.4 through 5.7, paying attention to the illustrations and captions. Permit student volunteers to demonstrate how inside and outside calipers and telescoping gages are used.

Assignments

1. Assign the Test Your Knowledge questions.
2. Assign Part IV in the *Modern Metalworking Workbook.*
3. Permit students to volunteer for the Research and Development activities at the end of the chapter.
4. Assign Part V of the *Modern Metalworking Workbook* for students who need practice in converting from SI Metric and US Conventional measurements, and vice versa.

Test Your Knowledge

1. 1. 5/32 A. 0.5 mm
   2. 17/32 B. 9.5 mm
   3. 1 1/16 C. 25.0 mm
   4. 1 19/32 D. 36.5 mm
   5. 1 29/32 E. 41.5 mm
   6. 2 9/32 F. 52.5 mm
   7. 2 21/32 G. 63.5 mm
   8. 2 31/32 H. 69.5 mm
   9. 3 5/32 I. 78.5 mm
   10. 3 15/32 J. 88.5 mm
2. A. 0.125″ G. 2.50 mm
   B. 0.250″ H. 5.00 mm
   C. 0.3125″ I. 6.54 mm
   D. 0.375″ J. 16.10 mm
   E. 0.4375″ K. 21.51 mm
   F. 0.500″ L. 25.67 mm
3. A. 8.683″ F. 5.008″
   B. 4.107″ G. 55.78 mm
   C. 7.500″ H. 73.34 mm
   D. 3.150″ I. 71.70 mm
   E. 8.793″ J. 24.84 mm
4. microinch
5. millimeters, microinches
6. Lines or divisions that represent points of measurement on a measuring tool.
7. 1/1000 or 0.001 and 1/10,000 or 0.0001; 0.01, 0.002.
8. An outside micrometer measures external diameters and thicknesses. An inside micrometer measures internal diameters.
9. thimble
10. In reverse order.
11. (c) can be used to make both inside and

outside measurements over a number of sizes

12. 1/1000 or 0.001, 0.01
13. (d) universal Vernier bevel protractor
14. hole diameters
15. external diameters
16. Jo
17. Any order: balanced, continuous.
18. A special instrument that uses air pressure to check internal and external diameters.
19. (a) Optical flats
20. A tool used to determine the pitch or number of threads per inch on a threaded section.
21. Fillet and radius gages can be used to check convex and concave radii on corners or against shoulders, and they can be used as templates for grinding form cutting tools.
22. (b) telescoping gage
23. A tool used to measure openings that are too small to be measured with a telescoping gage.
24. 1/64″, 0.4 mm
25. outside, inside

WORKBOOK

1. millionth (0.000001″)
2. 1/64″, 0.5 mm
3. (a) 6″ steel rule
   (b) Rule with adjustable hook
   (c) Narrow rule
4. (a) 3/16″
   (b) 9/16″
   (c) 15/16″
   (d) 1 1/4″
   (e) 1 11/16″
   (f) 2 1/16″
   (g) 2 5/16″
   (h) 2 11/16″
   (i) 3 3/16″
   (j) 3 7/16″
   (k) 3 5/8″
   (l) 3 13/16″
   (m) 1/32″
   (n) 13/32″
   (o) 23/32″
   (p) 1 3/32″
   (q) 1 17/32″
   (r) 1 27/32″
   (s) 2 5/32″
   (t) 2 15/32″
   (u) 2 25/32″
   (v) 2 31/32″
   (w) 3 9/32″
   (x) 3 21/32″
   (y) 4 7/32″
5. (a) 306.0 mm
   (b) 295.0 mm
   (c) 289.0 mm
   (d) 281.0 mm
   (e) 267.0 mm
   (f) 258.0 mm
   (g) 245.0 mm
   (h) 234.0 mm
   (i) 222.0 mm
   (j) 208.0 mm
   (k) 2.50 mm
   (l) 11.5 mm
   (m) 20.5 mm
   (n) 29.5 mm
   (o) 36.5 mm
   (p) 45.5 mm
   (q) 55.5 mm
   (r) 63.5 mm
   (s) 71.5 mm
   (t) 83.5 mm
   (u) 93.5 mm
   (v) 99.5 mm
   (w) 106.5 mm
6. (a) 0.856″
   (b) 0.663″
   (c) 0.817″
   (d) 0.532″
   (e) 0.748″
   (f) 0.142″
7. (a) ~~1.39 mm~~ 1.89
   (b) 19.51 mm
   (c) 0.56 mm
   (d) 14.61 mm
8. (a) Inside micrometer
   (b) Micrometer depth gage
   (c) Screw-thread micrometer
9. The measuring range can be increased by changing to longer spindles. The micrometer scale is read from right to left.
10. Screw-thread micrometers
11. Vernier caliper
12. Dial
13. (a) 0.743
    (b) 4.157
    (c) 6.991
    (d) 12.108
    (e) 8.475
    (f) 11.708

(g) 5.057
(h) 3.343
14. (a) Double end cylindrical plug gage
(b) Ring gages
(c) Adjustable snap gage
15. (a) are within specified tolerances
16. external
17. Evaluate individually.
18. (d) All of the above.
19. (a) Dial indicator
(b) Thickness (feeler) gages
20. (a) Screw pitch gages
(b) Fillet and radius gage
(c) Telescoping gage
21. fillet and radius
22. (b) telescoping gage
23. 0.32 cm
24. 23.8 mm
25. 0.95 cm
26. 11.9 mm
27. 50.8 mm
28. 0.97″
29. 0.02 m
30. 1.59 cm
31. 20.24 mm
32. 0.65 cm
33. 12.7 mm
34. 3.18 mm
35. 2.22 cm
36. 96.84 mm
37. 28.71 L
38. 1.36 kg
39. 0.64″
40. 16.1 km

Inch-Based Rule Pretest (4-1)

1. 1/16″
2. 1/2″
3. 1 1/8″
4. 1 13/16″
5. 2 7/16″
6. 2 15/16″
7. 3 11/16″
8. 4 3/16″
9. 4 3/4″
10. 5 5/16″
11. 5 7/8″
12. 3/32″
13. 13/32″
14. 1 5/32″
15. 1 25/32″
16. 2 11/32″
17. 2 29/32″
18. 3 19/32″
19. 4 7/32″
20. 4 17/32″
21. 5 1/32″
22. 5 9/32″
23. 5 21/32″
24. 5 31/32″

Metric Rule Pretest (4-2)

1. 4.0 mm
2. 12.0 mm
3. 21.0 mm
4. 29.0 mm
5. 42.0 mm
6. 53.0 mm
7. 65.0 mm
8. 76.0 mm
9. 87.0 mm
10. 90.0 mm
11. 2.5 mm
12. 13.5 mm
13. 23.5 mm
14. 40.5 mm
15. 49.5 mm
16. 60.5 mm
17. 71.5 mm
18. 87.0 mm
19. 94.5 mm
20. 101.5 mm

Chapter 5 Quiz—Part I

1. 5.0 mm
2. 26.0 mm
3. 51.0 mm
4. 69.0 mm
5. 90.0 mm
6. 11.5 mm
7. 35.0 mm
8. 55.5 mm
9. 76.0 mm
10. 97.5 mm
11. 5/16″
12. 1 3/8″
13. 3 1/8″
14. 4 3/8″
15. 5 9/16″
16. 15/32″
17. 1 23/32″

18. 2 15/32″
19. 3 27/32″
20. 5 13/32″

Chapter 5 Quiz—Part II

1. 0.125″
2. 0.339″
3. 0.481″
4. 0.062″
5. 0.033″
6. 0.198″
7. 0.435″
8. 0.358″
9. 0.260″
10. 0.126″
11. 6.78 mm
12. 13.22 mm

Chapter 5 Quiz—Part III

1. 4.107″
2. 7.500″
3. 73.34 mm
4. 55.78″
5. 5.008″
6. 71.70 mm
7. 8.793″
8. 3.150″
9. 24.84 mm
10. 8.683″

Notes

U.S. Conventional and Metric Rules

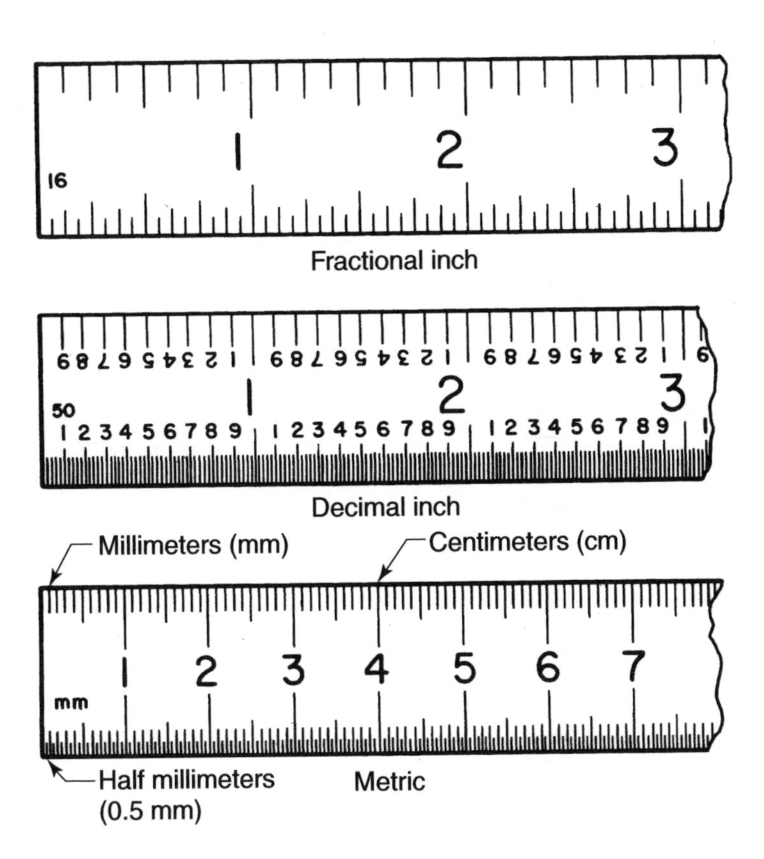

Inch-Based Rule

1 2 3 4 5 6 7 8 9 10 11

16

1 2 3 4 5 6

32

12 13 14 15 16 17 18 19 20 21 22 23 24

1. ________	9. ________	17. ________
2. ________	10. ________	18. ________
3. ________	11. ________	19. ________
4. ________	12. ________	20. ________
5. ________	13. ________	21. ________
6. ________	14. ________	22. ________
7. ________	15. ________	23. ________
8. ________	16. ________	24. ________

Name: ________

5-2

Metric Rule

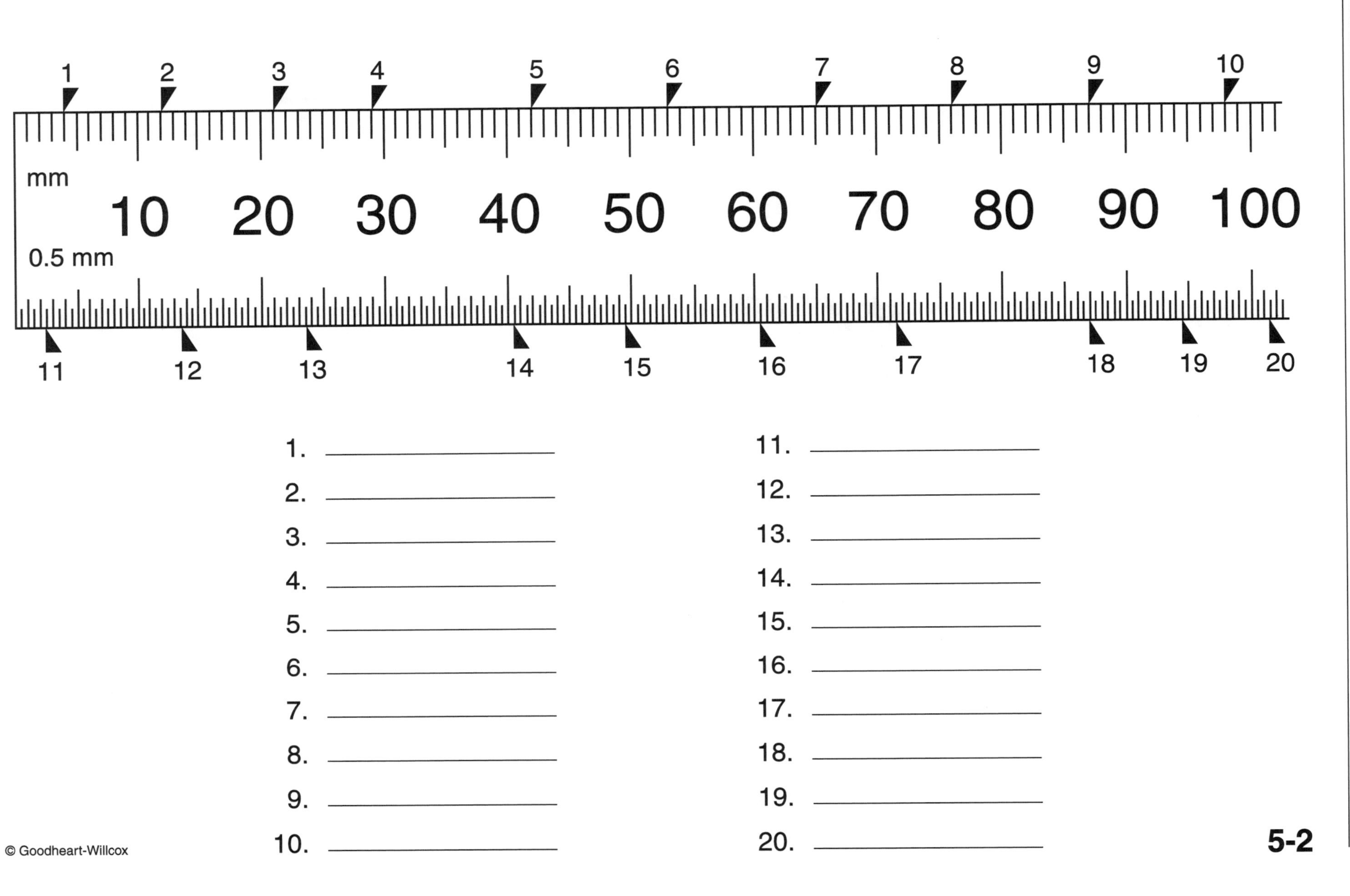

1. ______
2. ______
3. ______
4. ______
5. ______
6. ______
7. ______
8. ______
9. ______
10. ______
11. ______
12. ______
13. ______
14. ______
15. ______
16. ______
17. ______
18. ______
19. ______
20. ______

5-2

Inch-Based Micrometer

The reading is composed of:

4 Large graduations or 4 × 0.100	= 0.400
2 Small graduations or 2 × 0.025	= 0.050
8 Graduations on the thimble or 8 × 0.001	= 0.008
	0.458″

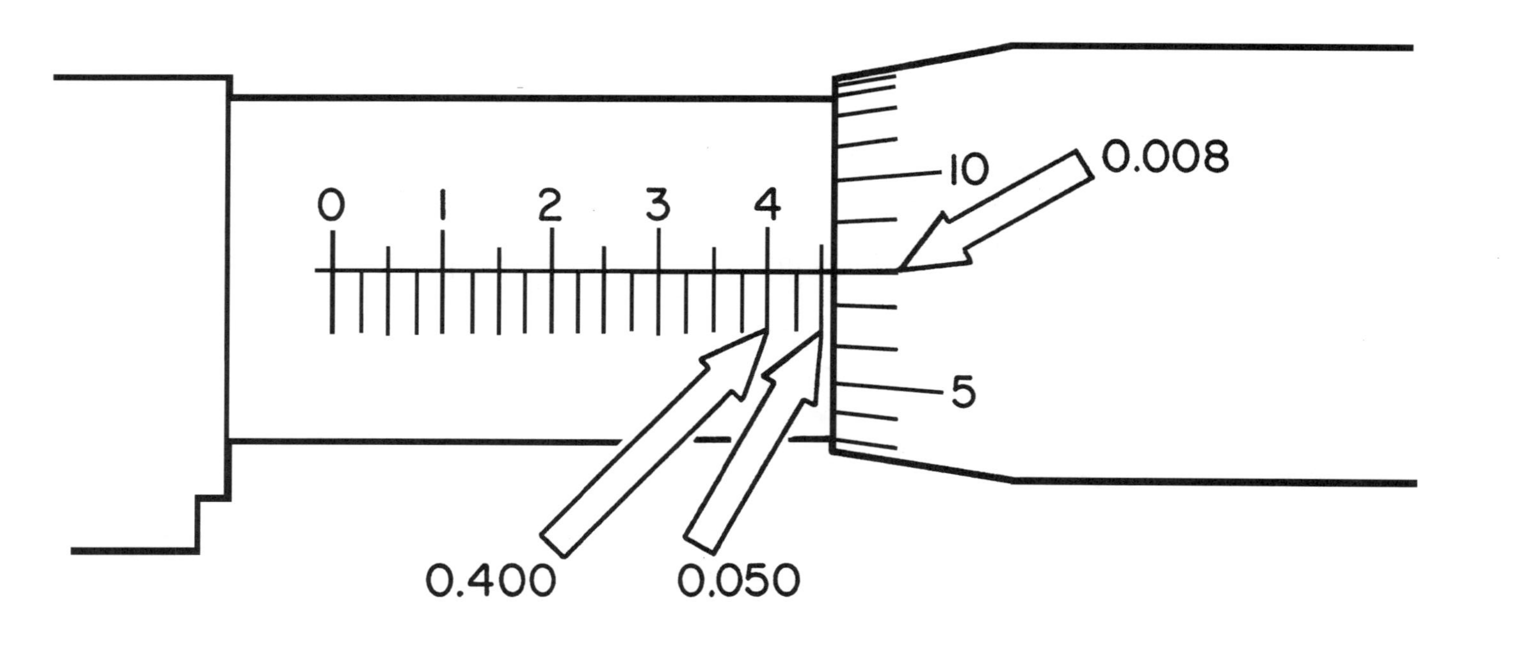

5-4

Inch-Based Vernier Micrometer

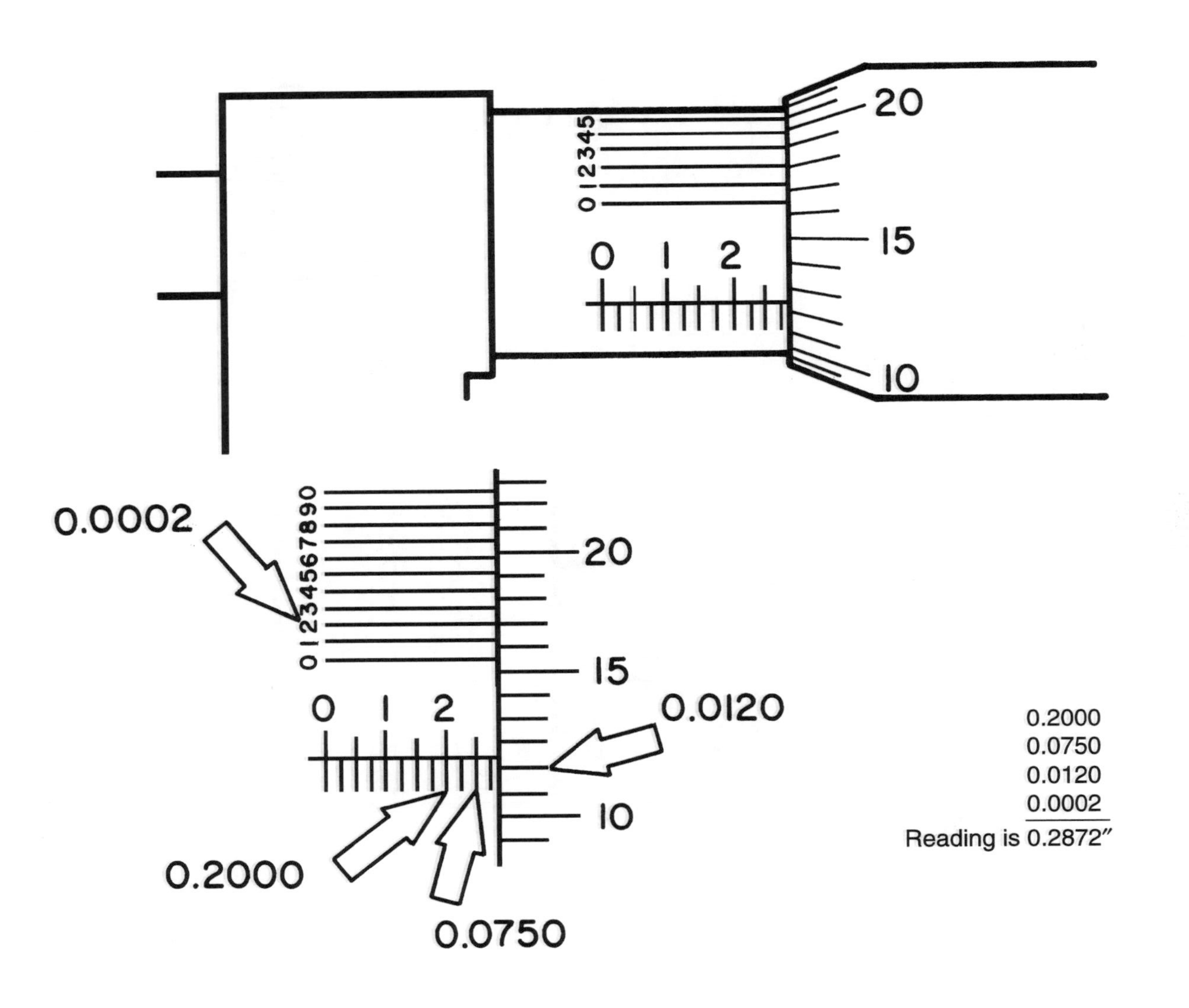

Metric Micrometer

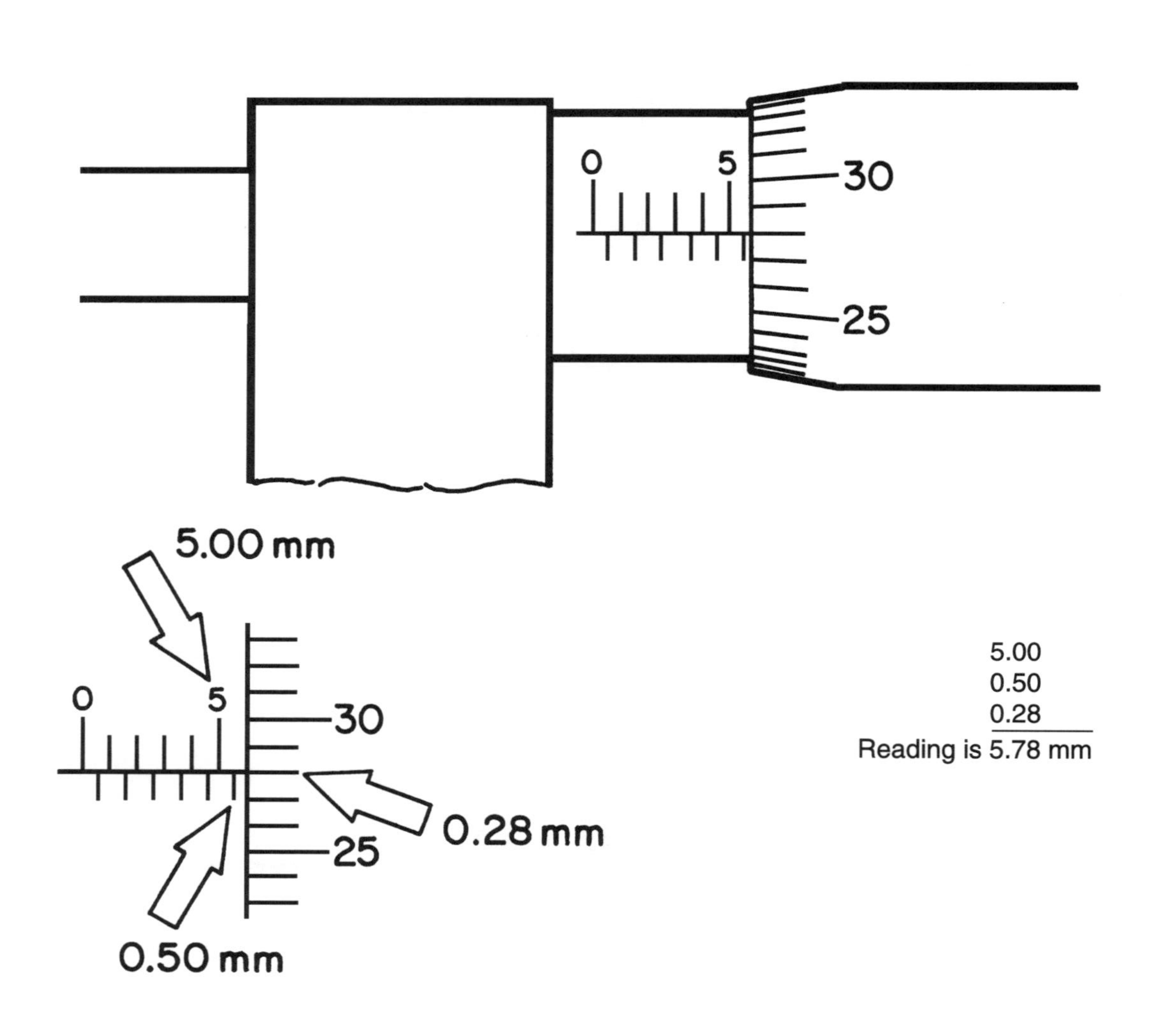

Metric-Based Vernier Micrometer

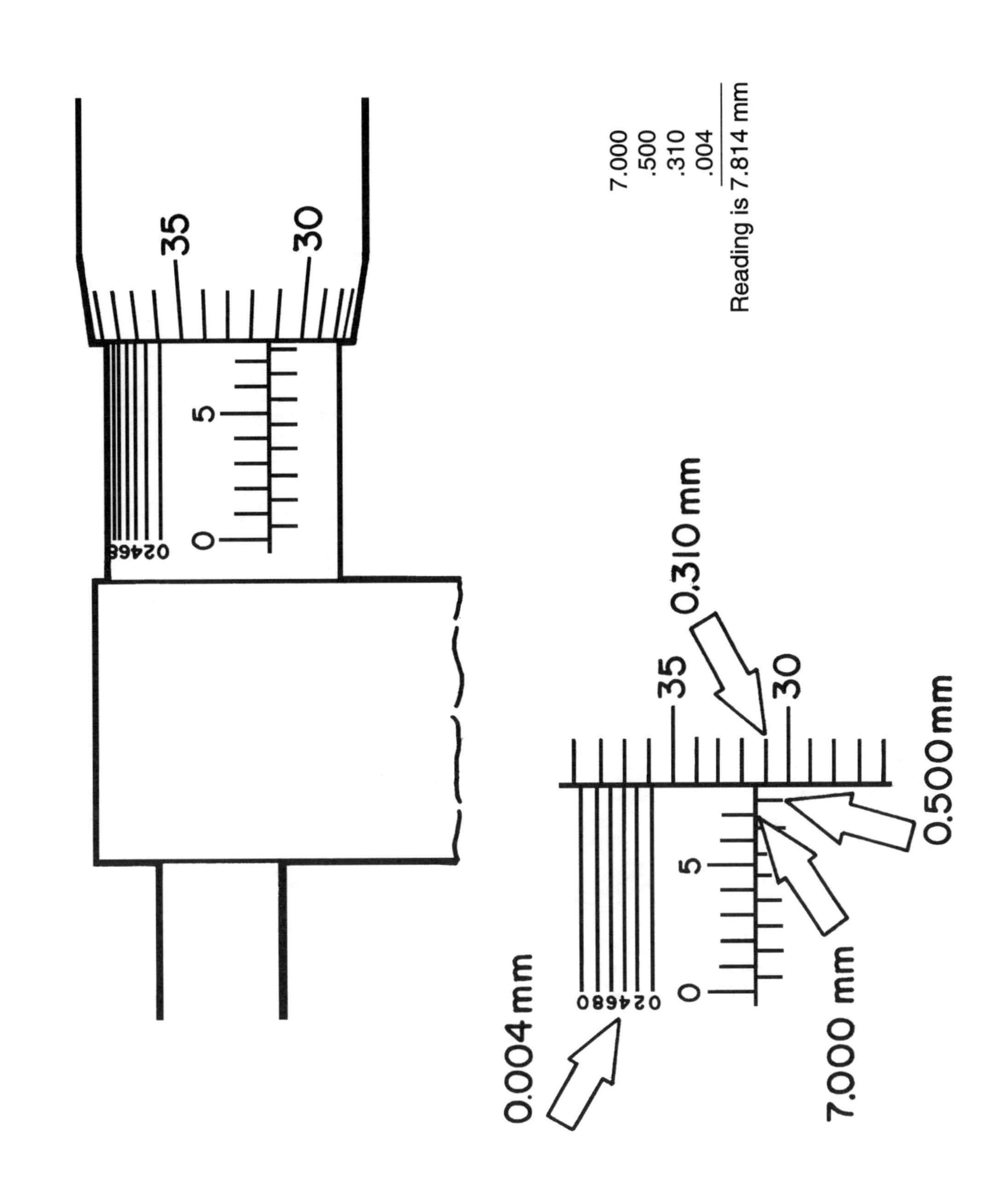

25-Division Inch-Based Vernier Caliper

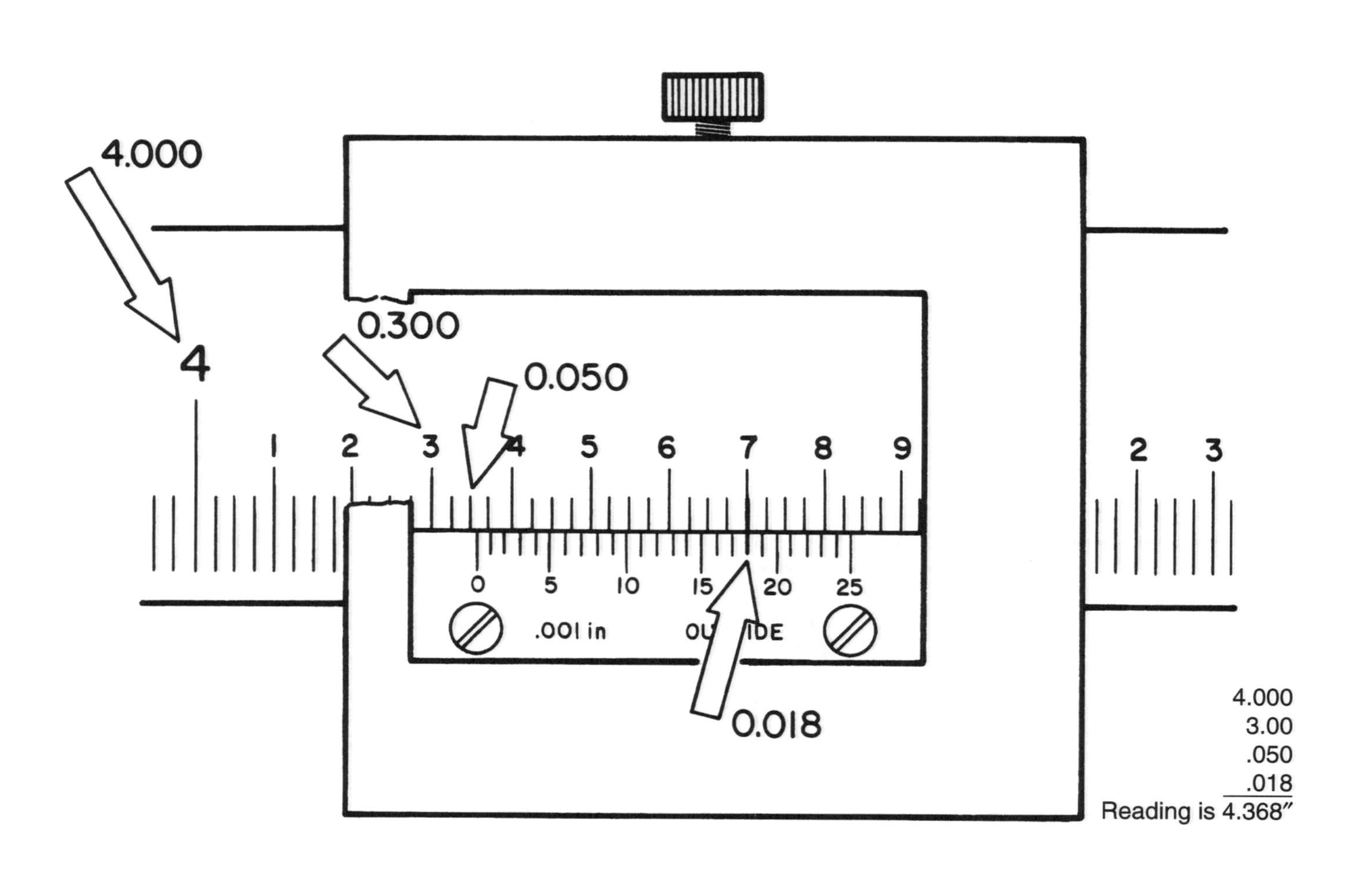

50-Division Inch-Based Vernier Caliper

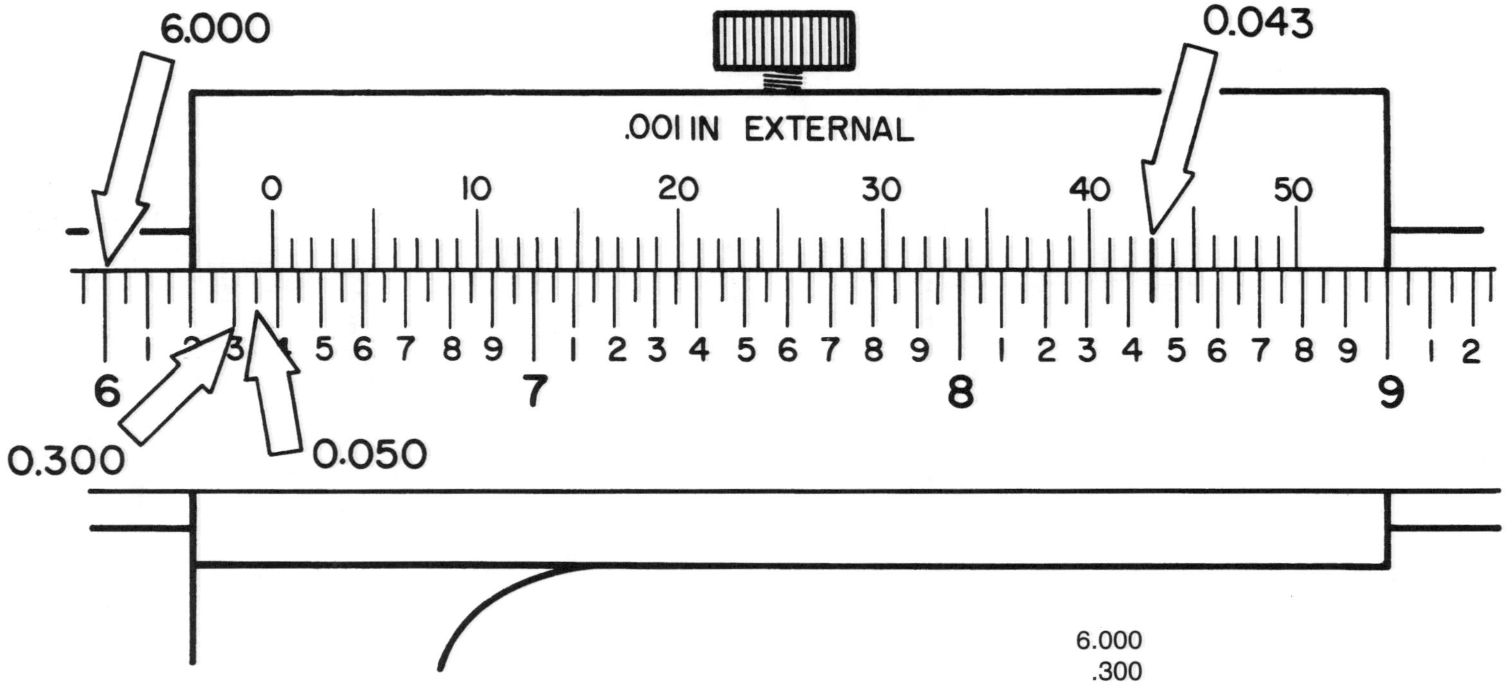

6.000

0.043

.001 IN EXTERNAL

0 10 20 30 40 50

6 7 8 9

0.300

0.050

6.000
.300
.050
.043
Reading is 6.393″

5-8

25-Division Metric Vernier Caliper

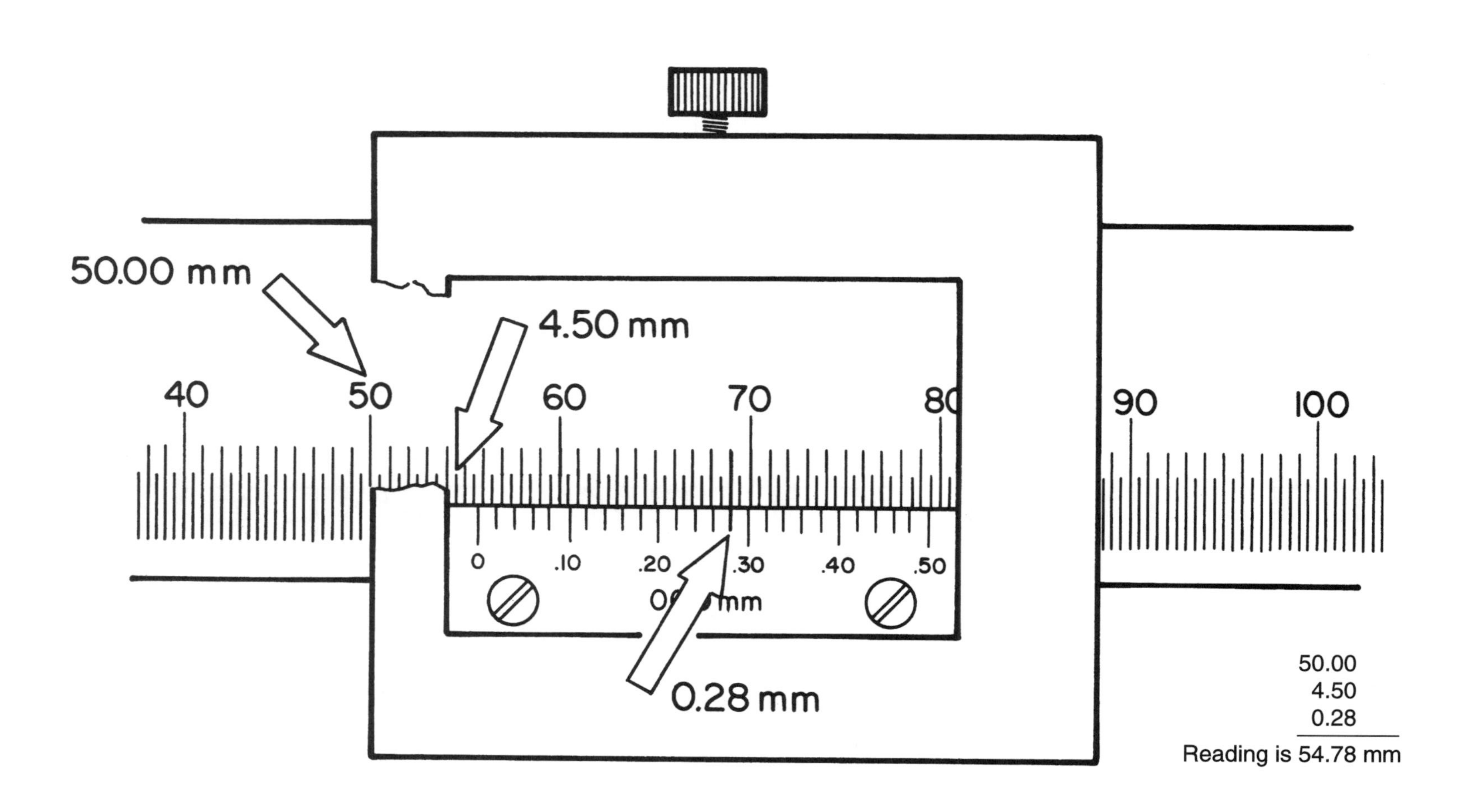

5-9

5-10

50-Division Metric Vernier Caliper

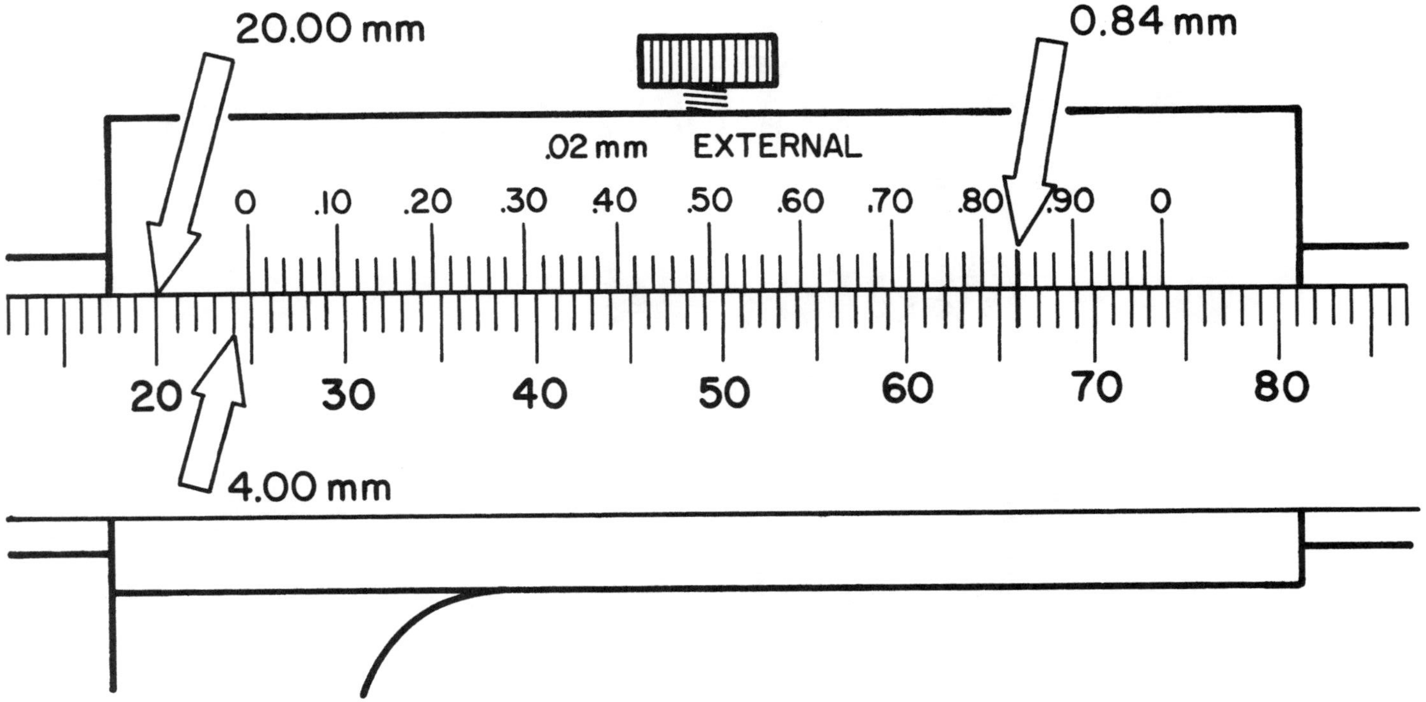

20.00 mm

0.84 mm

.02 mm EXTERNAL

0 .10 .20 .30 .40 .50 .60 .70 .80 .90 0

20 30 40 50 60 70 80

4.00 mm

20.00
4.00
0.84

Reading is 24.84 mm

Universal Bevel Protractor

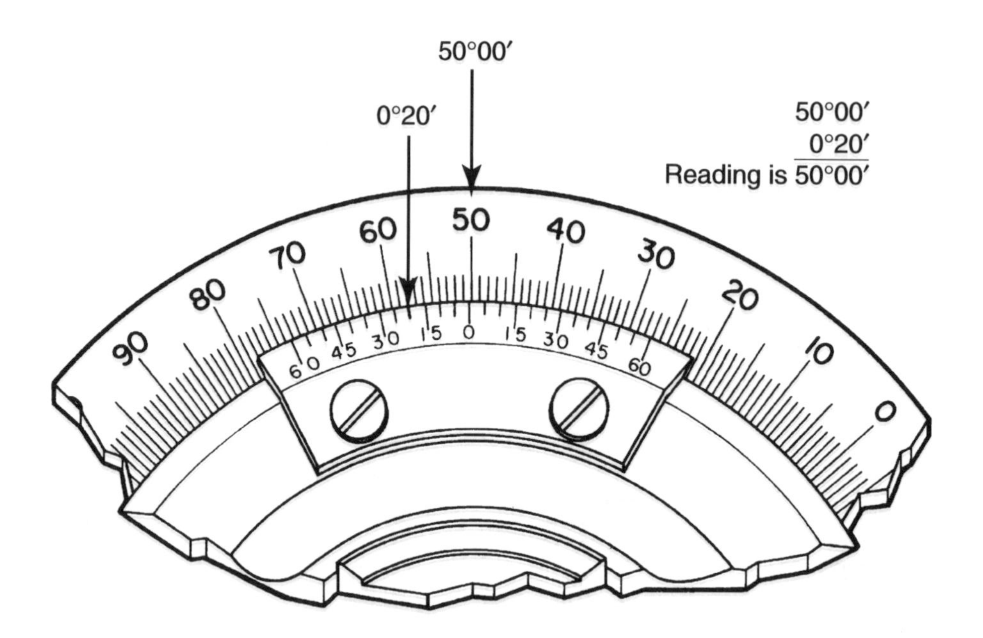

To read the protractor, note the number of degrees that can be read up to the "0" on the Vernier plate. To this, add the number of minutes indicated by the line beyond the "0" on the Vernier plate that aligns exactly with a line on the dial.

The "0" is past the 50° mark, and the Vernier scale aligns at the 20′ mark. Therefore, the measurement is 50°20′.

Chapter 5 Quiz—Part I
Measurement

Name: ______________________ **Date:** ______________ **Period:** ______________

Make the readings from the rules. Place the answers in the proper blanks.

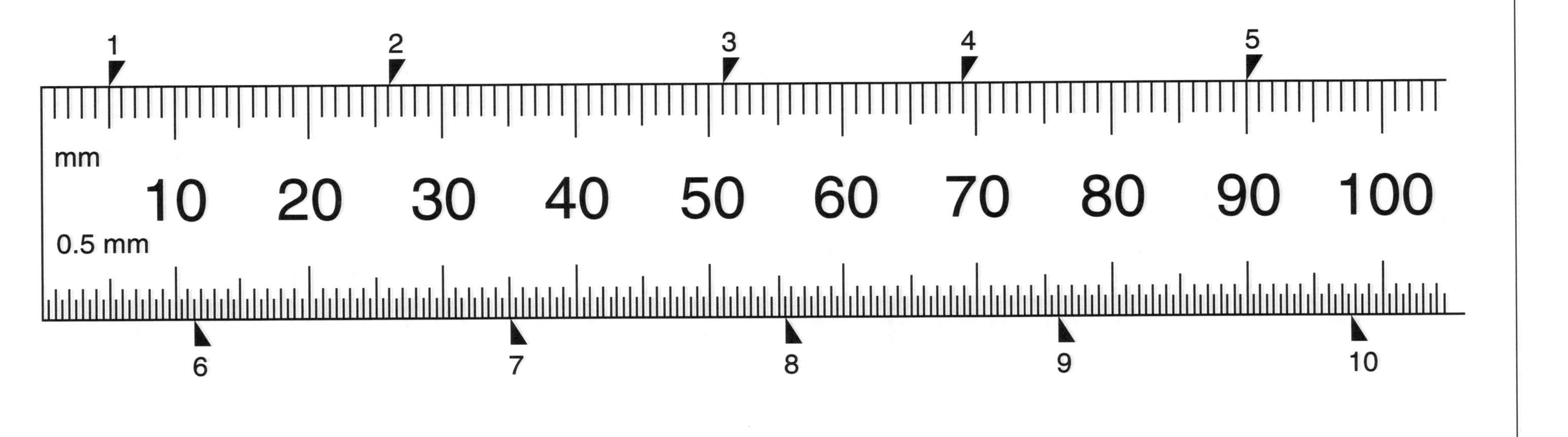

1. ______________________
2. ______________________
3. ______________________
4. ______________________
5. ______________________

6. ______________________
7. ______________________
8. ______________________
9. ______________________
10. ______________________

5-12
(Continued)

Chapter 5 Quiz—Part I *(Continued)*

Measurement

Name: ______________________ **Date:** ____________ **Period:** ____________

Make the readings from the rules. Place the answers in the proper blanks.

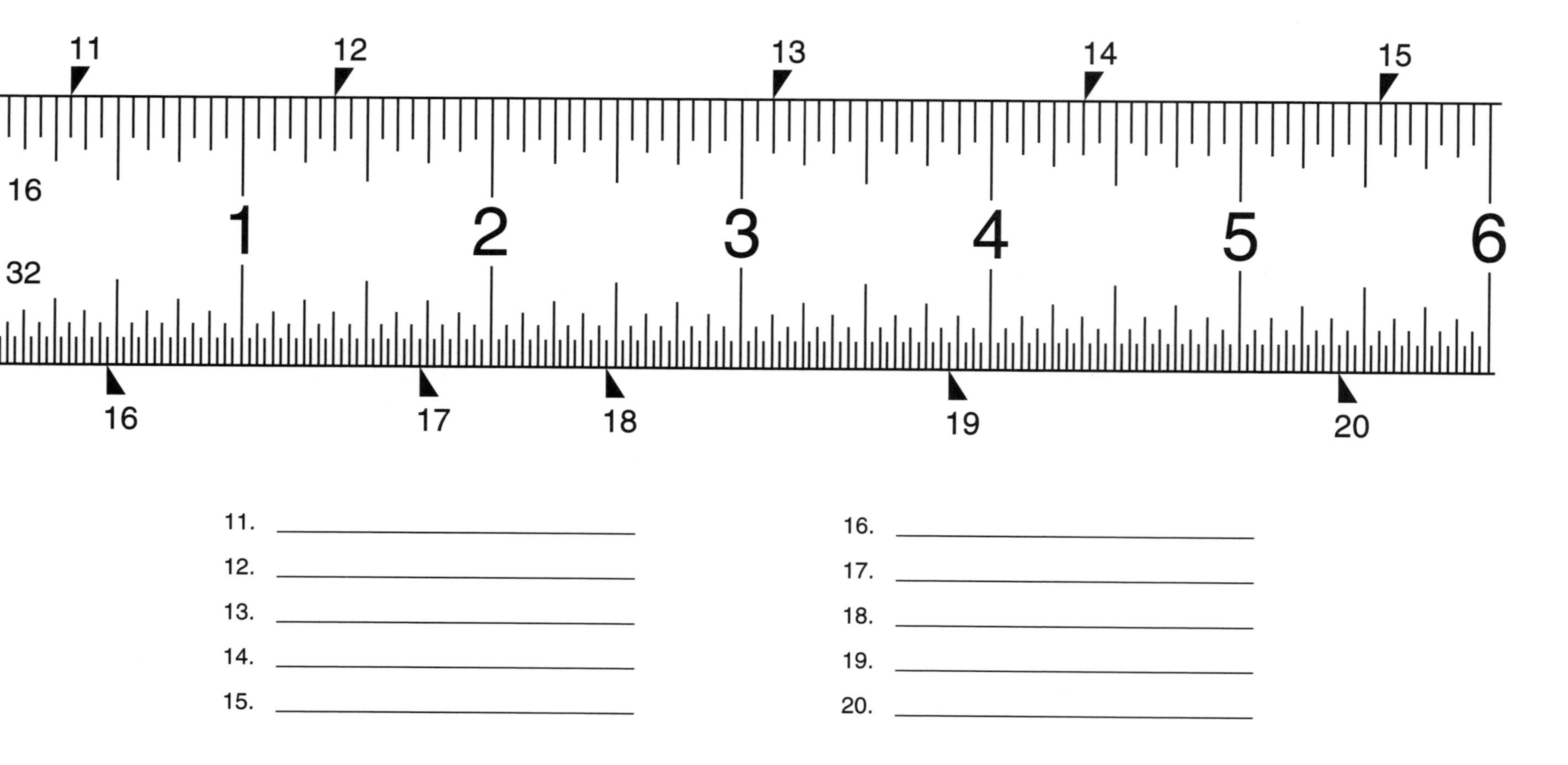

11. ______________________
12. ______________________
13. ______________________
14. ______________________
15. ______________________
16. ______________________
17. ______________________
18. ______________________
19. ______________________
20. ______________________

Chapter 5 Quiz—Part II

Measurement

Name: ______________________ **Date:** ____________ **Period:** ____________

Make readings from the micrometers shown below and place answers in the proper blanks.

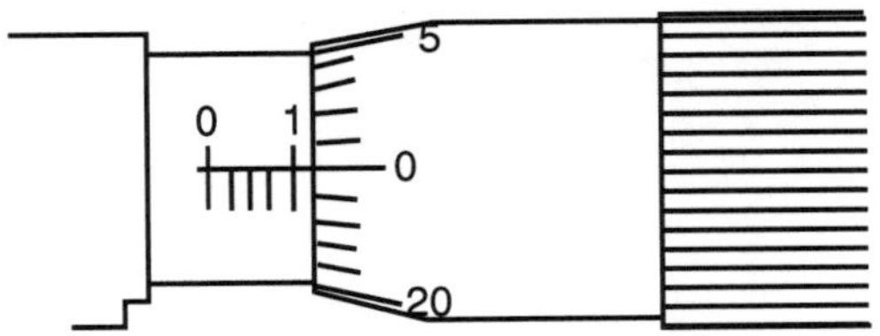

A

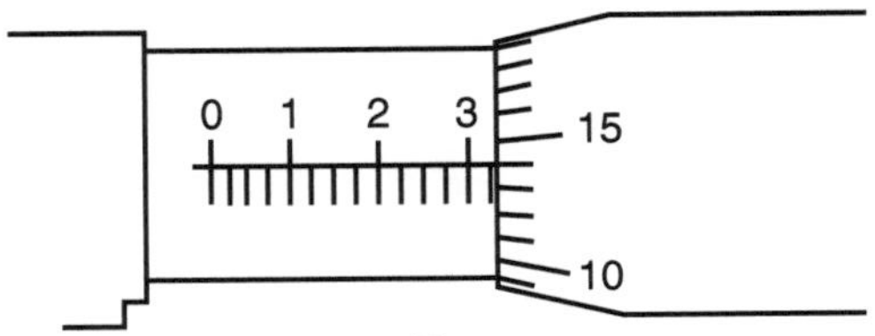

B

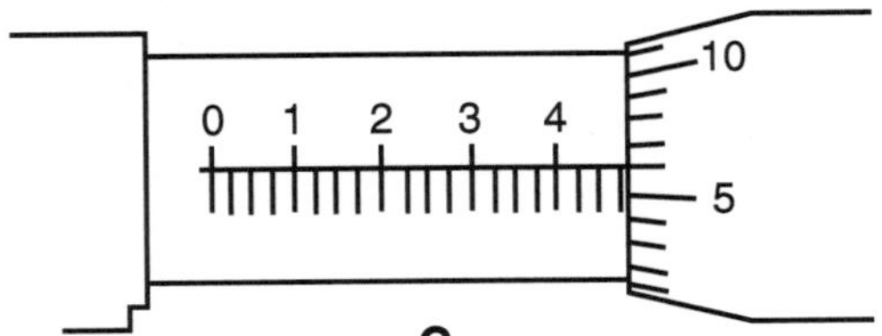

C

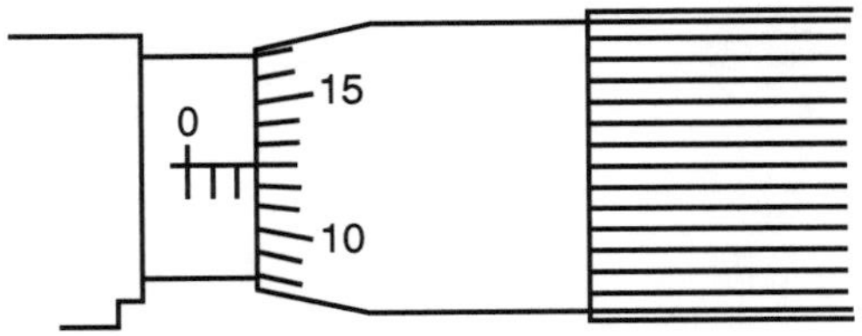

D

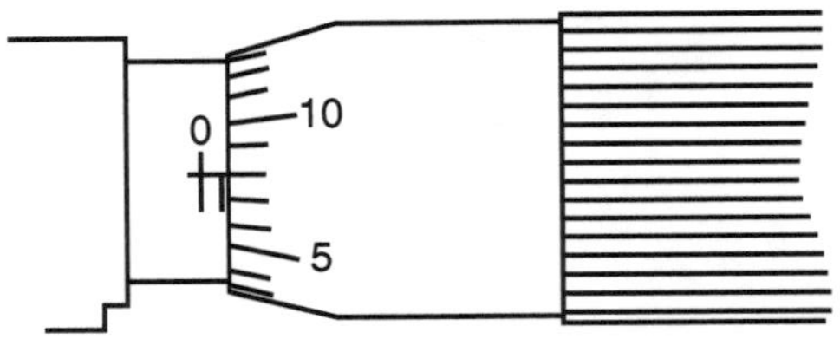

E

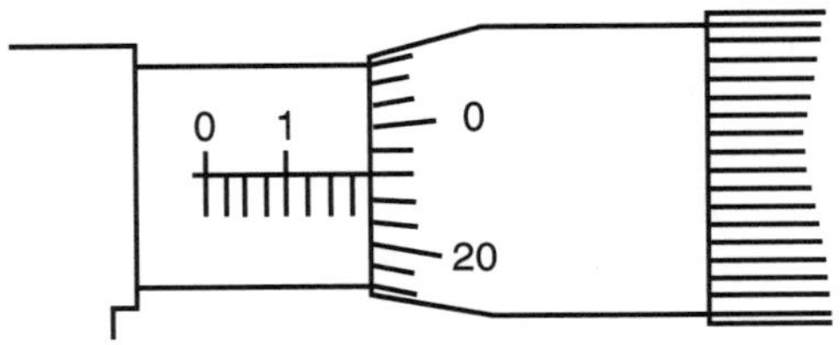

F

A. ______________________

B. ______________________

C. ______________________

D. ______________________

E. ______________________

F. ______________________

5-13
(Continued)

Chapter 5 Quiz—Part II *(Continued)*

Measurement

Name: ______________________ **Date:** ____________ **Period:** ____________

Make readings from the micrometers shown below and place answers in the proper blanks.

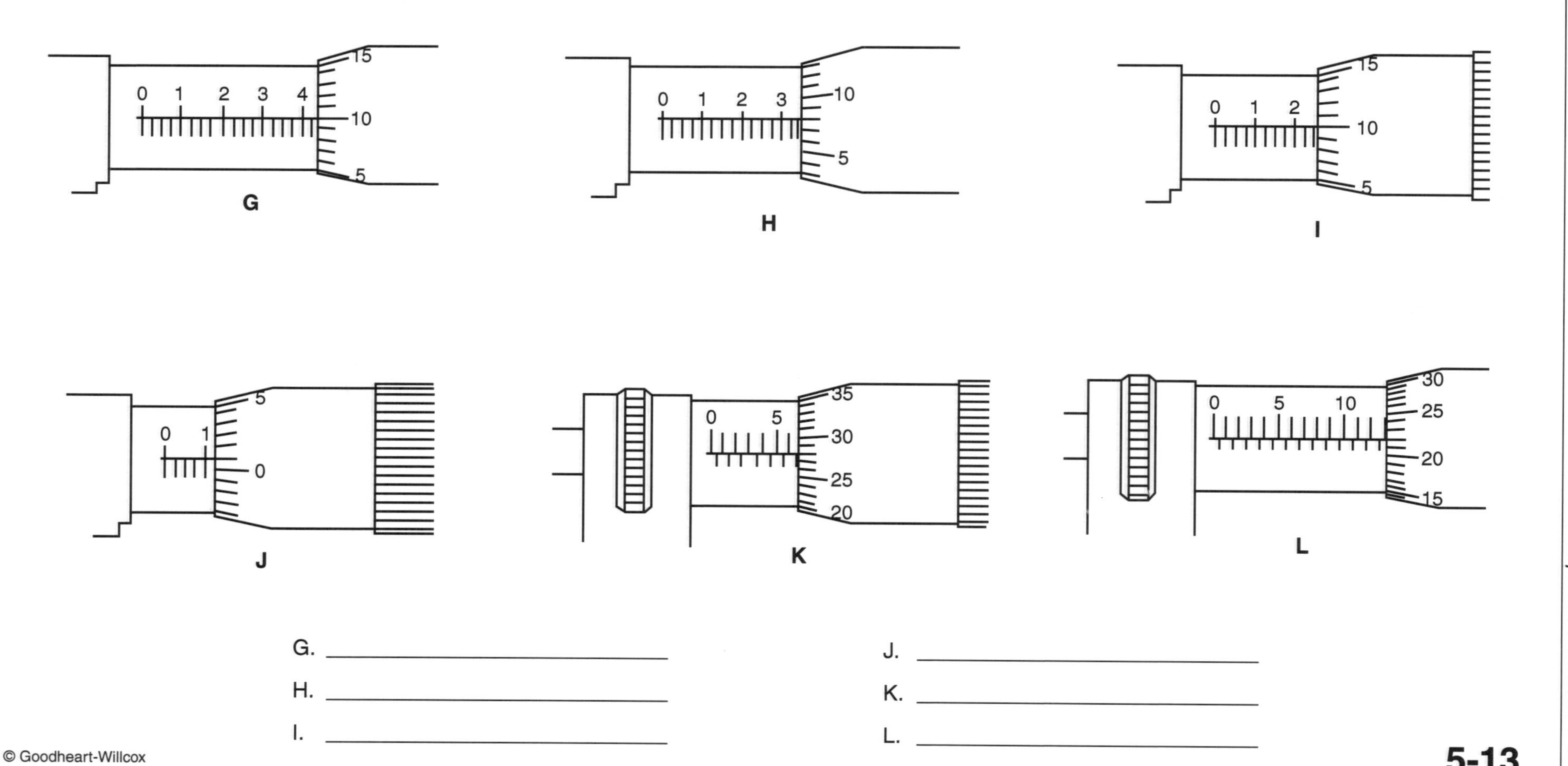

G. ______________________

H. ______________________

I. ______________________

J. ______________________

K. ______________________

L. ______________________

Chapter 5 Quiz—Part III

Measurement

Name: ____________________ **Date:** ____________ **Period:** ____________

Make readings from the Verniers shown below. Place your answers in the proper blanks.

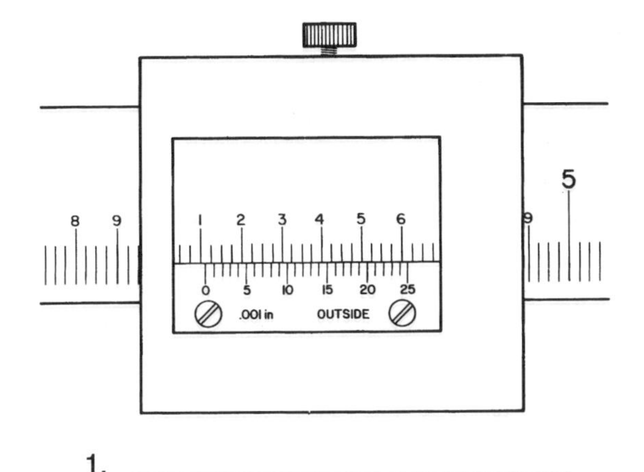

1. ____________________

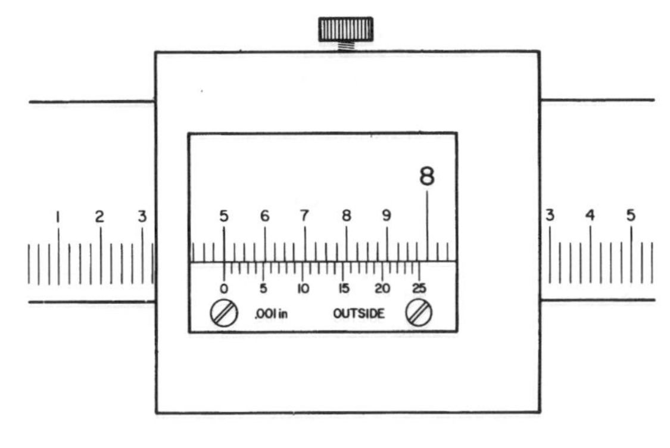

2. ____________________

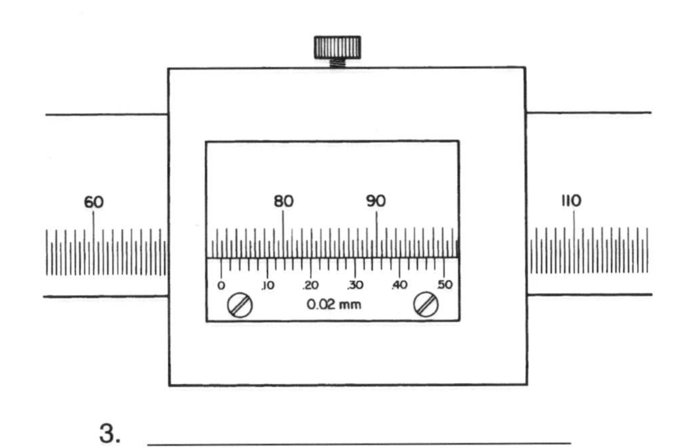

3. ____________________

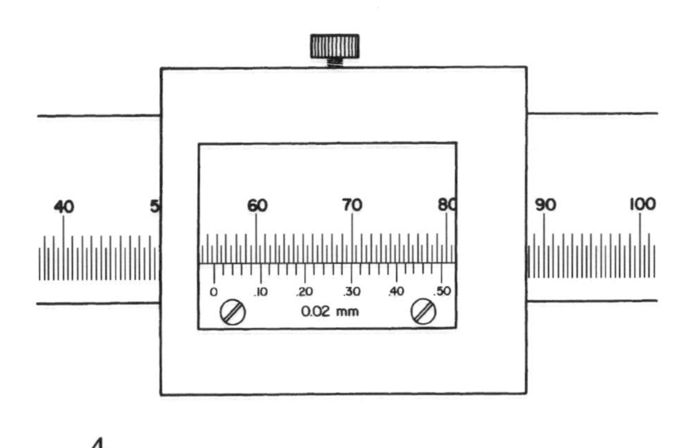

4. ____________________

.001 IN. EXTERNAL

0 10 20 30 40 50

6 7 8 9 1 2 3 4 5 6 7 8 9 1 2 3 4 5 6 7 8 9 1 2 3 4 5 6 7 8

5 6 7

5. ____________________

Chapter 5 Quiz—Part III *(Continued)*

Measurement

Name: ______________________ **Date:** ____________ **Period:** ____________

Make readings from the Verniers shown below. Place your answers in the proper blanks.

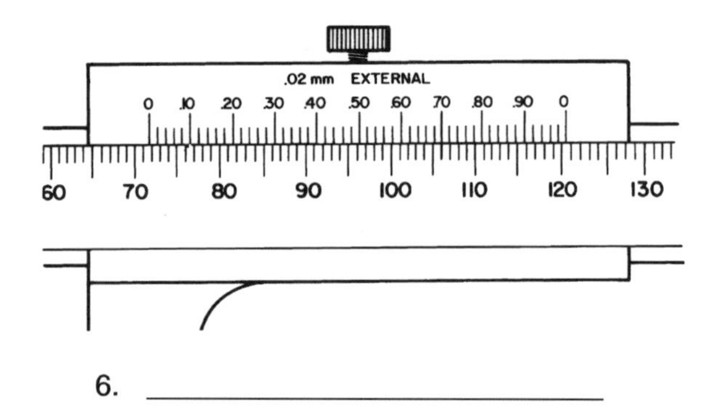

6. ______________________

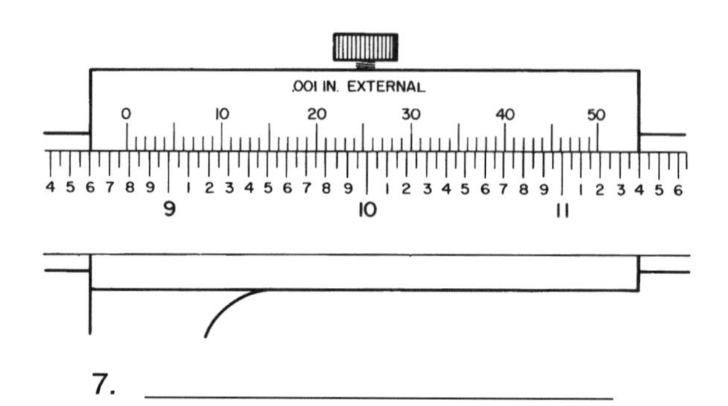

7. ______________________

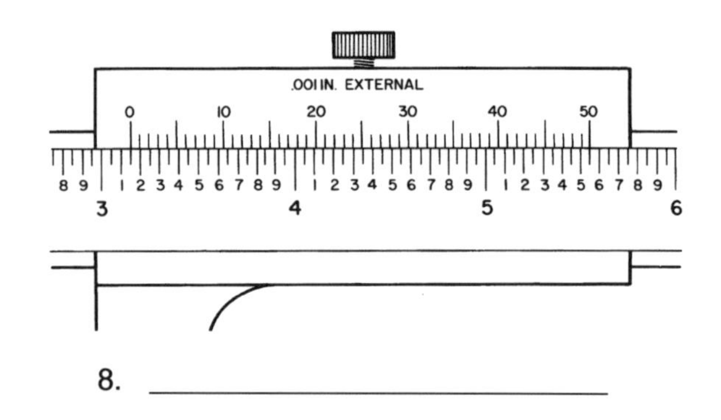

8. ______________________

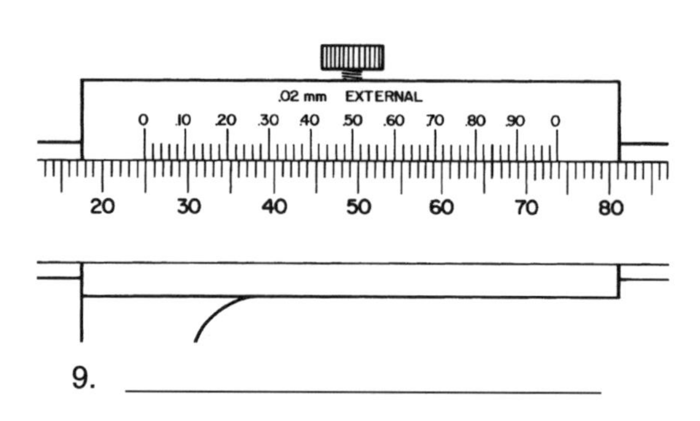

9. ______________________

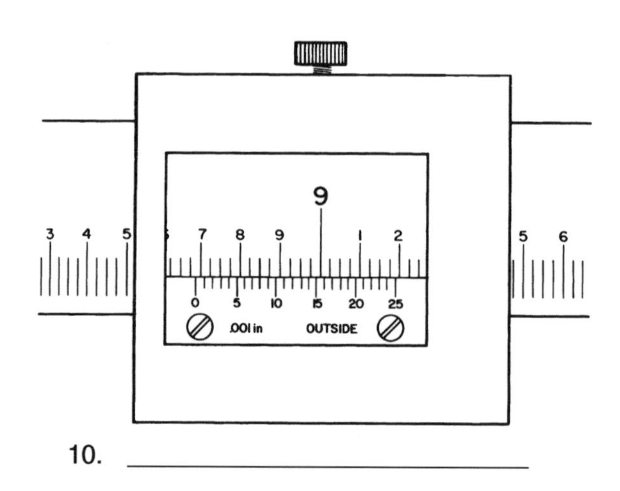

10. ______________________

Layout Work 6

Learning Objectives

After studying this chapter, students should be able to:

- Explain the purpose of a layout and how it is used to prepare metal for machining.
- Identify common layout tools and use them safely.
- Make simple layouts.
- List and observe common safety precautions used in layout work.

Chapter Resources

Text, pages 97–108
Test Your Knowledge, page 107
Research and Development, page 107
Workbook, pages 39–42
Instructor's Manual
Answer keys for:
Test Your Knowledge Questions
Workbook
Chapter Quiz
Reproducible Masters:
6-1 Typical Layout Problem
6-2 Steps in Making a Typical Layout
6-3 Chapter 6 Quiz

Guide for Lesson Planning

Prepare for the lesson by having the following materials available:

- Sections of clean metal to demonstrate layout techniques.
- Layout dye, scribers, hermaphrodite caliper, divider, surface gage, selection of squares, combination set, hammer, punches and a surface plate or flat section of metal.

Class Discussion

Have students read and study the chapter. Discuss and demonstrate the layout tools and equipment they will be using. This should include:

- How to use layout dye.
- Safe use of layout tools and equipment.
- The proper way to use a square and combination set.
- Sequence in laying out a typical job.
- Safety rules to be observed when doing layout work.

Assignments

1. Assign the Test Your Knowledge questions at the end of the chapter.
2. Assign Chapter 6 in the *Modern Metalworking Workbook.*
3. Assign the chapter quiz. Copy and distribute Reproducible Master 6-3.
4. Allow students to volunteer for the Research and Development activities at the end of the chapter.

Test Your Knowledge

1. Layout is a metalworking term that describes the locating and marking of lines, circles, arcs, and points for drilling holes in metal.
2. (b) tell the machinist where to machine
3. dye
4. A pointed tool commonly used to draw layout lines on metal.
5. (b) divider
6. Any order: base, spindle, scriber.
7. A hardened steel square has a fixed blade that is used to lay out extremely accurate lines. A double square uses interchangeable blades that are designed for other layout tasks.

8. (c) draw circles and arcs
9. center head
10. (b) Vernier protractor
11. A precise, flat surface used to inspect the flatness of layout work surfaces.
12. (a) V-blocks
13. steel, granite
14. Lines that provide a base location for scribing other lines on a layout.
15. prick, center

Workbook

1. Layout
2. layout dye
3. Scriber. It is used to mark layout lines on metal.
4. dividers
5. trammel
6. Hermaphrodite caliper. It is used to make lines and locate points.
7. (a) Surface gage
   (b) Double square
   (c) Combination set
8. To check the accuracy of 90° (square) angles, lay out lines at right angles to a given edge, or parallel to another surface. Some simple machine setups can be made quickly and easily with the square.
9. Any three of the following: used as a standard square, a miter square, a depth gage, or level; to locate the center of round stock; to measure or lay out angles.
10. Protractor depth gage. It can be used to check 30°, 45°, and 60° angles and measure slot depths.
11. surface plate
12. (c) V-blocks
13. straightedge
14. Evaluate student sketches individually.
15. Evaluate student responses individually. Refer to Section 6.6.

Chapter Quiz

1. the locating and marking out lines, circles, arcs, and points for drilling holes
2. (c) Both a and b.
3. In order: scriber, divider, trammel.
4. surface plate
5. V-blocks
6. square
7. The answers may be in any order.
   (a) Rule. For making measurements.
   (b) Square head. Used with rule to serve as T–square and miter square.
   (c) Center head. Fitted to rule it is used to quickly locate the center of round stock.
   (d) Bevel protractor. Used with rule to measure and layout angles.
8. straight edge
9. They provide a base location for scribing other lines on a layout.
10. Doing so can cause severe puncture wounds.

Typical Layout Problem

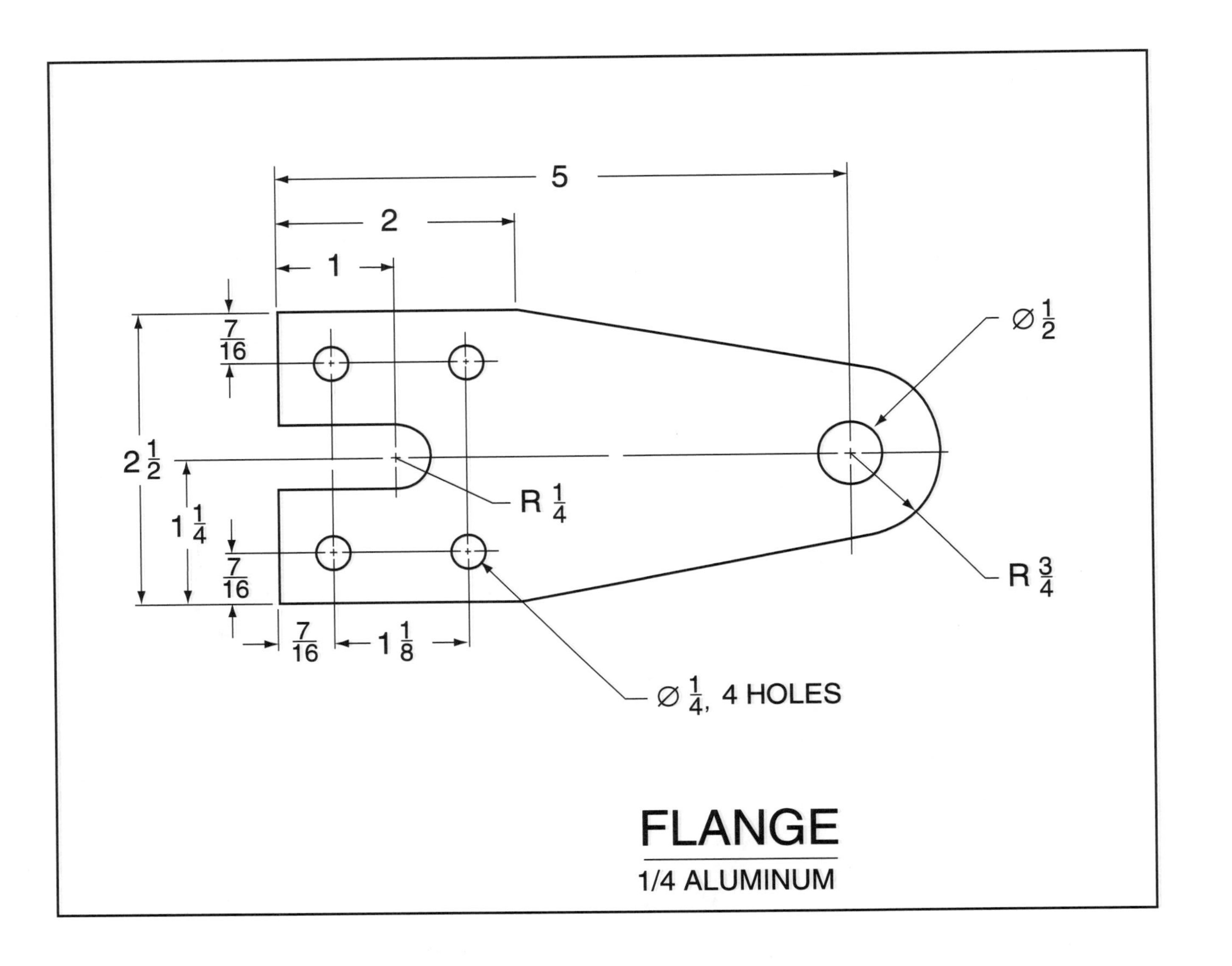

Steps in Making a Typical Layout

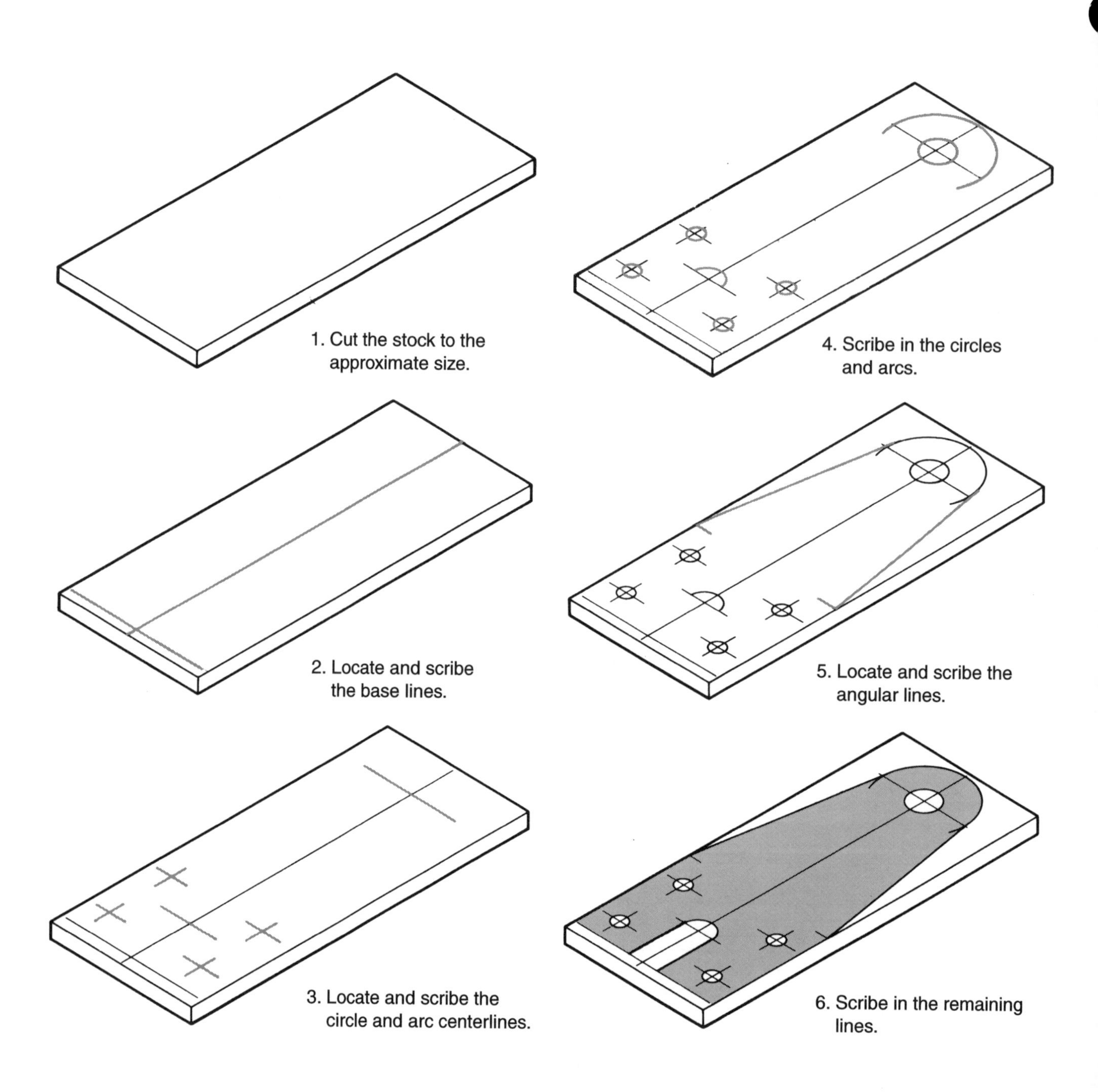

Chapter 6 Quiz

Layout Work

Name: ______________________ **Date:**______________ **Period:**______________

1. *Layout* is the metalworking term used to describe ____
____
____.

2. Layout dye is used to _____. 2.______________
   (a) provide contrast between the metal and layout lines
   (b) make it easier to see the layout lines
   (c) Both a and b.
   (d) Neither a nor b.

3. The _____ is used to draw straight lines. Circles and arcs are made with the _____. Circles and arcs too large to be made with the above tool are drawn with a(n) _____. 3.______________
____
____

4. The precise flat surface needed for accurate layout work is called a(n) _____. 4.______________

5. _____ are often used to support round work for layout and inspection. 5.______________

6. The accuracy of a 90° angle is checked with a(n) _____. 6.______________

7. Name the parts of a combination set. How is each part used in layout work?
   (a) ____
   ____
   (b) ____
   ____
   (c) ____
   ____
   (d) ____
   ____

8. Long flat surfaces can be checked with a(n) _____. 8.______________

9. What purpose do reference lines serve? ____
____

10. Why is it poor practice to carry an open scriber or divider in your pocket?
____
____
____

7 Hand Tools

Learning Objectives

After studying this chapter, students should be able to:

- Identify basic hand tools used in metalworking.
- Select the right tool(s) for a given job.
- Demonstrate the safe and correct use of hand tools.
- Explain how to maintain tools properly.

Chapter Resources

Text, pages 109–122
  Test Your Knowledge, pages 121–122
  Research and Development, page 122
Workbook, pages 43–48
Instructor's Manual
  Answer keys for:
    Test Your Knowledge Questions
    Workbook
    Chapter Quiz
  Reproducible Masters:
    7-1 Chapter Quiz

Guide for Lesson Planning

Due to the amount of material covered, it would be advisable to divide this chapter into two segments. Assign one or two students, or have them volunteer, to present each tool group—vises, clamps, pliers, etc. Have them demonstrate and explain the proper use and care for the tools on which they are providing instruction.

Class Discussion

Have students read and study tool assignment before each demonstration. Make the following preparations before each presentation:

- Necessary tools in safe working condition.
- Additional tools available for demonstration.
- Demonstration area clearly visible to all students.
- Materials available for demonstration.
- All safety precautions taken.

A review of the demonstration will provide an opportunity to answer any questions students may have. Hand tool safety precautions can be emphasized at this time.

Assignments

1. Assign the Test Your Knowledge questions.
2. Assign Chapter 7 in the *Modern Metalworking Workbook.*
3. Assign the chapter quiz. Copy and distribute Reproducible Master 7-1.
4. Permit students to volunteer for the Research and Development activities at the end of the chapter.

Test Your Knowledge

1. (b) permit clamping long work in a vertical position
2. By the width of the jaws.
3. parallel
4. slip-joint
5. (b) flush to the surface
6. (c) Side-cutting
7. Tongue and groove
8. Needle-nose
9. adjustable clamping
10. wrench
11. threaded fasteners
12. foot-pounds, newton-meters
13. To tighten a threaded fastener or part and give it maximum holding power while limiting excessive stress caused by over-tightening.
14. adjustable
15. pull
16. round

17. (b) completely surrounds the fastener and does not normally slip
18. Socket wrenches consist of a detachable tool or handle that fits many sizes of sockets.
19. Special wrenches with drive heads or lugs designed to turn flush-type and recessed-type threaded fittings.
20. Allen
21. Phillips
22. (e) Standard
23. (a) Stubby
24. (f) Heavy-duty
25. (c) Electrician's
26. (g) Pozidriv®
27. (d) Ratchet
28. machinist's ball-peen
29. A soft-face hammer or mallet.
30. The faces are very hard and the blow might cause a chip to break off and fly out at high speed.

Workbook

1. Swivel base vise
2. (d) Needle-nose pliers
3. (a) Vise
4. (b) Combination pliers
5. (e) Diagonal pliers
6. (c) Side-cutting pliers
7. (a) Diagonal pliers
   (b) Side-cutting pliers
   (c) Straight-nose pliers
   (d) Round-nose pliers
8. (a) Adjustable wrench
   (b) Open-end wrench
   (c) Socket wrench
9. (k) Ball-peen hammer
10. (e) Open-end wrench
11. (h) Socket wrench
12. (a) Wrenches
13. (b) Torque-limiting wrench
14. (c) Adjustable wrench
15. (d) Pipe wrench
16. (f) Box wrench
17. (l) Dead blow hammer
18. (j) Spanner wrench
19. (g) Combination wrench
20. (i) Allen wrench
21. Evaluate individually. Refer to Section 7.3.9.
22. (a) Standard
    (b) Phillips
    (c) Clutch
    (d) Square
    (e) Trox
    (f) Hex
23. Evaluate individually. Refer to Section 7.4.2.
24. Evaluate individually. Refer to Section 7.5.3.
25. Evaluate student observances individually. Use the observances to discuss and correct all unsafe practices.

Chapter Quiz

1. (g) Vise
2. (r) Clamps
3. (a) Diagonal pliers
4. (d) Side-cutting pliers
5. (n) Round-nose pliers
6. (h) Needle-nose pliers
7. (u) Tongue-and-groove pliers
8. (i) Torque-limiting wrenches
9. (b) Adjustable wrench
10. (j) Box wrench
11. (v) Socket wrench
12. (f) Allen wrench
13. (l) Spanner wrench
14. (c) Pipe wrench
15. (t) Phillips screwdriver
16. (t) Ball-peen hammer
17. (m) Dead blow hammer
18. (k) Stubby screwdriver
19. (e) Heavy-duty screwdriver
20. (q) Pozidriv® screwdriver

Chapter 7 Quiz
Hand Tools

Name: ____________________ **Date:**____________ **Period:**____________

Match each phrase below with the correct term.

_________ 1. Clamping device employed to hold and position material while it is worked.

_________ 2. Used to hold parts together while being worked upon.

_________ 3. Designed to cut flush with work surface.

_________ 4. Useful for cutting heavier wire and pins.

_________ 5. Helpful when forming wire and light metal.

_________ 6. Handy when space is limited and for holding small work.

_________ 7. Type of pliers whose jaw opening is easily adjusted to the desired size.

_________ 8. Permits tightening a threaded fastener for maximum holding power without danger of fastener failing.

_________ 9. Can be adjusted to fit different size bolt heads and nuts.

_________10. Wrench opening completely surrounds bolt head or nut.

_________11. Box type and made to fit many types of handles.

_________12. Used with socket headed fasteners.

_________13. Designed to turn flush and recessed type threaded fasteners.

_________14. Wrench used to grip round stock.

_________15. Has “X” shaped tip.

_________16. Has hardened striking face.

_________17. Will not rebound like other hammers.

_________18. Is short and used where space is limited.

_________19. Has a square shank to permit additional force to be applied with a wrench.

_________20. Screwdriver tip similar in appearance to a Phillips type tip.

(a) Diagonal pliers
(b) Adjustable wrench
(c) Pipe wrench
(d) Side-cutting pliers
(e) Heavy-duty screwdriver
(f) Allen wrench
(g) Vise
(h) Needle-nose pliers
(i) Torque-limiting wrench
(j) Box wrench
(k) Stubby screwdriver
(l) Spanner wrench
(m) Dead blow hammer
(n) Round-nose pliers
(o) Raw hide mallet
(p) Ball-peen hammer
(q) Pozidriv® screwdriver
(r) Clamps
(s) Wrenches
(t) Phillips screwdriver
(u) Tongue-and-groove pliers
(v) Socket wrench

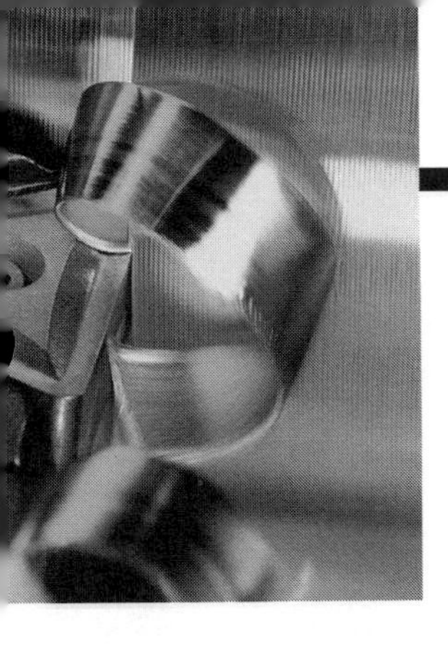

8 HAND TOOLS THAT CUT

LEARNING OBJECTIVES

After studying this chapter, students should be able to:

- Identify the basic hand tools that are commonly used to cut metal.
- Demonstrate the safe and proper use of hand tools that cut.
- Select the proper hand tool(s) for each job.
- Maintain and store tools properly.
- Use common safety practices for using metal-cutting tools.

CHAPTER RESOURCES

Text, pages 123–140
 Test Your Knowledge, pages 137–138
 Research and Development, pages 138–139
Workbook, pages 49–54
Instructor's Manual
 Answer keys for:
  Test Your Knowledge Questions
  Workbook
  Chapter Quiz
 Reproducible Master:
  8-1 Chapter 8 Quiz

GUIDE FOR LESSON PLANNING

Divide this chapter into the following four segments: 8.1 Chisels, 8.2 Hacksaws, 8.3 Files, and 8.4 Reaming Tools. Assign one or two students, or have them volunteer, to present each segment. They should demonstrate and explain the proper use and care for the tools on which they are providing instruction.

Make the following preparations before each presentation:

- Have necessary tools sharpened and in safe working condition.
- Additional tools available for student examination.
- Demonstration area clearly visible to all students.
- Material available for demonstration.
- All safety precautions taken.

CLASS DISCUSSION AND DEMONSTRATIONS

Each segment of the chapter should be read and studied before tools are demonstrated. When discussing the cutting tools, raise the following questions:

- Why is the cutting edge of a chisel for cutting on a flat plate slightly rounded and the cutting edge of a chisel used for shearing metal flat?
- Why do hacksaws have the blade positioned to cut on the forward stroke?
- Refer to Figure 8-3. Why is the preferred position of work, when cutting, better?
- Why should a new hacksaw blade *not* be used in a cut already started?
- Why are different cut files needed?
- Why a file should never be used without a handle?
- Why should the hole being reamed be slightly smaller than the reamer diameter?
- Why is a taper roughing reamer necessary? For what reason are the cutting edges on a taper roughing reamer notched?
- How to determine which file to use for a specific job.

A review of each demonstration will provide an opportunity to answer any questions the students may have. Hand cutting safety should be reemphasized at this time.

ASSIGNMENTS

1. Assign the Test Your Knowledge questions.
2. Assign Chapter 8 in the *Modern Metalworking Workbook.*
3. Assign the chapter quiz. Copy and distribute Reproducible Master 8-1.

4. Permit students to volunteer for the Research and Development activities at the end of the chapter.

Test Your Knowledge

1. Any order: flat, cape, round-nose, diamond-point.
2. straight
3. Rivet buster
4. A chisel used in this condition can cause cuts and other injuries because metal chips can break and fly off.
5. Answer may include any three of the following: wear safety goggles and erect a shield around the work; hold the chisel properly against the work; do not use a chisel with a mushroomed head; remove sharp edges from the work after making a cut.
6. forward
7. (b) The work will vibrate or "chatter" and cause the teeth to snap off.
8. set
9. (c) stick and bind, ruining the new blade
10. all-hard, hard
11. high-grade carbon steel
12. Any order: single-cut, double-cut, rasp, curved-tooth.
13. (a) general outline and cross section
14. Any order: rough, coarse, bastard, second-cut, smooth, dead smooth.
15. Evaluate responses individually.
16. file card, brush
17. (d) Rasps
18. safe edge
19. Single-cut files are normally used to produce a smooth surface finish. Their use requires light pressure. Double-cut files can remove metal much faster. They require heavier pressure, and they produce a rougher surface finish.
20. (c) machinist's
21. elbow
22. (b) Straight
23. handle
24. So it can be held in a tap wrench.
25. 0.005″ to 0.010″
26. When a hole must be cut to a few thousandths of an inch over nominal size for fitting purposes.
27. roughing
28. clockwise
29. Because it permits an even application of pressure.
30. Answer may include any four of the following: remove all burrs from reamed holes; never use your hands to remove chips and cutting fluid from the reamer; store reamers properly so that they do not make contact with one another; clamp the work solidly before starting to ream; never use compressed air to clean a reamed hole.

Workbook

1. Any order: flat, cape, round nose, diamond point. Evaluate sketches individually. Refer to Section 8.1.
2. slightly curved, straight
3. rivet buster
4. Evaluate individually. Refer to Section 8.1.1.
5. forward
6. chatter (vibrate)
7. The new blade will bind and stick and be ruined in the first few strokes. Start a new cut from the other side.
8. kerf
9. Sandwich it between two sections of wood.
10. Evaluate individually. Refer to Section 8.2.8.
11. (a) Single-cut
    (b) Double-cut
    (c) Rasp
    (d) Curved-tooth
12. pinning
13. file card, brush
14. Any order: length, kind, cut.
15. Evaluate student drawings individually. Refer to text Figure 8-26.
16. straight (cross)
17. draw
18. Evaluate individually. Refer to Section 8.3.7.
19. Evaluate individually. Refer to Section 8.4.1.
20. When a hole must be a few thousandths of an inch over nominal size for fitting purposes.
21. counterclockwise
22. Evaluate individually. Refer to Section 8.4.3.
23. Evaluate student work individually.
24. Evaluate student work individually.
25. Evaluate student work individually.

Chapter Quiz

1. (a) Diamond point chisel
   (b) Round nose chisel
   (c) Flat chisel
   (d) Cape chisel
2. slightly rounded
3. flat
4. forward
5. close
6. chatter, vibration
7. Continuing the cut in the same slot will cause the new blade to bind and stick and be ruined in the first few strokes.
8. Mount metal between two sections of wood.
9. (d) All of the above.
10. Single-cut
11. Double-cut
12. Rasp
13. Curved tooth

Notes

Chapter 8 Quiz

Hand Tools That Cut

Name: ______________________ **Date:**______________ **Period:**______________

1. Identify the cold chisels shown.

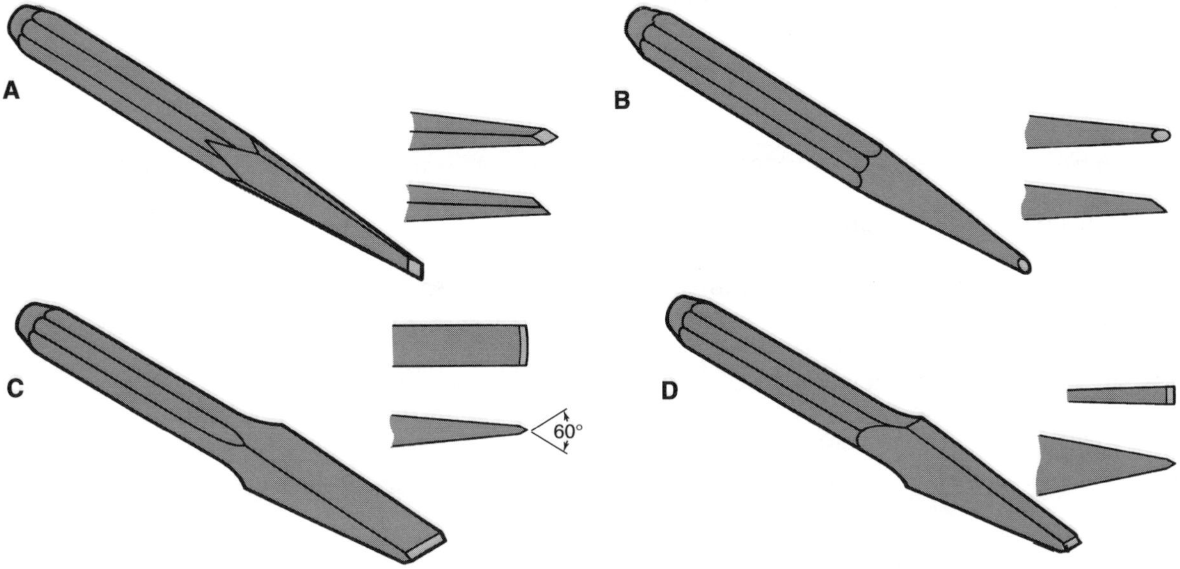

A B C D 60°

(a) ______________________

(b) ______________________

(c) ______________________

(d) ______________________

2. A chisel with a(n) _____ cutting edge will work best when cutting on a flat metal plate. 2.______________

3. If the chisel is to be used to shear metal held in a vise, it will work best if the cutting edge is _____. 3.______________

4. The blade should be fitted in a hacksaw so it will cut on the _____ stroke. 4.______________

5. When using a hacksaw, the work must be held securely, with the point to be cut as _____ to the vise as possible. 5.______________

6. If the above is done, _____ and _____ that dull the blade can be reduced or eliminated. 6.______________

____

7. When replacing a broken or dull blade, why is it necessary to begin the cut in a new place?

____

____

8. What is the recommended way to cut thin metal with a hacksaw? ______________________

____

Chapter 8 Quiz *(Continued)*

Name: ______________________ **Date:**__________ **Period:**__________

9. A reamer is used _____. 9.__________
   (a) because a drill does not produce a smooth or accurate enough hole for a precision fit
   (b) to produce a hole that has the required smoothness and accuracy
   (c) for final hole sizing
   (d) All of the above.
   (e) None of the above.

The cut of a file indicates the relative coarseness of the teeth. Identify the file cuts in the following descriptions.

10. Usually used to produce a smooth surface finish. 10.__________

11. Remove metal much faster than those in #10, but produce a less smooth finish. 11.__________

12. Used to work wood and soft materials. 12.__________

13. Used to file flat surfaces of aluminum and steel sheet. 13.__________

9 Hand Threading

Learning Objectives

After studying this chapter, students should be able to:

- Describe how threads are specified on drawings.
- Explain thread nomenclature.
- Select the proper tap(s) and tap wrench for each job.
- Determine the correct tap drill size for specified threads.
- Adjust a die for different classes of fits.
- Correct problems that may occur when hand threading.
- Use, clean, and store threading tools properly.
- Observe hand threading safety rules.

Chapter Resources

Text, pages 141–150
Test Your Knowledge, page 150
Workbook, pages 55–58
Instructor's Manual
Answer keys for:
Test Your Knowledge Questions
Workbook
Chapter Quiz
Reproducible Masters:
9-1 Methods Used to Depict Threads on Drawings
9-2 How Threads Are Specified on Drawings
9-3 Thread Nomenclature
9-4 Starting a Die
9-5 Cutting an External Thread to a Shoulder
9-6 Chapter Quiz

Guide for Lesson Planning

The items listed below are necessary to demonstrate hand threading:

- Examples of UNC and UNF threaded sections (bolts, nuts, threaded rods, etc.) of the same diameter.
- Different tap sets.
- Different sizes of tap wrenches.
- Material drilled for tapping.
- A selection of dies.
- Die holders.
- Stock for threading with a die.
- Cutting fluid.

Class Discussion and Demonstrations

The demonstration area should be clearly visible to all students. Have students read and study all or part of the chapter. Using Reproducible Masters 9-1 through 9-5 as overhead transparencies or handouts, discuss and demonstrate:

- How threads are depicted on drawings.
- How UNC and UNF threads of the same diameter differ.
- Thread nomenclature.
- Why there are taper, plug, and bottom taps.
- Tap drills and how to select them.
- Proper way to tap a hole.
- Why cutting fluids are needed.
- Removing broken taps.
- The proper way to cut external threads.
- Advantage of using an adjustable die instead of a solid die.
- How to correct problems encountered when cutting threads.
- Precautions to be taken when using a die. *Emphasize hand threading safety!*
- Hand threading safety.
- How to clean and store hand threading tools after use.

Assignments

1. Assign the Test Your Knowledge questions.

2. Assign Chapter 9 in the *Modern Metalworking Workbook.*
3. Assign the chapter quiz. Copy and distribute Reproducible Master 9-6.
4. Permit students to volunteer for the Research and Development activities at the end of the chapter.

Test Your Knowledge

1. tap
2. die
3. (b) has more threads per inch for a given diameter
4. Evaluate responses individually.
5. Evaluate responses individually.
6. In order: taper, plug, bottoming.
7. (a) Thread diameter in inches
   (b) Number of threads per inch
   (c) Unified National Coarse Thread Series
   (d) Class of fit (free fit)
8. (a) Metric thread 5.0 mm in diameter
   (b) Pitch of thread (0.8 mm)
   (c) Tolerance class designation
9. tap drill
10. T-handle, it allows a more sensitive feel when tapping.
11. Evaluate responses individually.

Workbook

1. The UNF has more threads per inch of length for a given size thread.
2. inside (internal)
3. Taper, plug, and bottom.
4. (a) taper
5. A special drill used to make a hole before tapping.
6. (a) #7 (0.210" DIA)
   (b) 29/64" DIA
   (c) 27/64" DIA
   (d) "I" (0.272" DIA)
   (e) "F" (0.275" DIA)
   (f) 5/16" DIA
7. T-handle tap wrench, Hand tap wrench
8. Evaluate individually. Refer to Section 9.4.6.
9. outside (external)
10. die stock
11. Evaluate individually. Refer to Section 9.5.3.
12. (d) None of the above. The shaft must be the same size as thread diameter.
13. Evaluate individually. Refer to Section 9.6.
14. Evaluate student work individually.
15. Evaluate student work individually.

Chapter Quiz

1. die
2. tap
3. UNC has fewer threads per inch than UNF threads of the same diameter.
4. Taper, plug, bottom
5. tap drill, smaller
6. 1/4 = Thread diameter
   28 = Threads per inch
   UNF = Unified National Fine Thread Series
   2 = Class of fit (free fit)
7. (b) the same as
8. Evaluate individually. Refer to Section 9.6 in the text.

Methods Used to Depict Threads on Drawings

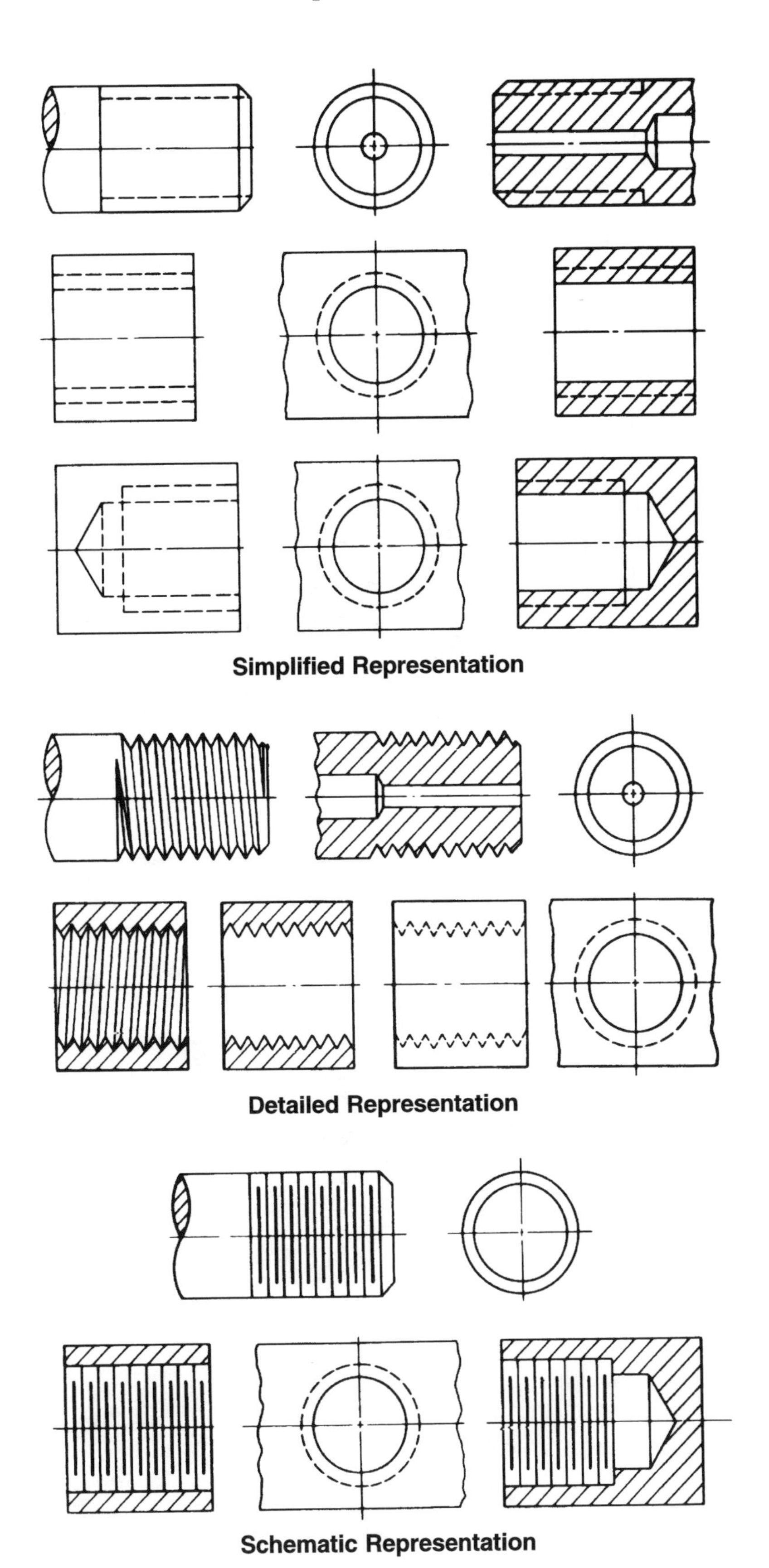

Simplified Representation

Detailed Representation

Schematic Representation

How Threads Are Specified on Drawings

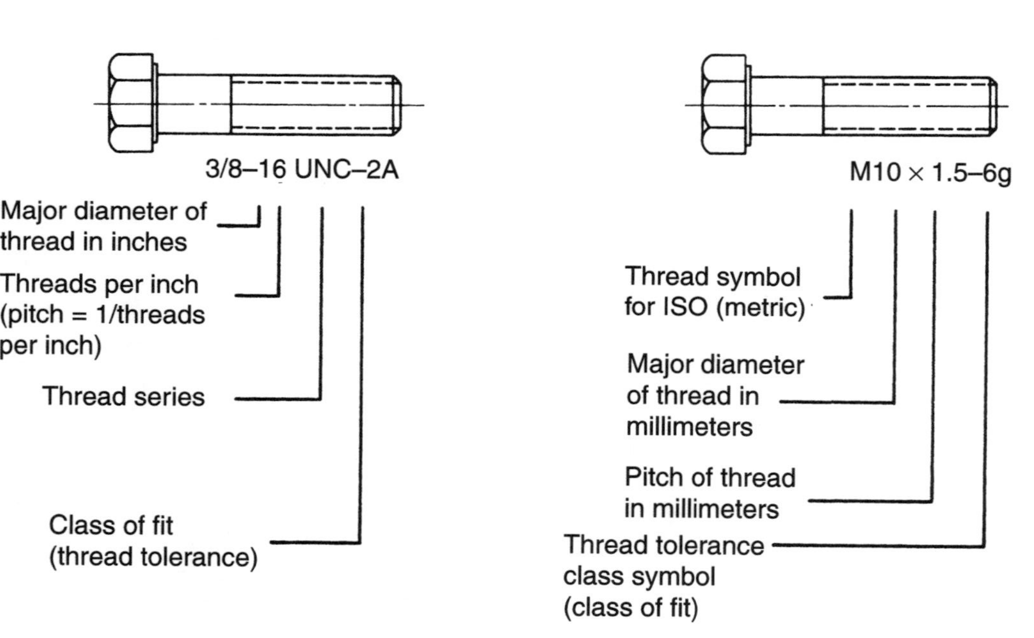

Thread Nomenclature

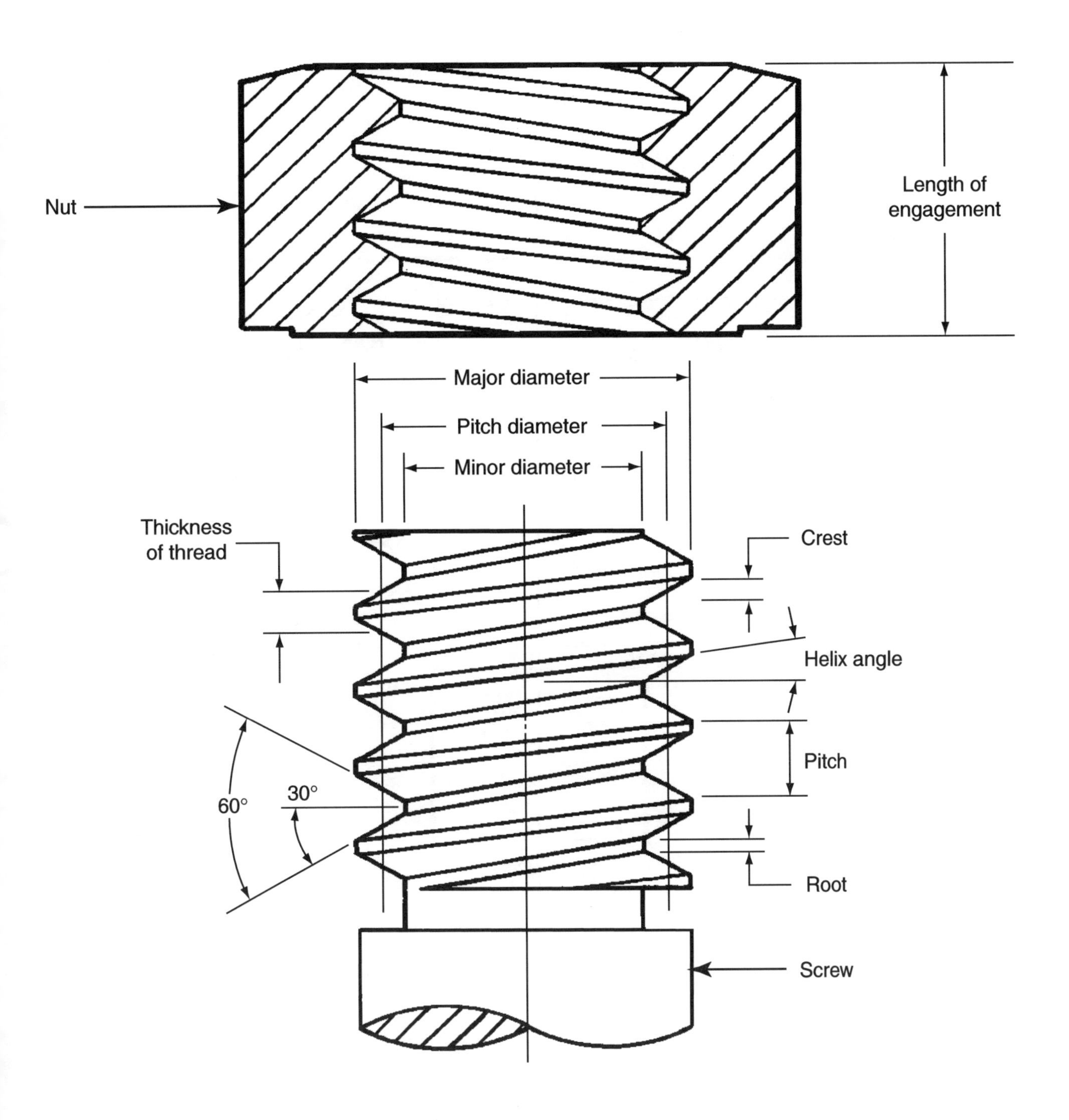

Starting a Die

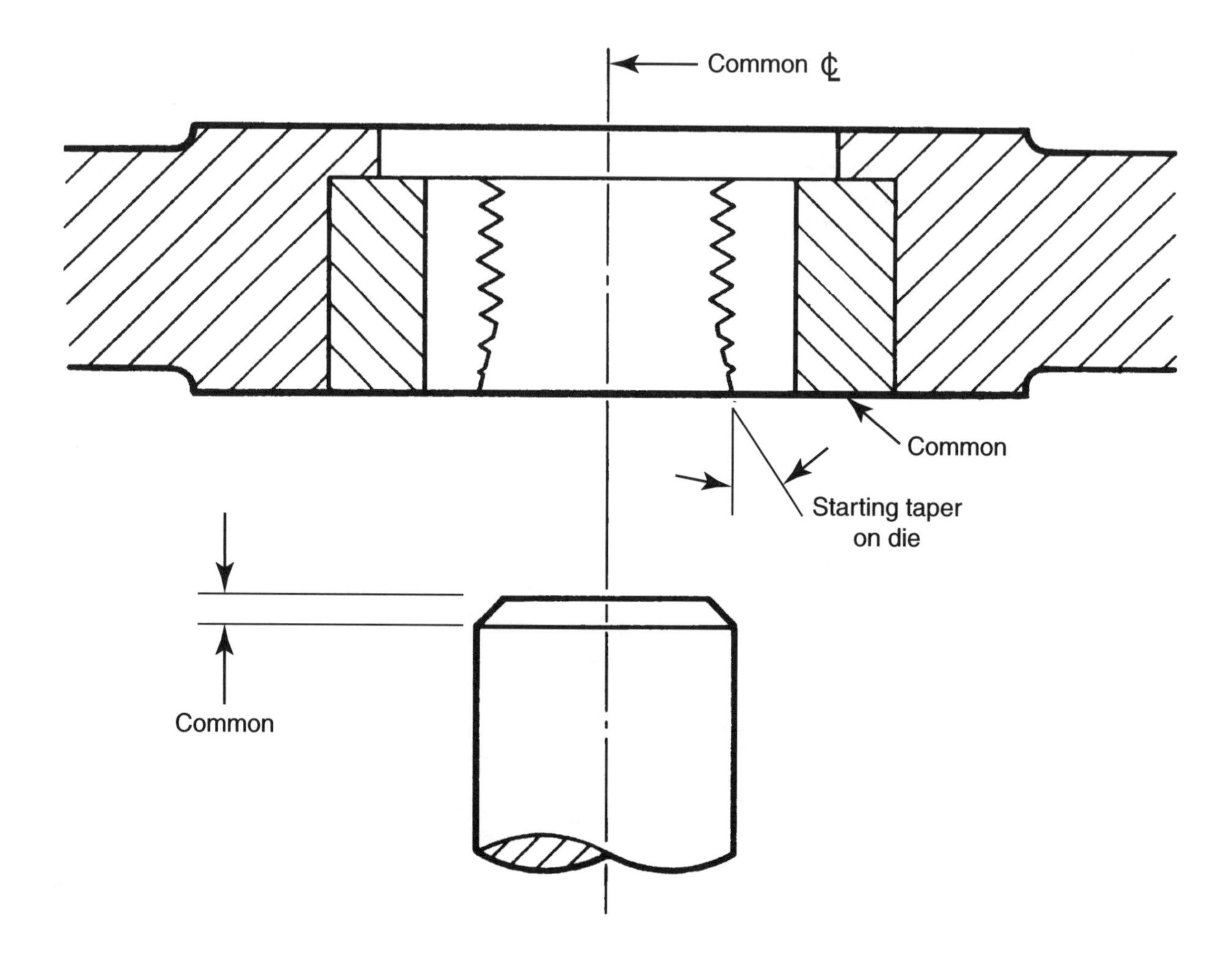

Cutting an External Thread to a Shoulder

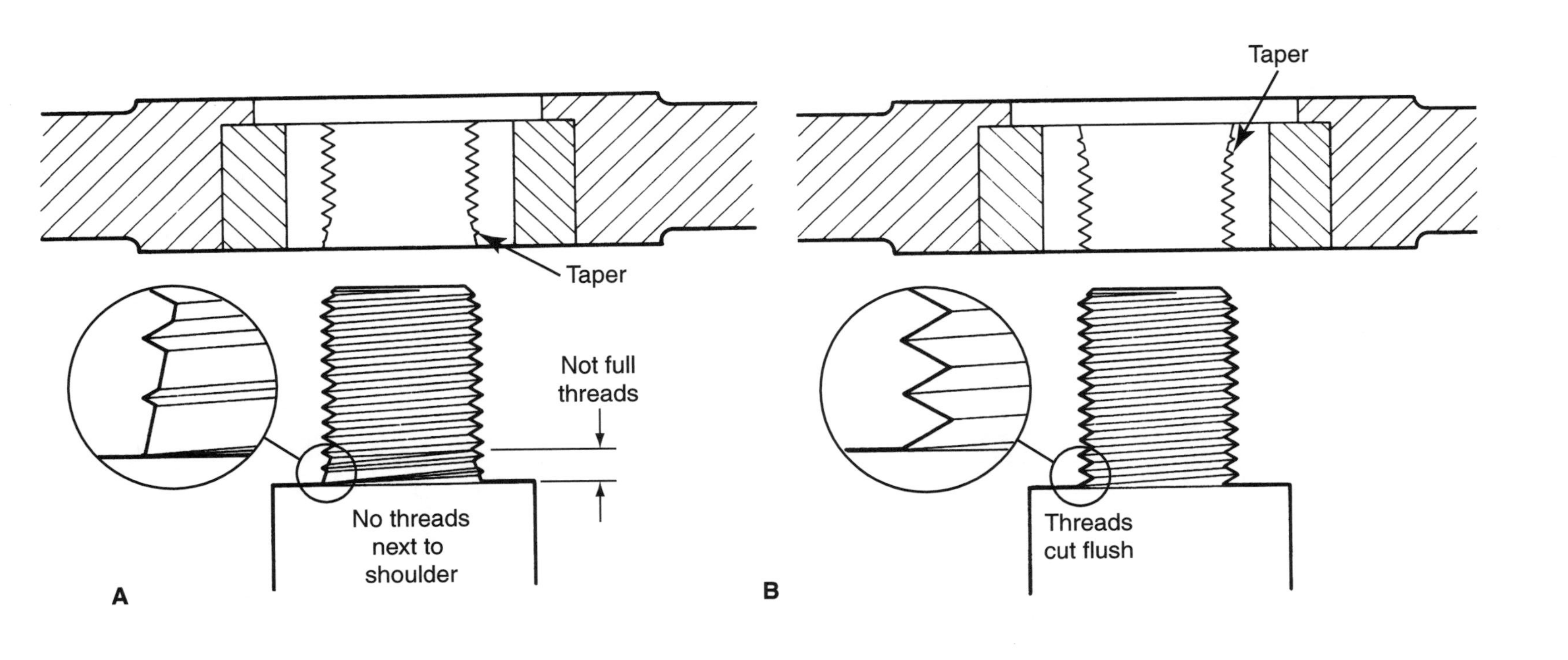

After die has been run down as far as possible, the die is reversed. When rotated down the shaft, it will cut threads almost flush with shoulder. A—Running die down normally. B—Reversing die to cut flush.

Chapter 9 Quiz

Hand Threading

Name: ______________________ **Date:**______________ **Period:**______________

1. External threads are cut with a(n) _____. 1.______________

2. Internal threads are cut with a(n) _____. 2.______________

3. How does a Unified coarse thread (UNC) differ from a Unified thread (UNF)?

____

____

____

____

4. Standard hand taps are made in sets of three and are known as _____, _____, and _____ taps. 4.______________

____

____

5. The drill used to make the hole prior to tapping is called a(n) _____ drill. It is always _____ than the thread diameter. 5.______________

____

6. Explain what the following thread information means. It would be shown on a drawing as 1/4-28UNF-2.

1/4 ______________________

28 ______________________

UNF ______________________

2 ______________________

7. When cutting external threads, stock diameter is always _____ the desired thread diameter. 7.______________
(a) larger than
(b) the same as
(c) smaller than

8. List three safety precautions that should be observed when cutting threads by hand.

1)____

____

2)____

____

3)____

____

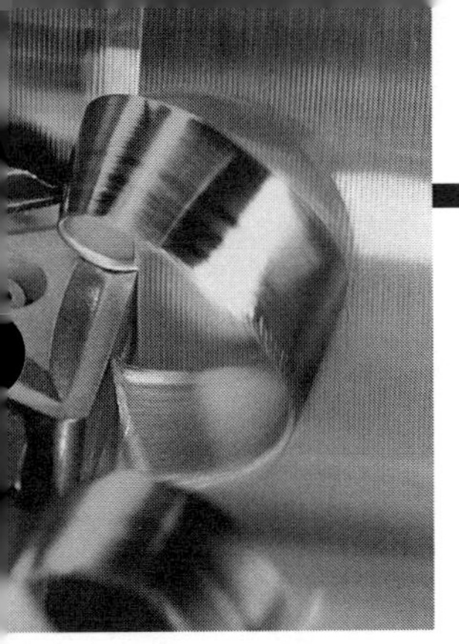

10 FASTENERS

LEARNING OBJECTIVES

After studying this chapter, students should be able to:

- Identify several types of fasteners.
- Discuss why inch-based fasteners are not interchangeable with metric-based fasteners.
- Explain how some fasteners are used.
- Select the proper fastening technique for each job.
- Demonstrate the application of permanent fasteners.
- Describe chemical fastening techniques.

CHAPTER RESOURCES

Text, pages 151–164
  Test Your Knowledge, pages 163–164
  Research and Development, page 164
Workbook, pages 59–62
Instructor's Manual
  Answer keys for:
    Test Your Knowledge Questions
    Workbook
    Chapter Quiz
  Reproducible Masters:
    10-1 Why Inch-Based and Metric Threads Are not Interchangeable
    10-2 Identifying Metric Fasteners
    10-3 Relative Strength of Hex Head Cap Screws
    10-4 Chapter Quiz

GUIDE FOR LESSON PLANNING

Refer students to the old mill illustration on Figure 37-30, page 586 of the text. Question them on why no metal fasteners (they were available when it was built) were used in its construction. No metal tools were used in its operation either.

The following items will aid in teaching this chapter:

- A selection of fasteners for students to examine.
- Examples showing how several types of fasteners are used.
- A selection of adhesives suitable for bonding metal.

CLASS DISCUSSION

Have students read and study all or part of the chapter. They should pay particular attention to the illustrations. Using Reproducible Masters 10-1 and 10-2, review the assignment and discuss:

- Why there are so many types of fasteners.
- The five basic uses of the screw thread.
- How threaded fasteners are measured.
- Why there is a need for metric fasteners.
- How to identify metric fasteners.
- Thread nomenclature.
- The various types of fasteners available in the shop/lab.

Questions that can be raised during class discussion:

- What unusual fasteners have students observed?
- Why is it not possible to substitute metric based fasteners for inch based fasteners?
- Where are metric fasteners used today?
- What problems are encountered when inch-based and metric-based fasteners are used on the same product? (Some automobiles use both types of fasteners at present.)
- Why are metric fasteners used today?
- Where would stainless steel fasteners be used?
- Why are the special fasteners used on aircraft so expensive?
- Name applications for different types of fasteners.
- Are students aware of any products where adhesives are used to join metal parts?

- A specially designed fastener is used to mount alloy wheels on some automobiles. They were developed to prevent the theft of the wheels. How do they differ from conventional fasteners?
- How are broken fasteners removed?

Assignments

1. Assign the Test Your Knowledge questions.
2. Assign Chapter 10 in the *Modern Metalworking Workbook.*
3. Assign the chapter quiz. Copy and distribute Reproducible Master 10-4.
4. Permit students to volunteer for the Research and Development activities at the end of the chapter.

Test Your Knowledge

1. Evaluate responses individually.
2. (a) Machine screws
   (b) Setscrew
   (c) Thread forming screw
   (d) Thread cutting screw
3. rivets
4. Lock
5. wing nut
6. (a) Acorn nut
7. (k) Machine screw
8. (f) Jam nut
9. (j) Machine bolt
10. (e) Drive screws
11. (m) Standard washer
12. (d) Dowel pins
13. (o) Threaded fasteners
14. (l) Retaining ring
15. (b) Blind rivet
16. (n) Thread forming screw
17. (c) Cotter pin
18. (g) Key
19. (i) Keyway
20. (h) Keyseat

Workbook

1. (h) Stud bolt
2. (c) Nut
3. (i) Drive screw
4. (f) Dowel pin
5. (e) Cotter pin
6. (j) Retaining ring
7. (b) Rivet
8. (k) Washer
9. (g) Adhesive
10. wedging action
11. penetrating oil
12. (c) a regular nut must be locked in place
13. (c) locked in place with a cotter pin or safety wire
14. Where frequent adjustment or frequent removal is necessary.
15. A special form of nut or internal thread. Inserts are designed to provide higher strength threads in soft metals and plastics. They are frequently used to replace damaged or stripped threads.
16. To provide an increased bearing surface for bolt heads and nuts. This distributes the load over a larger area. They also prevent surface marring.
17. (b) prevent bolts and nuts from loosening under vibration
18. (a) Square key
    (b) Keyway
    (c) Keyseat
19. Evaluate individually. Refer to Section 10.2.7.
20. Setscrews
21. Drive screws
22. Thread forming screw
23. Self-drilling screw
24. Flat washer
25. Split-ring lock washer

Chapter Quiz

1. (c) permit work to be assembled and disassembled without damage to the parts
2. Machine
3. Machine
4. Cap
5. stud
6. (d) All of the above.
7. square, hexagonal
8. jam, full size nut
9. wing
10. Washers
11. permanent
12. (c) the joint is accessible from one side only
13. cotter pin
14. key
15. keyseat, keyway

Why Inch-Based and Metric Threads Are not Interchangeable

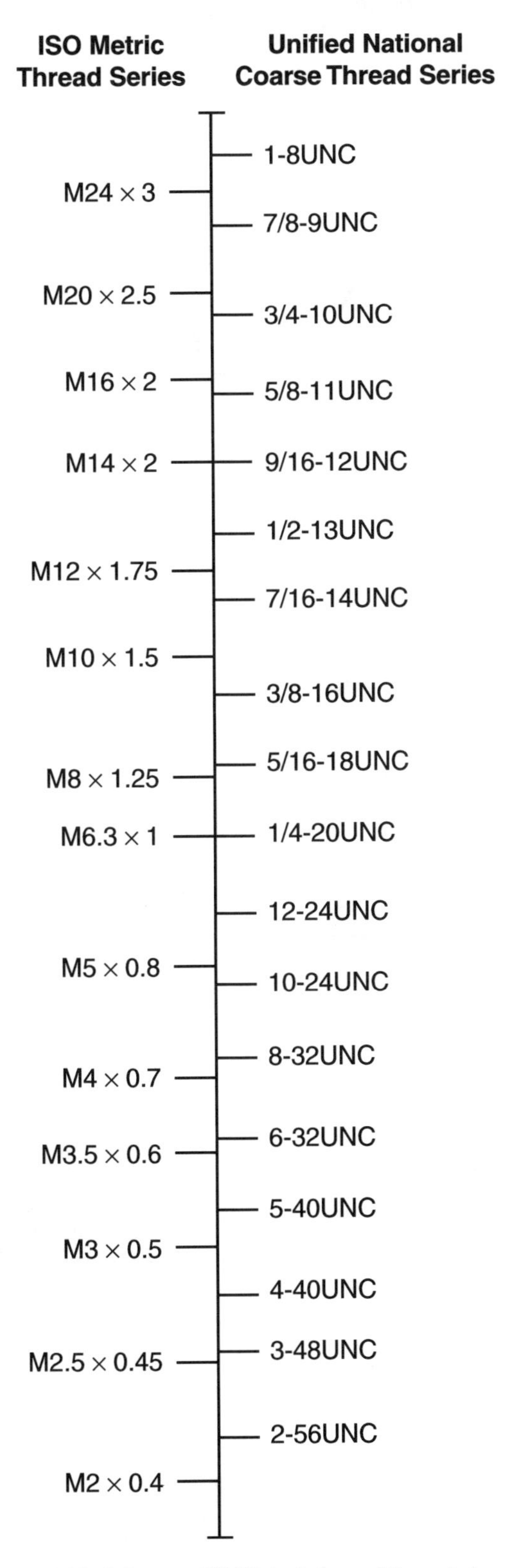

A comparison of ISO metric coarse and Unified Coarse (UNC) inch-based thread sizes. Even though several of them seem to be the same size, they are not interchangeable (one cannot be substituted for the other).

Identifying Metric Fasteners

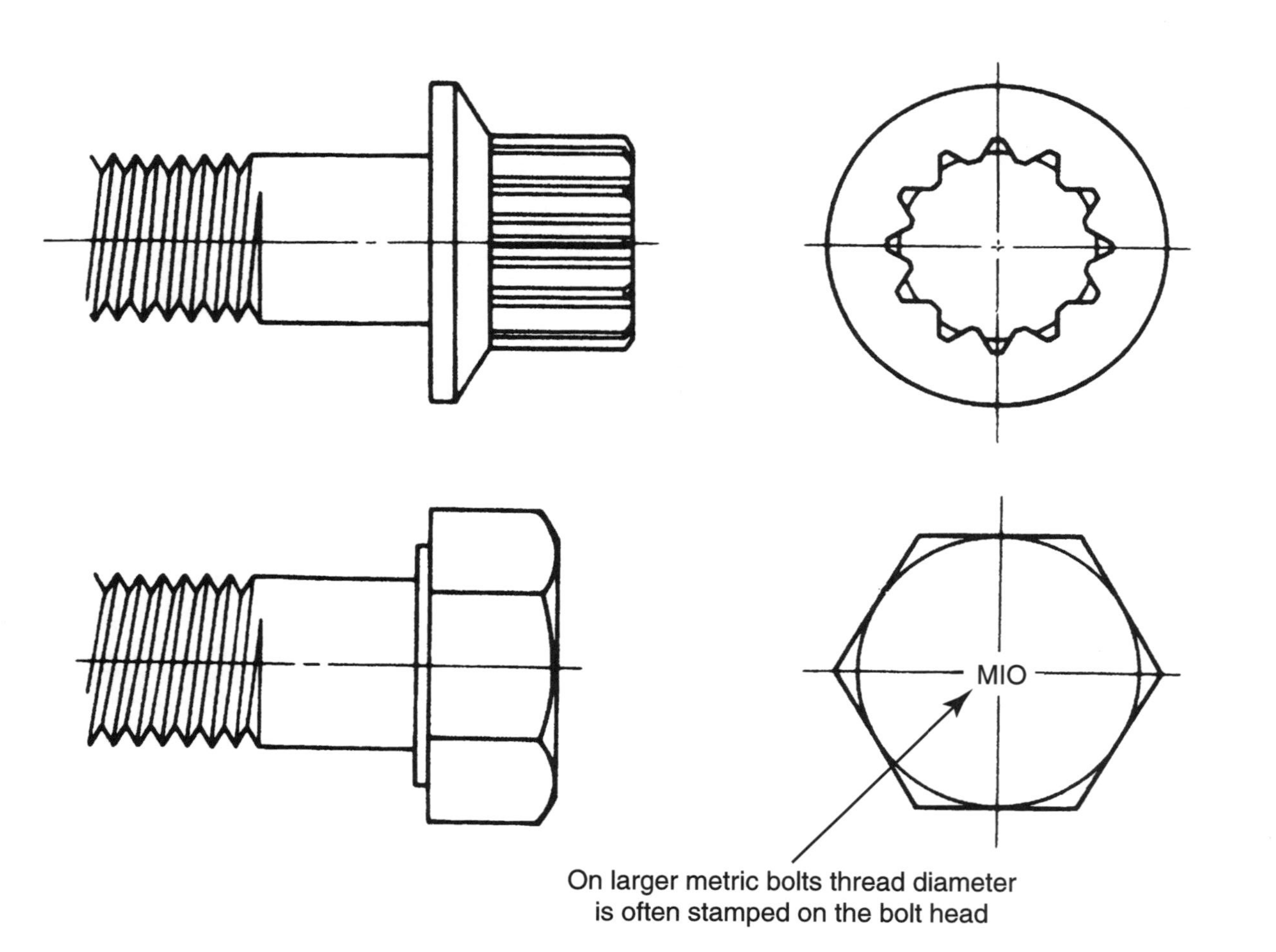

Metric fasteners are manufactured in the same variety of head shapes as inch-based fasteners. However, there is a problem in finding an easy way to distinguish between the two fastener types. Bottom—Some larger size hex-head metric fasteners have the size stamped on the head. Top—A twelve-spline flange head is under consideration for use on eight sizes of metric fasteners: 5, 6.3., 8, 10, 13, 14, 16, and 20 mm.

Relative Strength of Hex Head Cap Screws

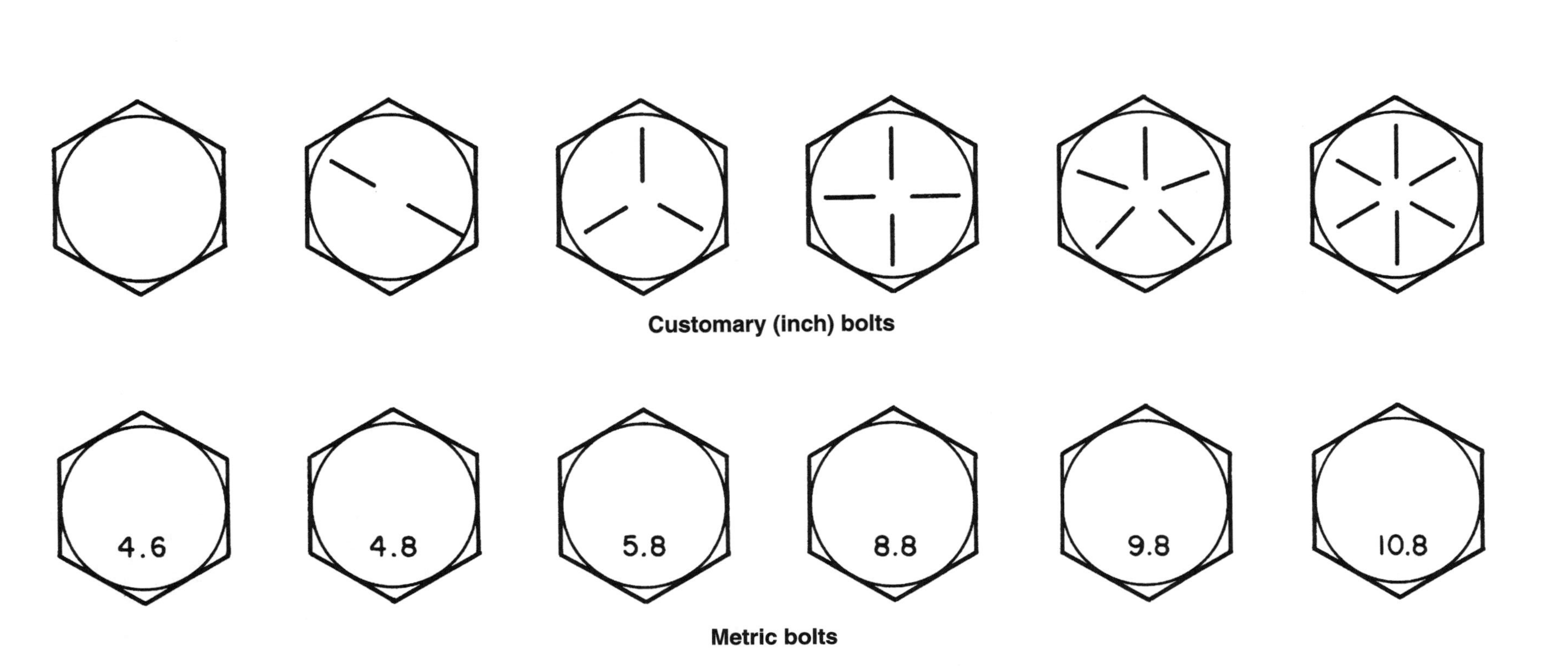

Identification marks (inch size) and class numbers (metric size) are used to indicate the relative strength of hex head cap screws. As identification marks increase in number, or class numbers become larger, increasing strength is indicated.

10-3

Chapter 10 Quiz
Fasteners

Name: ______________________ **Date:** ______________ **Period:** ______________

1. Threaded fasteners are often used because they _____. 1.__________
   (a) are standard size
   (b) are readily available
   (c) permit work to be assembled and disassembled without damage to the parts
   (d) All of the above.
   (e) None of the above.

2. _____ screws are widely used in general assembly work. They are made in a number of head styles. 2.__________

3. _____ bolts are used to assemble parts that do not require close tolerance fasteners. 3.__________

4. _____ screws are found in assemblies that require higher quality fastener with a more finished appearance. 4.__________

5. The _____ bolt is threaded on both ends. 5.__________

6. Setscrews are used to _____. 6.__________
   (a) prevent pulleys from slipping on shafts
   (b) hold collars in place on shafts
   (c) hold shafts in place on assemblies
   (d) All of the above.
   (e) None of the above.

7. Nuts for most threaded fasteners have external _____ and _____ shapes. 7.__________
   ____

8. The _____ nut is thinner than the standard nut, and is frequently used to lock a(n) _____ in place. 8.__________
   ____

9. The _____ nut provides for rapid loosening and tightening of the fastener without the need for a wrench. 9.__________

10. _____ permit a bolt or nut to be tightened without damaging the work surface. 10.__________

11. Rivets are used to make _____ assemblies. 11.__________

12. Blind rivets have been developed for applications where _____. 12.__________
    (a) no tool is available to put them in place
    (b) heavier rivets are required
    (c) the joint is accessible from one side only
    (d) All of the above.
    (e) None of the above.

13. The _____ is fitted into a hole drilled crosswise in a shaft to prevent parts from slipping on a shaft or falling off a shaft. 13.__________

14. A(n) _____ is a section of metal used to prevent a gear or pulley from slipping on a shaft. 14.__________

15. One-half of the above fits into a(n) _____ on the shaft and a(n) _____ in the hub of the gear or pulley 15.__________
    ____

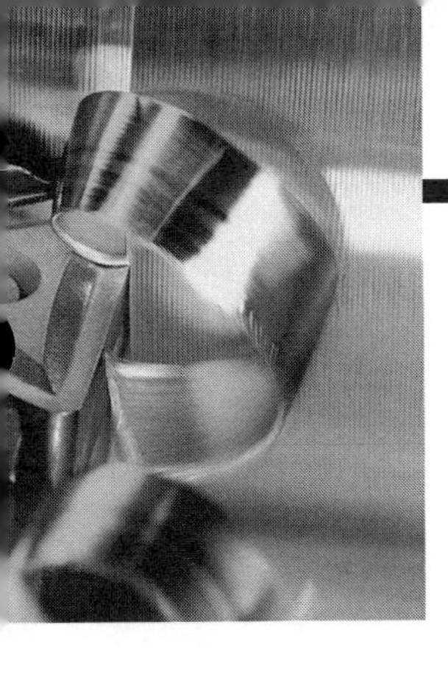

11 Art Metal

Learning Objectives

After studying this chapter, students should be able to:

- Identify the various areas of art metal.
- Demonstrate annealing and pickling.
- Use a jeweler's saw for sawing and piercing.
- Form bowls and trays by beating down and raising techniques.
- Identify art metal tools and equipment.
- Planish art metal projects.
- Explain how to prepare work for soldering.
- Apply art metal safety rules.

Chapter Resources

Text, pages 165–178
  Test Your Knowledge, page 177
  Research and Development, page 177
Workbook, pages 63–66.
Instructor's Manual
  Answer keys for:
    Test Your Knowledge Questions
    Workbook
    Chapter Quiz
  Reproducible Master:
    11-1 Chapter Quiz

Guide for Lesson Planning

This chapter will provide students with an opportunity to demonstrate their ability to design and produce objects of beauty and utility. As students develop their ideas they will learn that a simple solution to a design problem is usually the best solution. Remind them that there may be different solutions to design problems.

Introduce this chapter by having on hand samples of the metals that will be available and examples and/or illustrations of well designed and executed projects.

Have students read and study all or part of the chapter. Ask appropriate questions while discussing and demonstrating:

- Elements of good design.
- Reasons for annealing and pickling metals.
- Sawing and piercing.
- Raising metal forms. How to calculate blank diameter.
- Hammers, mallets, stakes, and anvil heads.
- Beating down.
- Planishing and the reason for doing it.
- Soldering.
- Polishing.
- Art metal safety.

Tools and equipment used in demonstrations must be in safe operating condition.

Assignments

1. Assign the Test Your Knowledge questions.
2. Assign Chapter 11 in the *Modern Metalworking Workbook.*
3. Assign the chapter quiz. Copy and distribute Reproducible Master 11-1.
4. Permit students volunteer for the Research and Development activities at the end of the chapter.

Test Your Knowledge

1. softening
2. Pickling
3. goggles
4. True
5. Aluminum
6. chasing
7. (a) to thrust back
8. Raising is the process of giving three-dimensional shape to flat sheet metal using hammers, mallets, stakes, and sandbags.
9. To mark the height of the finished piece.
10. (d) All of the above.

Workbook

1. (a) Annealing
2. (b) Pickling
3. (d) Repoussé
4. (e) Chasing
5. (g) Snarling iron
6. (f) Raising
7. (h) Stake
8. (j) Planishing
9. (i) Beating down
10. Pewter
11. To secure satisfactory joints and prevent the pieces from slipping.
12. By heating to a dull red then quenching in a water or pickling solution.
13. (b) the acid is poured into the water
14. (c) cut internal designs in metal
15. Any order: shaped hardwood block; sandbag and mallet; forming blocks; raising stake.
16. Evaluate student sketches individually.
17. Evaluate student sketches individually.
18. 14 5/8″
19. Evaluate student responses individually. Refer to Section 11.8.
20. Evaluate student designs individually.

Chapter Quiz

1. annealing
2. pickling
3. (c) internal decorative designs are cut into the metal
4. planishing
5. (d) All of the above.
6. thrust back

Chapter 11 Quiz

Art Metal

Name: ______________________ **Date:**______________ **Period:**______________

1. The process by which metal is softened by heating is called _____. 1.______________

2. The operation that removes oxides from metal by heating it and plunging it into a dilute solution of sulfuric acid is known as _____. 2.______________

3. Piercing is the operation in which _____. 3.______________
   (a) three-dimensional shape is given to the work
   (b) the object's outline is drawn on the metal
   (c) internal decorative designs are cut into the metal
   (d) All of the above.
   (e) None of the above.

4. The hammering operation that smoothes the surface of art metal work is known as _____. 4.______________

5. Raising is the process that _____. 5.______________
   (a) gives three-dimensional shape to flat sheet metal
   (b) makes use of hammers, stakes, and sandbags
   (c) sometimes employs a hardwood block with a depression
   (d) All of the above.
   (e) None of the above.

6. Repoussé is a French word meaning to _____. 6.______________

12 SHEET METAL

LEARNING OBJECTIVES

After studying this chapter, students should be able to:

- Explain the need for patterns and stretchouts.
- Use the different methods of pattern development.
- Cut and bend sheet metal using a number of tools.
- Identify and safely use a variety of sheet metal tools.
- Make hems, edges, and seams in sheet metal.
- Bend sheet metal into three-dimensional shapes using special machines.
- Join sheet metal sections with rivets.
- Apply sheet metal safety rules.

CHAPTER RESOURCES

Text, pages 179–200
Test Your Knowledge, page 198
Research and Development, page 199
Workbook, pages 67–74
Instructor's Manual
Answer keys for:
Test Your Knowledge Questions
Workbook
Chapter Quiz
Reproducible Masters:
12-1 Parallel-Line Development
12-2 Radial-Line Development
12-3 Chapter Quiz

GUIDE FOR LESSON PLANNING

Have a selection of sheet metal containers in which food and other products are packaged available for student examination. Carefully open some of them so they are in a two-dimensional shape. Also have on hand large size patterns of the stretchouts shown in Figures 12-3 through 12-9 in the textbook for students to assemble. This will help illustrate the concept of creating three-dimensional shapes from two-dimensional (flat) material.

Use Reproducible Masters 12-1 and 12-2 to show the techniques used to make parallel-line and radial-line developments.

CLASS DISCUSSION AND DEMONSTRATIONS

Assign all or part of the chapter to be read and studied. When reviewing and discussing the assignment do the following:

- Demonstrate the skills students will develop from each assignment.
- Emphasize sheet metal safety.
- Have students name sheet metal products found in the home and school. Was parallel-line or radial-line development used in their manufacture?
- Show how bending and folding sheet metal increases its rigidity.
- Illustrate the needs for hems, seams, and wire edges.
- Have students identify and name sheet metal tools and equipment.
- Ask appropriate questions on the ways to handle sheet metal safely and the safe use of the tools and equipment.

Allow sufficient time for students to demonstrate how well they have developed the skills of making simple patterns and cutting, forming, rolling, and joining sheet metal sections.

Demonstration tools should be in safe operating condition with all guards in place. Edge tools should be sharp.

ASSIGNMENTS

1. Assign the Test Your Knowledge questions.
2. Assign Chapter 12 in the *Modern Metalworking Workbook.*

3. Assign the chapter quiz. Copy and distribute Reproducible Master 12-3.
4. Permit students to volunteer for the Research and Development activities at the end of the chapter.

Test Your Knowledge

1. (b) full-size
2. stretchouts
3. prisms, cylinders
4. pyramids, cones
5. A transition piece is used to connect two different shaped openings.
6. strength, rigidity
7. soldering, riveting
8. snips (hand shears)
9. hollow, solid
10. Squaring shears
11. hems, seams
12. (b) forming roll
13. (a) narrow flange turned on the edge of a circular sheet metal section to form part of a joint
14. decorative
15. sheet metal

Workbook

1. A full-size drawing of the surfaces of an object, stretched out on a single plane.
2. Parallel-line development
3. Radial-line development
4. It is used to connect two differently shaped openings. For instance, a circular opening into a square opening.
5. (g) Squaring shears
6. (f) Hollow punch
7. (d) Circular snips
8. (c) Seams
9. (e) Combination snips
10. (b) Wired edge
11. (i) Bar folder
12. (h) Metal stakes
13. Evaluate student sketches individually. Refer to Section 12.2.1.
14. Evaluate student responses individually. Refer to Section 12.7.
15. Evaluate individually. Refer to text Figure 12-3A.
16. Evaluate individually. Refer to text Figure 12-3B.
17. Evaluate individually. Refer to text Figure 12-4A.
18. Evaluate individually. Refer to text Figure 12-4B.
19. Evaluate individually. Refer to text Figure 12-5.
20. Evaluate individually.

Chapter Quiz

1. bending, forming
2. A full-size drawing of the surface of an object stretched out on a single plane.
3. Parallel
4. Radial
5. transition piece
6. hems
7. seams
8. bar folder
9. (f) Stake
10. (e) Bar folder
11. (a) Burr
12. (g) Beading
13. (b) Crimping
14. (d) Hand groover
15. (c) Slip roll forming machine

Parallel-Line Development

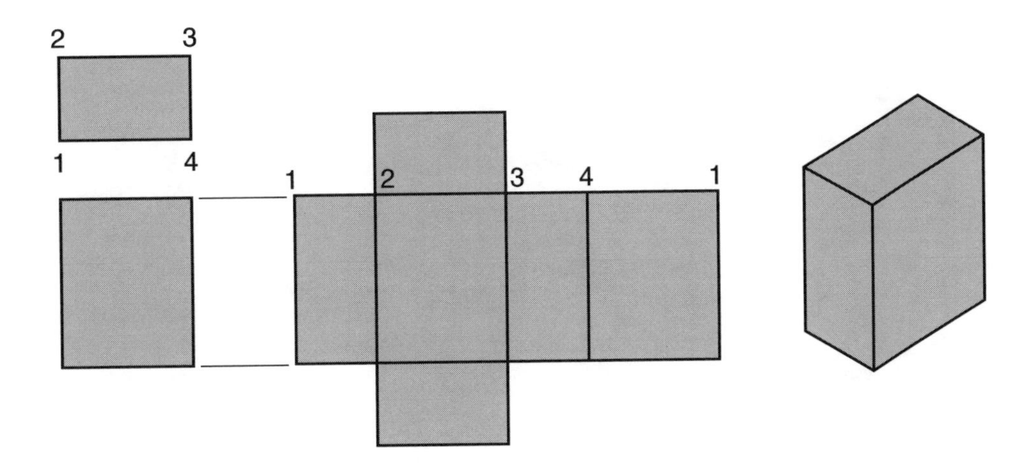

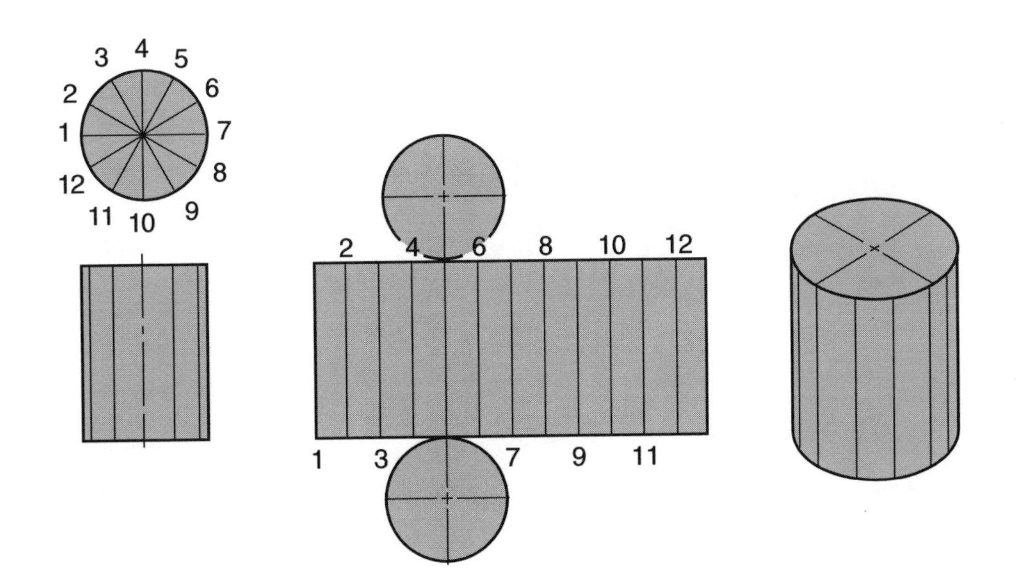

Radial-Line Development

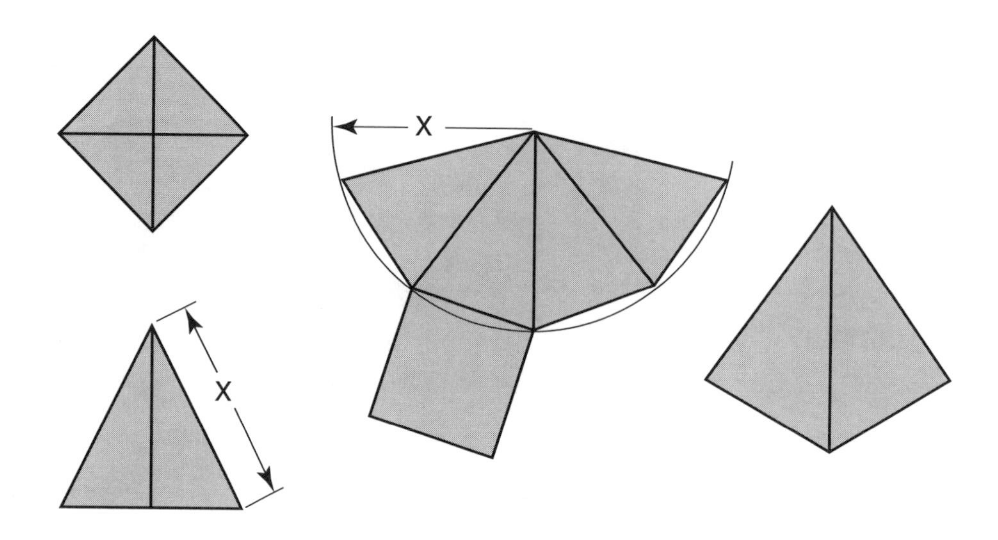

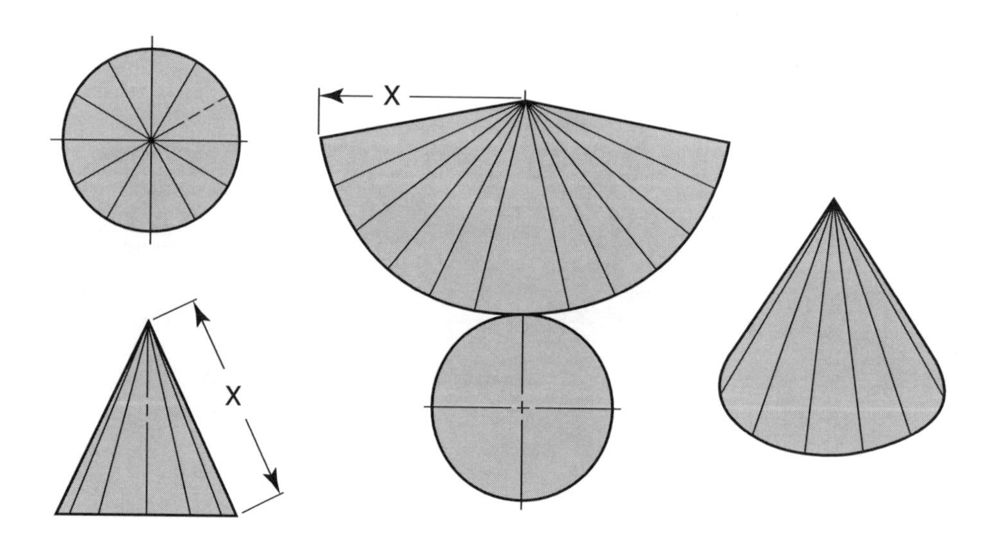

Chapter 12 Quiz

Sheet Metal

Name: ______________________ **Date:**______________ **Period:**______________

1. Sheet metal is given three-dimensional shape and rigidity by _____ and _____. 1.______________ ______________

2. What is a pattern or stretchout? ____

3. _____ line development is a technique used to make patterns of prisms and cylinders. 3.______________

4. _____ line development is used to make patterns of cones and pyramids. 4.______________

5. In sheet metal work, a(n) _____ is used to connect a circular section to a square or rectangular sheet metal section. 5.______________

6. Lips of sheet metal objects are strengthened with _____. 6.______________

7. _____ make it possible to join sheet metal sections. They are usually finished by soldering and/or riveting. 7.______________

8. The _____ is designed to bend sheet metal to form edges and also to prepare the metal for wire edges. 8.______________

Match each word or phrase with the correct sentence.

_________ 9. Used to make a variety of bends in sheet metal by hand.

_________ 10. Machine designed to bend sheet metal to form edges and prepare seams.

_________ 11. A narrow flange turned on the circular section that is to be attached to the end of a circular section.

_________ 12. Used to give additional rigidity to cylindrical sheet metal objects.

_________ 13. Reduces the diameter of a cylindrical section to permit it to be slipped into the next section.

_________ 14. Tool used to lock grooved seam joints.

_________ 15. Used to form circular sections.

(a) Burr
(b) Crimping
(c) Slip roll forming machine
(d) Hand groover
(e) Bar folder
(f) Stake
(g) Beading

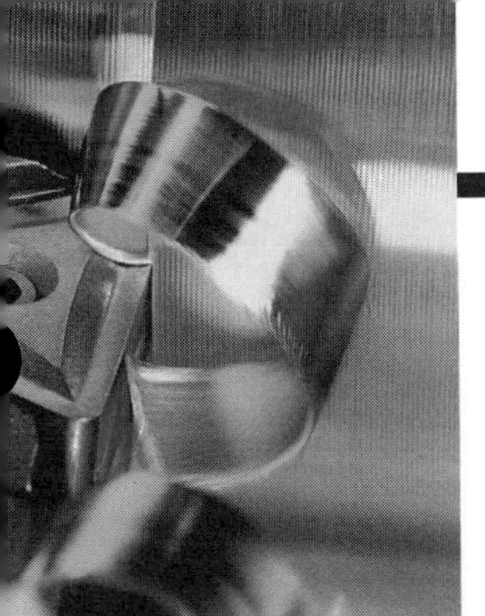

13 Soldering and Brazing

Learning Objectives

After studying this chapter, students should be able to:

- Explain how soldering differs from brazing.
- Demonstrate the use of soldering and brazing equipment.
- Describe the conditions required to produce a sound solder joint.
- Cite how solders are classified.
- Explain why flux is necessary.
- Tin a soldering copper.
- Solder a sheet metal joint.
- Prepare a joint for hard soldering.
- Demonstrate hard soldering.
- Follow soldering and brazing safety rules.

Chapter Resources

Text, pages 201–210
    Test Your Knowledge Questions, page 209
    Research and Development, page 209
Workbook, pages 75–76
Instructor's Manual
    Answer keys for:
        Test Your Knowledge
        Workbook
        Chapter Quiz
    Reproducible Master:
        13-1 Chapter Quiz

Guide for Lesson Planning

Prepare an assortment of materials for student examination. These items should include properly soldered sheet metal sections, solder in various shapes, fluxes, soldering coppers, and materials necessary for soldering.

Class Discussion and Demonstration

Have students read and study all or part of the chapter. Review the assignment with them. Discuss and demonstrate the following:

- Joint preparation for soldering.
- Solders and precautions taken when handling them.
- How solders are classified.
- Fluxes and why they are used.
- Soldering devices.
- How to tin a soldering copper and the reason for doing so.
- How to solder a lap joint.
- How to prepare a joint for hard soldering.
- Hard soldering techniques.
- Emphasize the safety precautions that must be observed when soldering.

The demonstration area must be arranged so all students can see and hear and there is adequate ventilation to remove generated fumes.

Allow adequate time for students to practice tinning soldering coppers and soldering joints.

Assignments

1. Assign the Test Your Knowledge questions.
2. Assign Chapter 13 in the *Modern Metalworking Workbook.*
3. Assign the chapter quiz. Copy and distribute Reproducible Master 13-1.
4. Permit students to volunteer for the Research and Development activities at the end of the chapter.

Test Your Knowledge

1. (b) with a nonferrous metal filler
2. To add strength to the soldered joint.
3. Any order: the correct solder alloy must be used; the proper flux must be applied; an

adequate source of heat must be available; the surfaces to be soldered must be clean.
4. Tin
5. 50-50
6. Flux removes the oxides and prevents more oxides from forming while the metal is heated to soldering temperature. The application of flux also lowers the surface tension of the molten solder, enabling it to cover the area and to alloy with the work.
7. corrosive, noncorrosive
8. corrosive
9. The soldering copper, electric soldering copper, and the gas torch.
10. Tinning is coating the metal surface with a thin layer or film of solder. A soldering copper is tinned so molten solder will adhere to it and make the solder easier to handle.
11. sweat
12. Easy flow, medium flow, and hard flow.
13. Any three of the following: cleans the surface to be joined; protects the surfaces from oxidation while the joint is being heated; dissolves any existing oxides; assists the flow of the brazing alloy.
14. In order: cleaning the metals; establishing a good fit and proper clearance; applying the proper flux; properly assembling and supporting the pieces; heating and flowing the alloy; and cleaning the assembly.

Workbook

1. A method of joining metals with a nonferrous metal filler without having to heat the base metals to a point where they melt.
2. tin, lead
3. 50% tin and 50% lead
4. 60% tin and 40% lead
5. (d) All of the above.
6. Corrosive and noncorrosive
7. corrosive
8. noncorrosive
9. Any order: soldering copper, electric soldering copper, gas torch.
10. adhere
11. Evaluate individually. Refer to Section 13.2.
12. Easy flow, medium flow, and hard flow.
13. Evaluate individually. Refer to Section 13.6.
14. A neutral flame.
15. Evaluate student answers individually. Refer to Section 13.8.
16. Evaluate student work individually.
17. Evaluate student work individually.
18. Evaluate student work individually.
19. Evaluate student work individually.
20. Evaluate student work individually.

Chapter Quiz

1. Any order: correct solder must be used; proper flux applied; adequate heat; clean surface.
2. tin, lead
3. tin, lead
4. (d) All of the above.
5. So molten solder will adhere to it. This makes the solder easier to handle.
6. Soft solders are typically tin-lead alloys. They melt about 420°F (215°C). Hard solders are silver-alloy solders used at temperatures above 800°F (426°C).
7. In order: good fit, clean metal, proper fluxing, proper fitting and support, sufficient heat to cause solder to flow, final cleaning.
8. Evaluate individually. Refer to Section 13.8 in the text.

Chapter 13 Quiz

Soldering and Brazing

Name: ______________________ **Date:** ______________ **Period:** ______________

1. Before two metal sections can be joined by soldering, four conditions must be met. List them.

____

____

____

____

2. Solders are alloys of _____ and _____. 2. ______________ ______________

3. A 60-40 solder indicates that it is 60% _____ and 40% _____. 3. ______________ ______________

4. Flux is required when soldering because _____. 4. ______________
   (a) a dirty or oxidized surface cannot be soldered
   (b) it cleans the metal surfaces being soldered
   (c) oxides form when metal is exposed to the atmosphere and must be removed to secure a solid soldered joint
   (d) All of the above.
   (e) None of the above.

5. Why must a soldering copper be tinned? ______________________________

____

____

____

6. There are two major differences between hard solder and soft solder. What are they?

____

____

7. Strong silver brazed joints require six steps. List them.

____

____

____

____

____

____

8. List three safety precautions that should be observed when soldering.
   1) ____
   2) ____
   3) ____

14 Sand Casting

Learning Objectives

After studying this chapter, students should be able to:

- Explain the sand casting process.
- Demonstrate the correct way to make a sand casting.
- Describe simple patterns, split patterns, and match plate patterns.
- Prepare and use a simple core.
- Heat and pour molten metal safely.
- Use a pyrometer.
- Follow safe casting procedures.

Chapter Resources

Text, pages 211–228
Test Your Knowledge, page 228
Research and Development, page 228
Workbook, pages 77–80
Instructor's Manual
Answer keys for:
Test Your Knowledge Questions
Workbook
Chapter Quiz
Reproducible Masters:
14-1 Parts of a Sand Mold
14-2 Shrink Rule
14-3 Chapter Quiz

Guide for Lesson Planning

Tour the foundry area. Show location of tools and equipment. Have finished mold (open), mold that has been poured with the casting exposed, casting that has been removed from mold, and selection of patterns available for student examination.

Class Discussion and Demonstrations

Have class read and study all or part of the chapter. Review the assignment with them and explain and demonstrate the sequence for making a simple mold.

Ask appropriate questions while discussing the following:

- How sand molds are made in industry.
- The correct way to make a sand mold.
- Parts of the mold.
- Correct way to remove a pattern from a mold.
- Cores and their uses.
- Using a pyrometer to check molten metal temperature.
- Procedure for heating and pouring molten metals. (Emphasize safety equipment that must be worn and care that must be taken when working with molten metal.)
- Foundry safety.

Assignments

1. Assign the Test Your Knowledge questions.
2. Assign Chapter 14 in the *Modern Metalworking Workbook.*
3. Assign the chapter quiz. Copy and distribute Reproducible Master 14-3.
3. Permit students to volunteer for the Research and Development activities at the end of the chapter.

Test Your Knowledge

1. (o) Crucible
2. (m) Riddle
3. (a) Flask
4. (d) Green sand mold
5. (e) Mold
6. (c) Drag

7. (g) Mulling machine
8. (n) Bench rammer
9. (f) Gating system
10. (b) Cope
11. (k) Simple pattern
12. (l) Draft
13. (j) Split pattern
14. (h) Parting line
15. (i) Core
16. Only once. Because the mold must be broken apart to remove the casting.
17. It is used to pack sand around the pattern.
18. A match plate is a plate of metal on which the pattern and gating system is split along the parting line, then mounted back to back to form a single piece. It is used for long production runs.
19. Because metal shrinks as it cools.
20. When openings in the casting are required.
21. It permits the pattern to be removed from the sand without damaging the mold.
22. strikeoff bar
23. Any order: withstand heat, hold the shape of the mold while the metal is being poured, be porous.
24. It removes impurities from the molten metal.

Workbook

1. The opening or cavity into which the molten metal is poured to produce the required casting.
2. green sand
3. A plate of metal on which the pattern and gating system is split along the parting line and mounted back to back to form a single piece.
4. gating
5. When a hole or cavity is required in the casting.
6. One
7. Two or more.
8. To allow for the dimensional changes as the casting cools.
9. Draft permits the pattern to be withdrawn from the sand without damaging the mold.
10. (a) Feeder
    (b) Sprue
    (c) Core
    (d) Casting
    (e) Runner
    (f) Gate
11. Evaluate student drawings individually.
12. (a) Flask
    (b) Gate cutter
    (c) Riser pin
    (d) Sprue pin
    (e) Draw screw
    (f) Bench rammer
    (g) Stick and oval
    (h) Molder's bellows
13. riddle
14. bench rammer
15. crucible
16. pyrometer
17. Evaluate individually. Refer to Section 14.4.
18. Evaluate student work individually.
19. Evaluate student work individually.
20. Evaluate student work individually.

Chapter Quiz

1. (a) Feeder
   (b) Sprue
   (c) Core
   (d) Casting
   (e) Runner
   (f) Gate
2. mold
3. pattern, draft
4. cores
5. Pattern must be made slightly oversize as metal shrinks as it cools. The shrink rule measurements allow for this shrinkage.
6. crucible
7. pyrometer

Parts of a Sand Mold

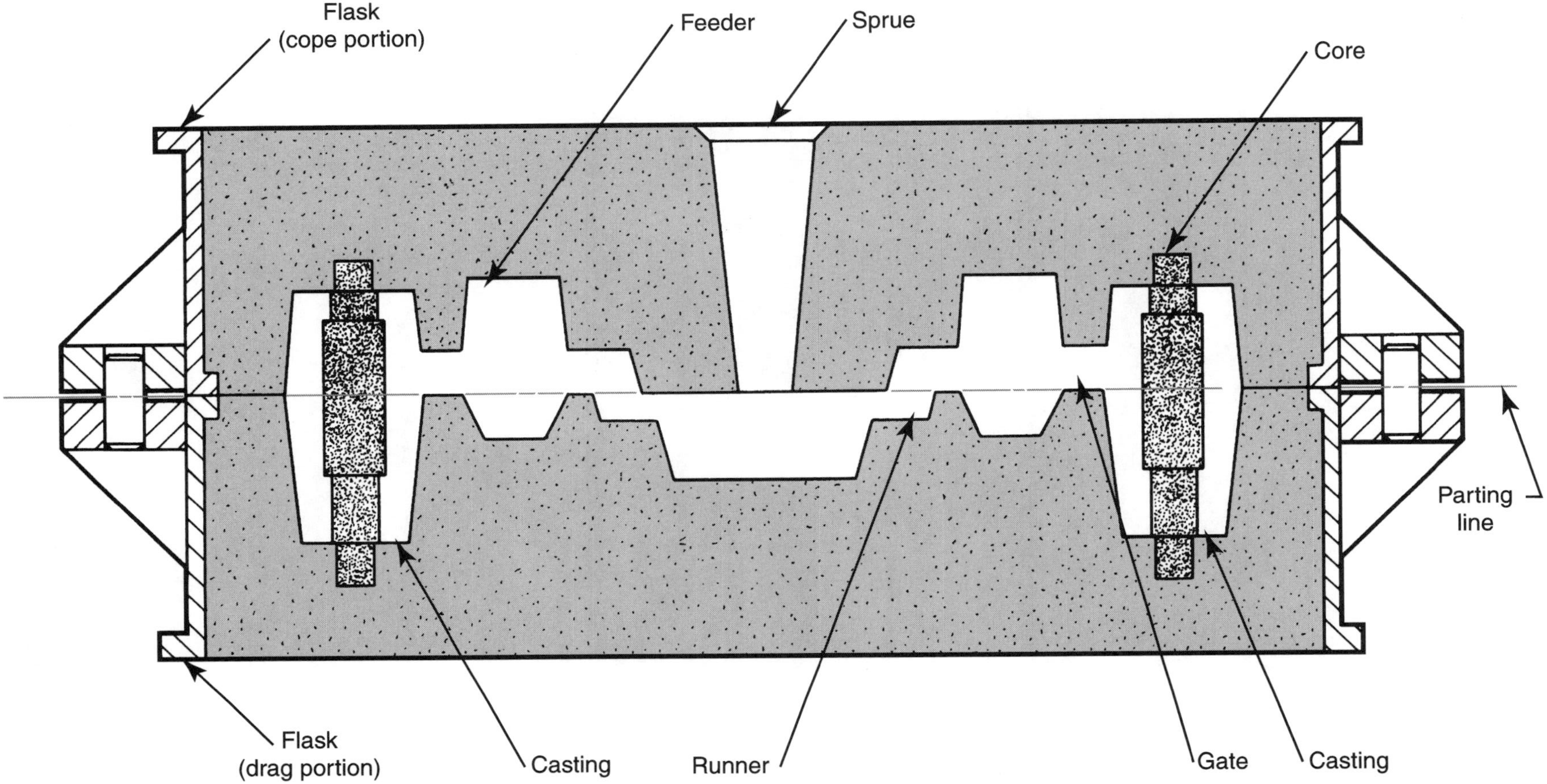

Flask
(cope portion)
Feeder
Sprue
Core
Parting
line
Flask
(drag portion)
Casting
Runner
Gate
Casting

14-1

Shrink Rule

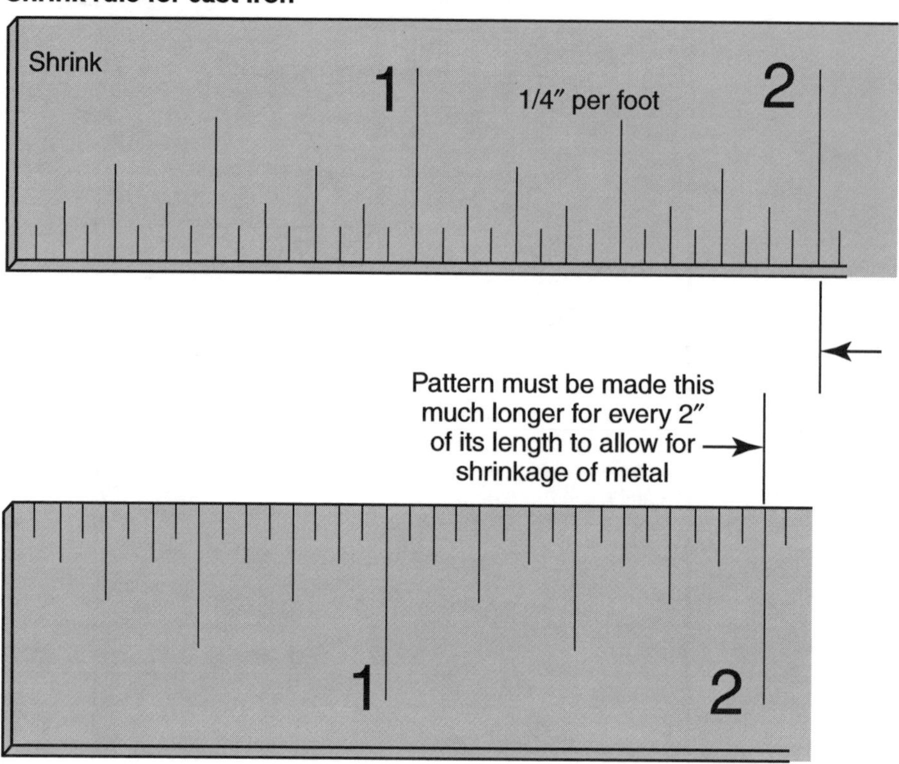
Shrink rule for cast iron
Shrink
1
1/4″ per foot
2
Pattern must be made this
much longer for every 2″
of its length to allow for
shrinkage of metal
1
2
Standard rule

Chapter 14 Quiz
Sand Casting

Name: ______________________ **Date:** ______________ **Period:** ______________

1. Name the parts of the two-part sand mold.

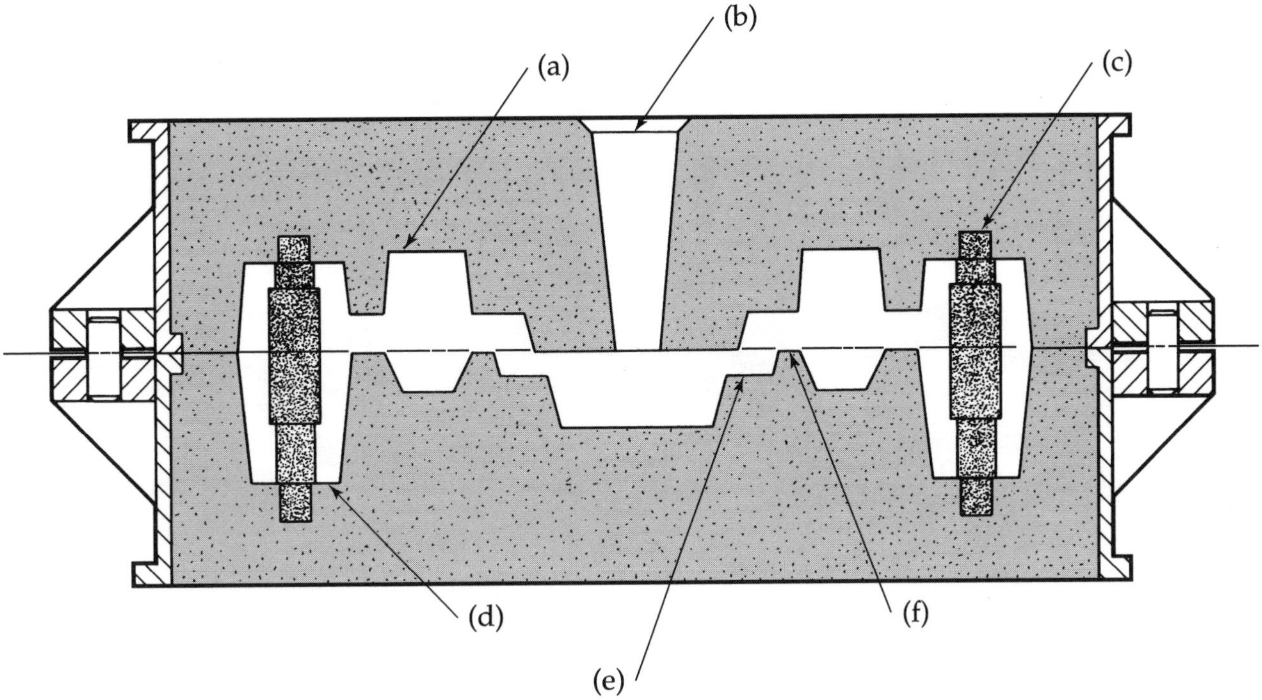

(b)
(a)
(c)
(d)
(f)
(e)

(a) ______________________ (d) ______________________

(b) ______________________ (e) ______________________

(c) ______________________ (f) ______________________

2. A(n) _____ is a cavity in the sand the shape and size of the object being cast. 2.______________

3. The cavity is made in the sand with a(n) _____ which must have _____ if it is to be removed from the sand without damaging the mold cavity. 3.______________ ______________

4. Openings or hollow spaces in a casting are made with _____. 4.______________

5. Why is a shrink rule used in place of a conventional rule when making patterns?

____

____

____

6. Metal is melted in a container called a(n) _____. 6.______________

7. The temperature of the molten metal is checked with a(n) _____. 7.______________

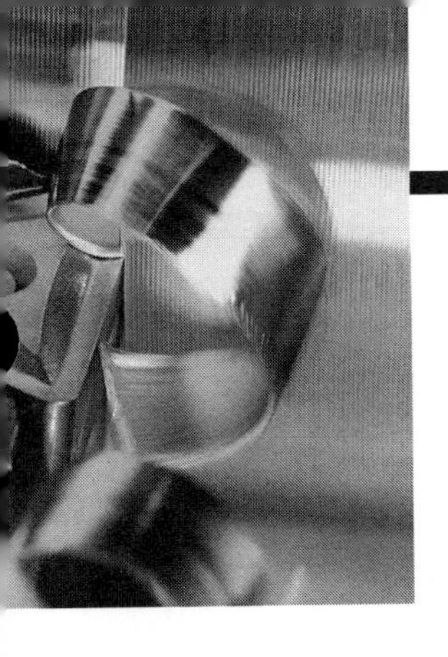

Metal Casting Techniques

Learning Objectives

After studying this chapter, students should be able to:

- Explain various casting techniques.
- List the casting sequences for some of the techniques described in this chapter.
- Describe various rapid prototyping techniques and their importance to the metal casting industry.

Chapter Resources

Text, pages 229–248
  Test Your Knowledge, pages 245–246
  Research and Development, pages 246–247
Workbook, pages 81–84
Instructor's Manual
  Answer keys for:
    Test Your Knowledge Questions
    Workbook
    Chapter Quiz
  Reproducible Masters:
    15-1 Plunger Die Casting Machine
    15-2 Air Injection Die Casting Machine
    15-3 Cold Chamber Die Casting Machine
    15-4 Centrifugal Casting
    15-5 How Stereolithography Works
    15-6 Laminated Object Manufacturing (LOM)
    15-7 Chapter Quiz

Guide for Lesson Planning

In most school metal shops/labs it is only possible to demonstrate lost wax casting, plaster mold casting, and permanent mold casting (making toy soldiers, fishing sinkers, etc.). To illustrate the other casting techniques, it will be necessary, therefore, to have examples of various other casting techniques available for student examination.

Class Discussion and Demonstration

Have students read and study all or part of the chapter. Using Reproducible Masters 15-1 through 15-6 as transparencies or handouts, review the assignment and ask appropriate questions on the following:

- Steps required to make a finished casting.
- Die casting techniques.
- Where are die castings used?
- How die casting machines operate.
- Permanent mold casting.
- Advantages of using the permanent mold casting process.
- Centrifugal casting.
- Investment (lost wax) casting.
- Plaster mold casting. Make sure to emphasize the dangers involved if the mold is not completely dry.
- Shell molding.
- How the stereolithography process operates.
- How the Laminated Object Manufacturing (LOM) process operates.
- Advantages of using rapid prototyping techniques when designing castings.
- Allow students to examine casting samples. Have them explain the differences they observe in the castings.

Note: Investigate the possibility of having a local dental laboratory demonstrate the investment casting process.

Assignments

1. Assign the Test Your Knowledge questions.
2. Assign Chapter 15 in the *Modern Metalworking Workbook.*
3. Assign the chapter quiz. Copy and distribute Reproducible Master 15-7.
4. Permit students to volunteer for the Research and Development activities at the end of the chapter.

Test Your Knowledge Questions

1. Refer to Section 15.1 of the text.
2. Preparation of a mold. Melting the metal. Putting the molten metal into the mold. Removing the casting from the mold. Cleaning and trimming the casting.
3. pressure
4. (a) sequence followed to make a casting
5. mold
6. (c) molten metal cannot be injected into the mold until it has been locked closed
7. Any order: plunger, air injection, cold chamber.
8. The thin fin of metal formed at the parting line of a forging or casting. Flash is formed where a small portion of metal is forced out between the edges of the die.
9. (b) do not have to be destroyed to remove the casting
10. fishing sinker, toy soldier
11. centrifugal
12. slush
13. lost wax
14. wax, plastic
15. dental, medical
16. Evaluate responses individually.
17. (d) All of the above.
18. thermosetting resin
19. roll over
20. immediately, an indefinite period
21. lost foam
22. The moisture will create steam when it comes into contact with the molten metal and cause the mold to explode.
23. (d) None of the above.
24. Chilling effect is the process of moisture cooling the molten metal. This makes it possible to cast thinner sections than with the sand or permanent mold methods.
25. Student answers will vary. Evaluate responses individually.

Workbook

1. Evaluate individually. Refer to Section 15.2.1.
2. (c) hardened steel
3. Any order: aluminum, copper, magnesium, lead, tin, zinc.
4. Permanent molds do not have to be destroyed to remove the casting.
5. The molten metal is poured into a rapidly rotating mold. Centrifugal force presses the molten metal against the mold wall, producing a casting with a fine grain metal structure and superior surface finish.
6. (c) long enough to form a thin shell of metal in the mold
7. Evaluate individually. Refer to Section 15.2.5.
8. Heat disposable wax or plastic.
9. thin shells
10. The pattern is made from expandable polystyrene beads. The beads are injected into a die and are bonded together using a heat source, usually steam.
11. Any moisture can be turned into steam when the molten metal is poured into the mold, and cause the mold to explode.
12. Evaluate individually. Refer to Section 15.4.1.
13. Evaluate individually. Refer to Section 15.4.2.
14. Evaluate individually. Refer to Section 15.5.
15. Review samples individually.

Chapter Quiz

1. (a) Preparation of mold
   (b) Melting metal
   (c) Pouring molten metal into mold
   (d) Removing casting from mold
   (e) Cleaning and trimming casting
2. (e) Die casting
3. (c) Permanent mold casting
4. (h) Slush molding
5. (g) Centrifugal casting
6. (a) Investment casting
7. (d) Plaster mold casting
8. (f) Shell molding
9. (b) Stereolithography

Plunger Die Casting Machine

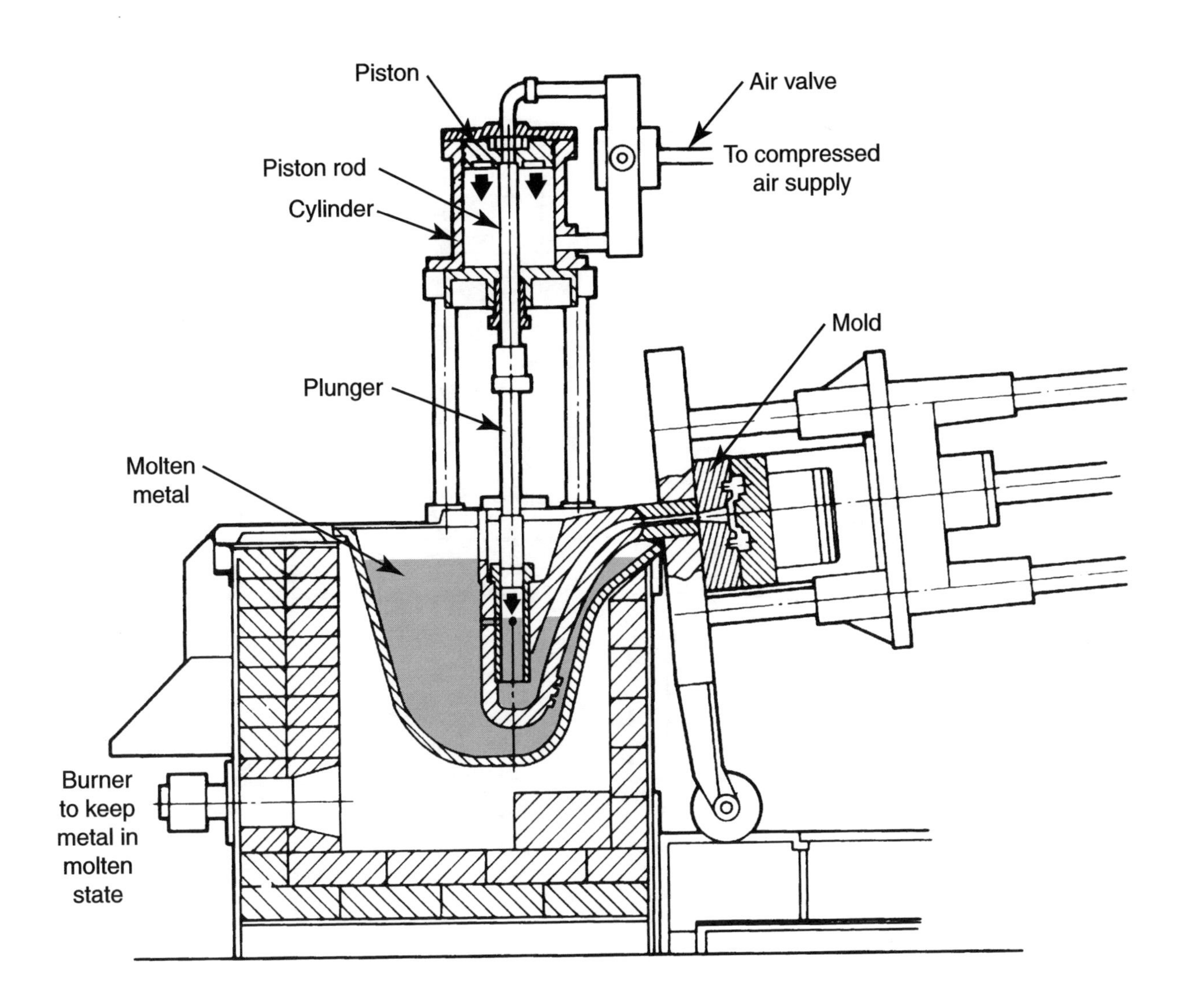

Air Injection Die Casting Machine

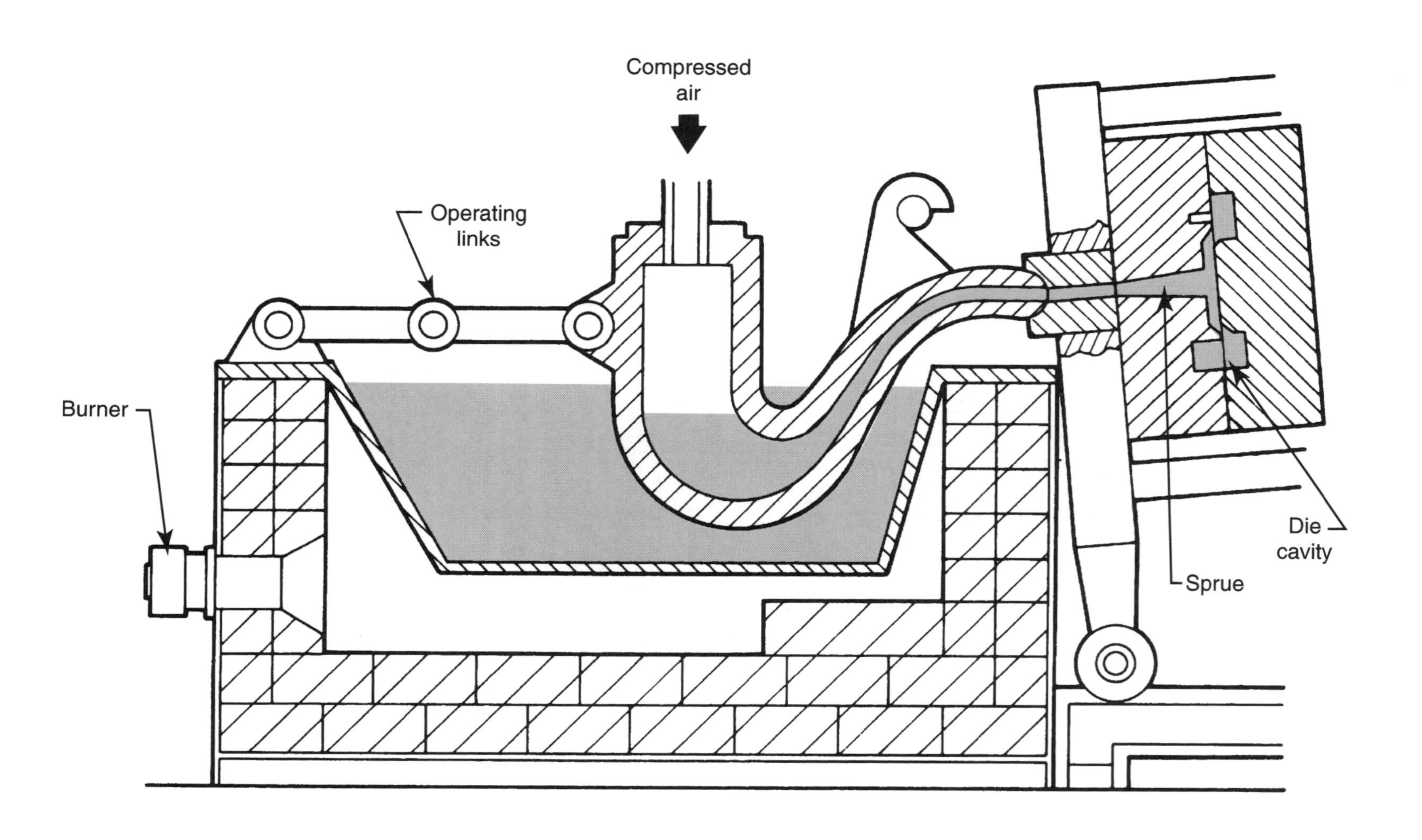

15-2

Cold Chamber Die Casting Machine

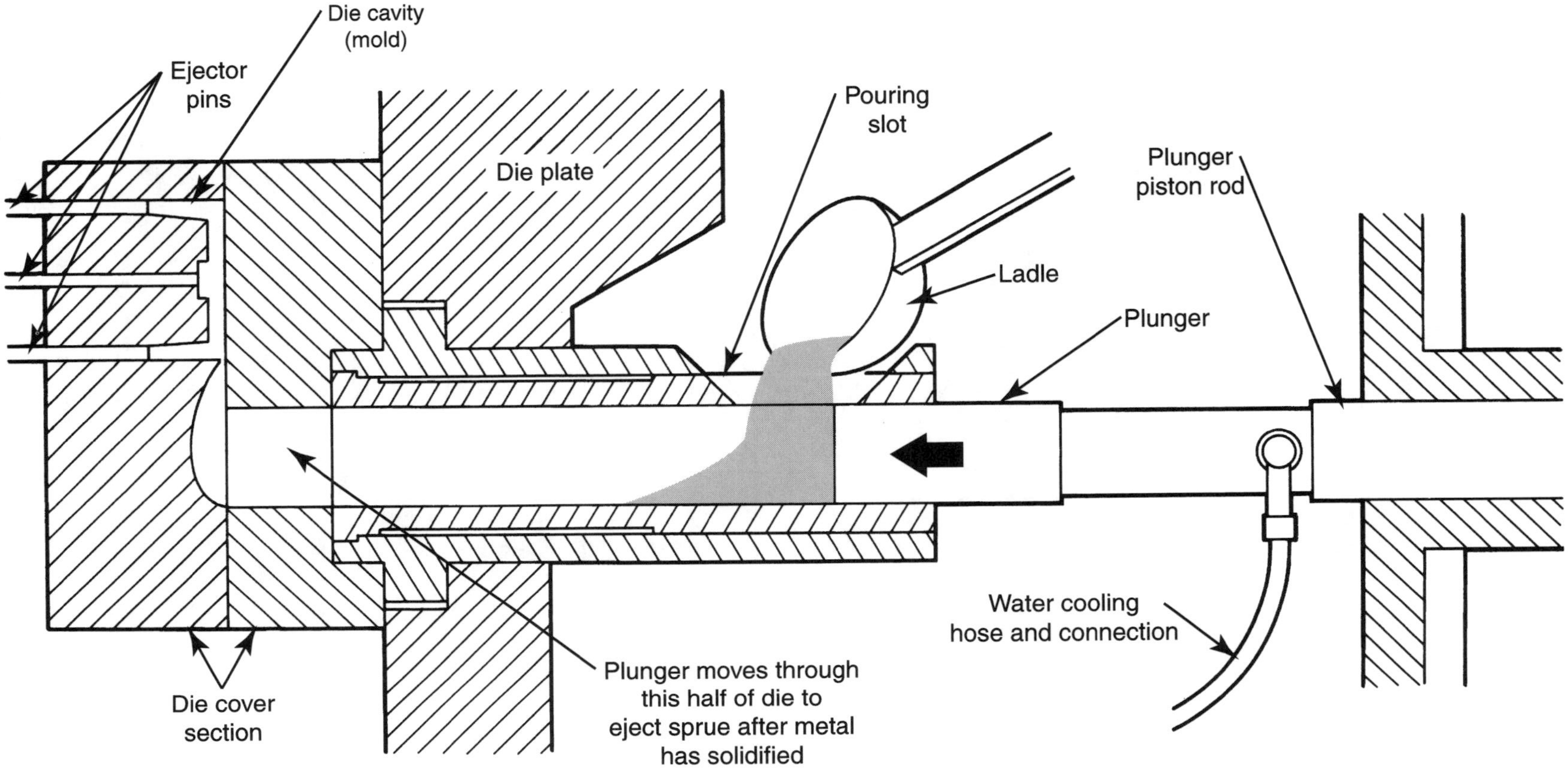

15-3

Centrifugal Casting

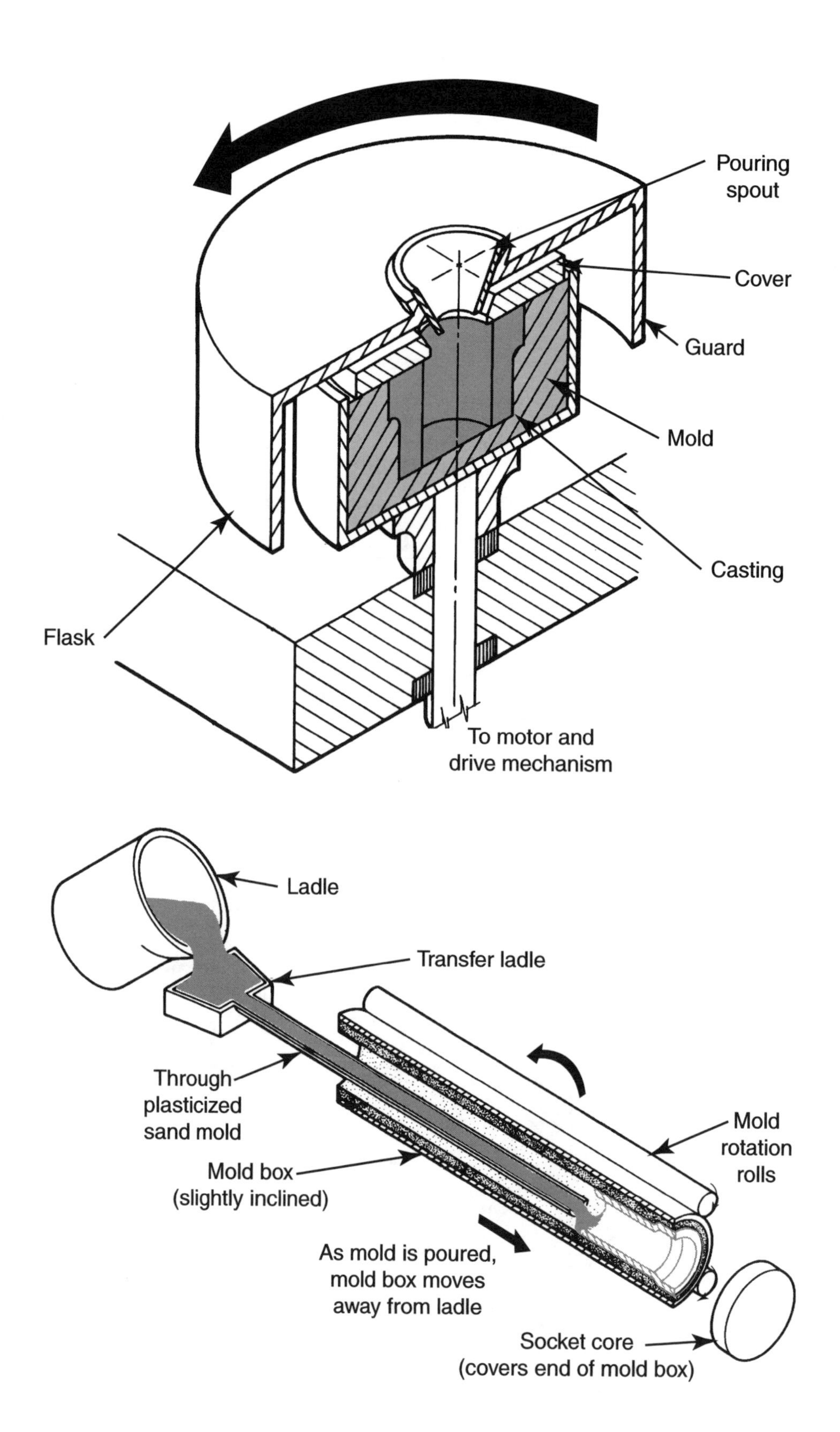

How Stereolithography Works

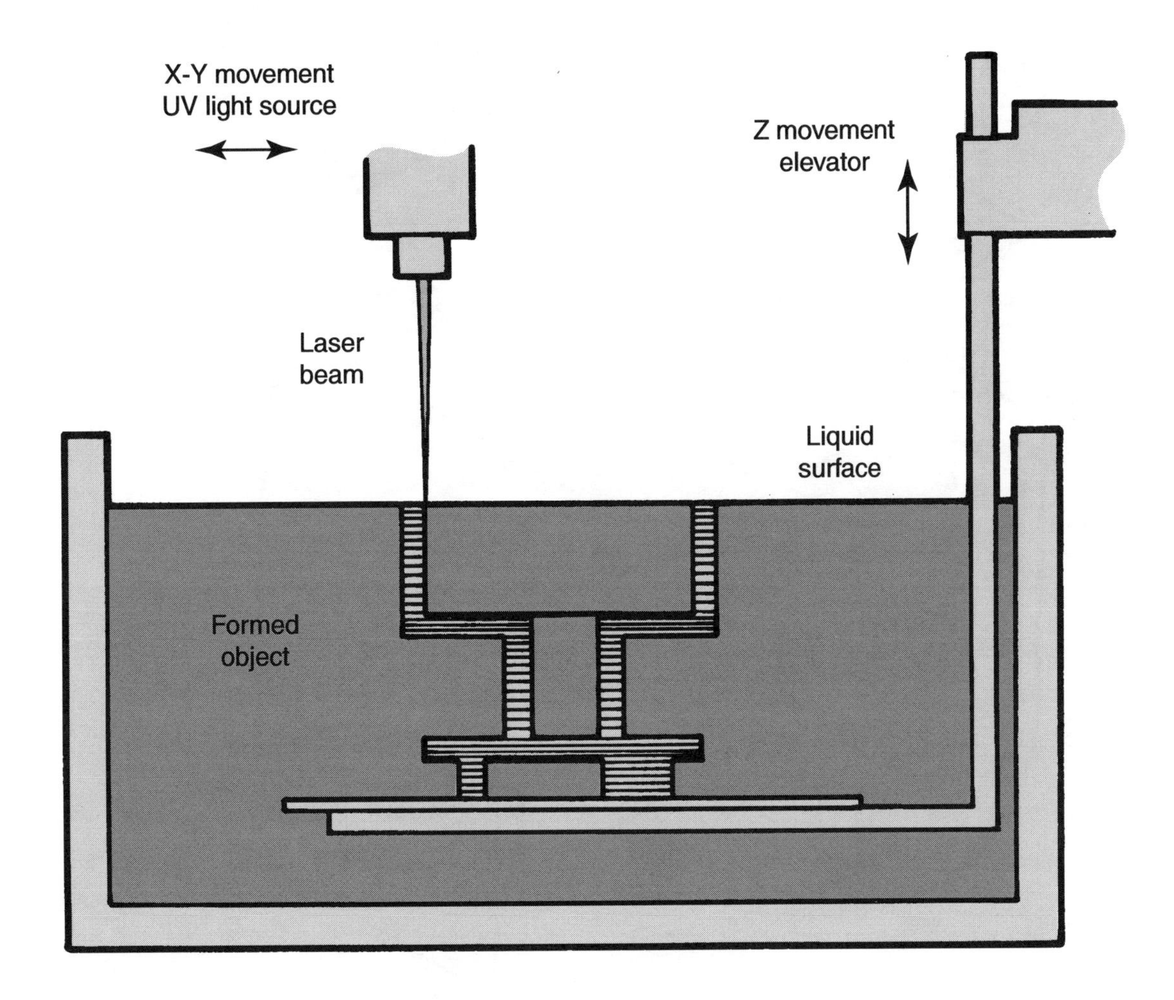

Laminated Object Manufacturing (LOM)

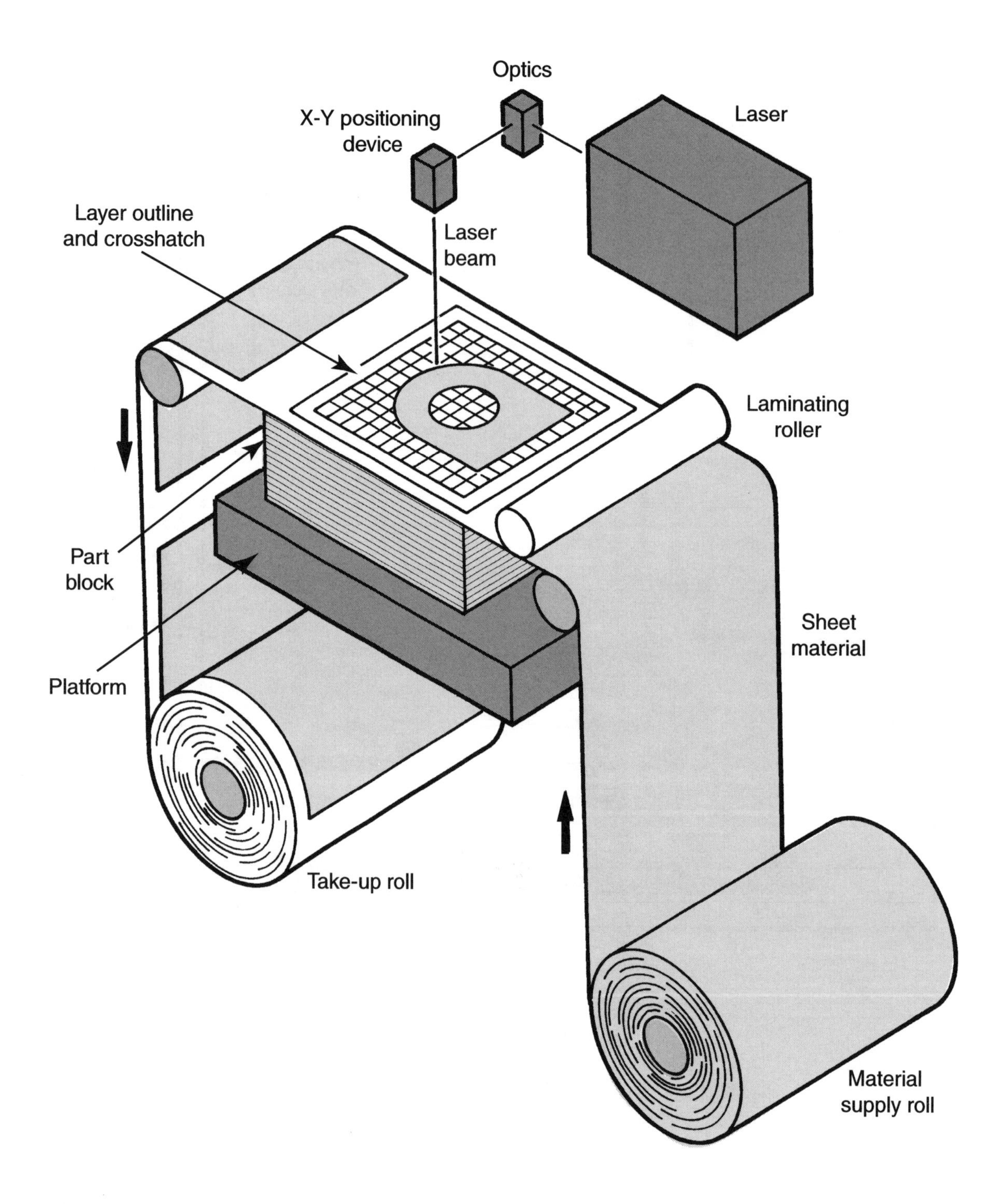

Chapter 15 Quiz

Metal Casting Techniques

Name: ______________________ **Date:**__________ **Period:**__________

1. List the five steps in making a casting.

   (a) ____

   ____

   (b) ____

   ____

   (c) ____

   ____

   (d) ____

   ____

   (e) ____

   ____

Match the word or phrase with the correct sentence.

_________ 2. Molten metal is forced into a die under pressure.

_________ 3. Casting technique using a metal mold that does not have to be destroyed to remove the casting.

_________ 4. Molten metal is poured into a metal mold and left long enough to form a thin shell. The remaining molten metal is poured out leaving a hollow casting.

_________ 5. Mold is rotated while metal is poured in.

_________ 6. Mold is made by coating a wax or plastic pattern with a refractory ceramic material. Mold is heated until the pattern is melted or burned out.

_________ 7. Casting technique used to make molds used in the manufacture of tires.

_________ 8. Molds made in the form of thin shells.

_________ 9. Technique employed to make prototypes of complex castings to check their designs.

(a) Investment casting
(b) Stereolithography
(c) Permanent mold casting
(d) Plaster mold casting
(e) Die casting
(f) Shell molding
(g) Centrifugal casting
(h) Slush molding

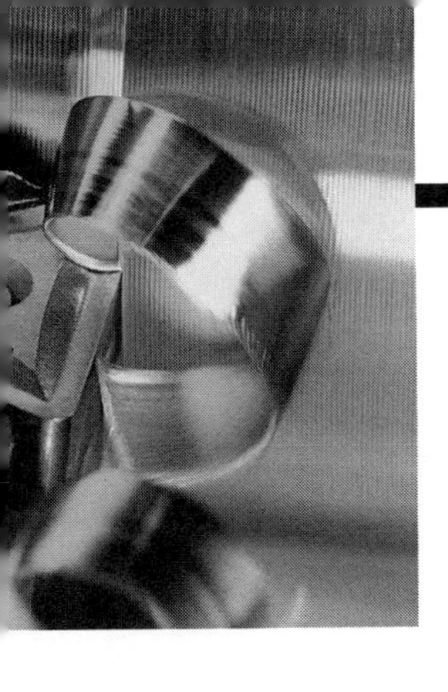

16

Wrought Metal

Learning Objectives

After studying this chapter, students should be able to:

- Explain the meaning of the term *wrought metal.*
- Identify the tools used in wrought metal work.
- Demonstrate the safe use of the tools used in wrought metal work.
- Bend metal rods, bars, and flats with hand tools.
- Form metal rods, bars, and flats with special equipment.
- Assemble wrought metal sections.
- Apply various finishes to wrought metal products.

Chapter Resources

Text, pages 249–260
  Test Your Knowledge, page 260
  Research and Development, page 260
Workbook, pages 85–88
Instructor's Manual
  Answer keys for:
    Test Your Knowledge Questions
    Workbook
    Chapter Quiz
  Reproducible Masters:
    16-1 Planning and Bending Sequence
    16-2 Drawing a Scroll
    16-3 Chapter Quiz

Guide for Lesson Planning

Use examples of wrought iron projects and products to introduce the chapter. Allow students to examine them.

Class Discussion and Demonstrations

Have students read and study all or part of the chapter. Review the assignment with them and discuss the following:

- Uses of wrought metal work.
- Tools, equipment, and materials available.
- Reasons twisted sections are used.
- Bending jigs and their uses.
- Need for full-size patterns.
- Types of scroll ends and why they are used.
- Bending machines.
- Assembling wrought ironwork.
- Finishing wrought ironwork.

Demonstrate the following operations. Be sure all students can observe and hear the demonstration.

- Techniques for bending angles.
- How to determine length of section to be bent.
- Problems encountered if bends are not carefully planned.
- Safe use of rod parter.
- Making twisted sections.
- Using bending jigs.
- Operating bending machines.
- Safety precautions that must be observed when working in wrought metal work.

Assignments

1. Assign the Test Your Knowledge questions.
2. Assign Chapter 16 in the *Modern Metalworking Workbook.*
3. Assign the chapter quiz. Copy and distribute Reproducible Master 16-3.
4. Permit students to volunteer for the Research and Development activities at the end of the chapter.

Test Your Knowledge

1. Metals that have been bent or formed into shape with various tools.
2. ornamental ironwork, bench metal
3. 1/4 (6.5 mm)
4. one-half
5. (d) All of the above.
6. full-size
7. decorative purposes
8. Evaluate individually. Refer to Section 16.1.3.
9. A modified machinist's vise to which forming rolls have been adapted.
10. (b) applying a flat black lacquer or paint

Workbook

1. blacksmith
2. Thickness of metal. One-half of its thickness must be added to the length for each bend. If several bends must be made in the same piece, the bending sequence must be carefully planned.
3. Acute
4. To check the curves to be made during the bending operation.
5. (d) All of the above.
6. A modified machinist's vise to which forming rolls have been adapted.
7. Rivets, welding
8. Evaluate individually. Refer to Section 16.4.
9. Evaluate student drawings individually. Refer to text Figure 16-14.
10. Evaluate student drawings individually. Refer to text Figure 16-15.

Chapter Quiz

1. carbon
2. cold
3. (e) None of the above.
4. greater, less
5. For additional stiffness and for decorative purposes.
6. Evaluate individually. Refer to Section 16.4 in the text.

Planning and Bending Sequence

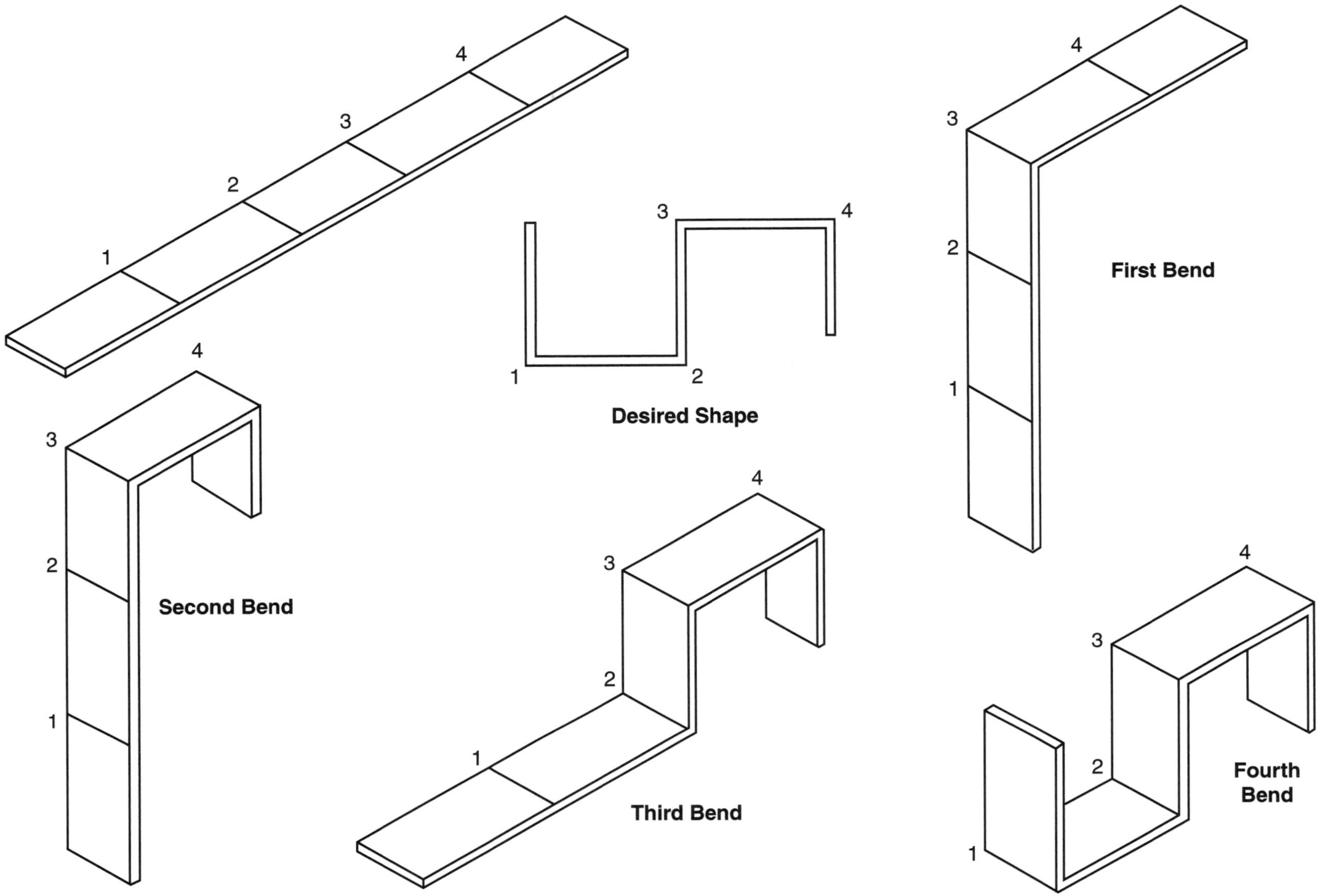

1
2
3
4

3 4
1 2
Desired Shape

4
3
2
1
First Bend

4
3
2
1
Second Bend

4
3
2
1
Third Bend

4
3
2
1
Fourth Bend

16-1

Drawing a Scroll

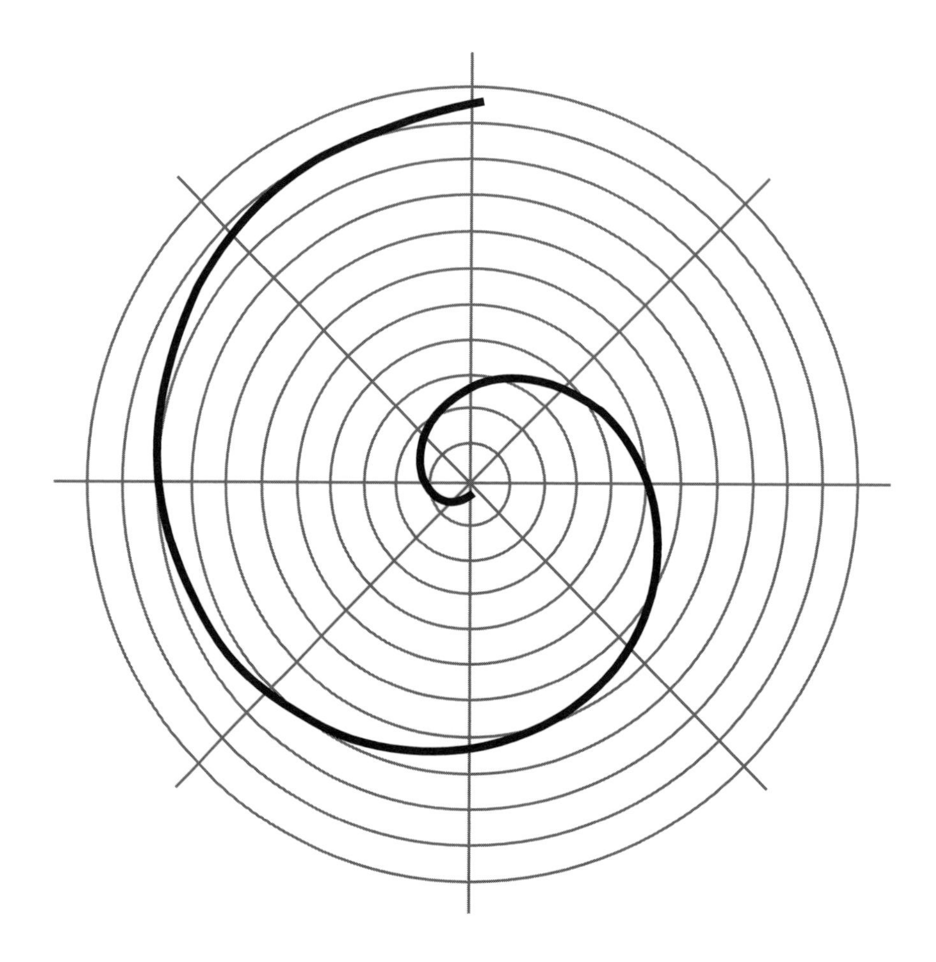

Chapter 16 Quiz

Wrought Metal

Name: ______________________________ **Date:**______________ **Period:**______________

Carefully read each question before attempting to answer it.

1. Wrought iron is almost pure iron containing very little _____. 1.______________
2. Metal up to 1/4″ thick can be bent _____. 2.______________
3. In working wrought metal _____. 3.______________
   (a) the length of the metal is increased slightly with each bend
   (b) subtract one-half the metal thickness to the length of the piece for each bend
   (c) the bending sequence is inconsequential
   (d) All of the above.
   (e) None of the above.
4. An obtuse angle is _____ than 90°; an acute angle is _____ than 90°. 4.______________

   ____
5. Why are wrought metal sections twisted at times?

   ____

   ____
6. List four safety rules that should be observed when working wrought metal.

   1)____

   ____

   2)____

   ____

   3)____

   ____

   4)____

   ____

17 FORGING

LEARNING OBJECTIVES

After studying this chapter, students should be able to:

- Identify the tools used in hand forging.
- Demonstrate several forging techniques.
- Bend, draw out, and upset metal by hand forging.
- Practice hand forging safety rules.
- Explain industrial forging processes.

CHAPTER RESOURCES

Text, pages 261–274
  Test Your Knowledge, pages 273–274
  Research and Development, page 274
Workbook, pages 89–92
Instructor's Manual
  Answer keys for:
    Test Your Knowledge Questions
    Workbook
    Chapter Quiz
  Reproducible Masters:
    17-1 Why a Forged Piece is Stronger
    17-2 Drop Forging
    17-3 Press Forging
    17-4 Roll Forging
    17-5 Upset Forging
    17-6 Intermittent Roll Forging
    17-7 Cross Forging
    17-8 The Basic Cold Forming Processes
    17-9 Chapter Quiz

GUIDE FOR LESSON PLANNING

Introduce chapter by having students examine similar forged and cast tools. Question them on which type, forged or cast, is the stronger. Which type do they think is more expensive? What are their reasons for determining which is the more costly?

CLASS DISCUSSION AND DEMONSTRATION

Have students read and study all or part of the chapter. Using the appropriate Reproducible Masters, review the assignment and discuss and demonstrate:

- Forging tools.
- Forging furnace and its operation.
- Drawing out metal sections.
- Bending metal sections.
- Upsetting or bulging metal sections.
- Safe forging practices.
- Industrial forging processes.

ASSIGNMENTS

1. Assign the Test Your Knowledge questions.
2. Assign Chapter 17 in the *Modern Metalworking Workbook.*
3. Assign the chapter quiz. Copy and distribute Reproducible Master 17-9.
4. Permit students to volunteer for the Research and Development activities at the end of the chapter.

TEST YOUR KNOWLEDGE

1. Student answers will vary. Evaluate individually.
2. improves
3. heat metal; coal, gas
4. Evaluate individually. Refer to Figure 17-3.
5. They hold hot metal while it is being forged.
6. Evaluate individually. Refer to Section 17.2.
7. bright
8. (b) lengthened
9. (a) the opposite of drawing out
10. hand
11. shaped dies
12. gradual application of pressure

13. Evaluate individually. Refer to Section 17.4.3.
14. rotary
15. Cold forming is a forging technique that forms cold metal wire or rod into a desired shape using a series of dies. It is also known as cold heading and chipless machining.
16. Evaluate individually. Refer to Figure 17-30.
17. Evaluate individually. Refer to Section 17.4.6.

Workbook

1. The process of using pressure to shape metal. In some forging techniques, the metal is rotated to make it easier to shape.
2. physical characteristics
3. tongs
4. (a) Pritchel hole
   (b) Hardy hole
   (c) Face
   (d) Rounded edge
   (e) Chipping block
   (f) Horn
5. (a) Straight lip tongs
   (b) Cross peen hammer
   (c) Ball peen hammer
   (d) Hardy
6. Drawing out means the metal is stretched or lengthened by the forging operation.
7. The operation that thickens or bulges the piece and, at the same time, shortens it.
8. (d) Dies
9. (a) Open die forging
10. (b) Drop forge
11. (h) Flash
12. (c) Precision forging
13. (e) Roll forging
14. (g) Intraform
15. Evaluate student work individually.
16. Evaluate student work individually.

Chapter Quiz

1. (d) All of the above.
2. (a) stronger
3. (f) Forge
4. (h) Anvil
5. (a) Tongs
6. (j) Hammers
7. (g) Hardies
8. (n) Drawing out
9. (m) Upsetting
10. (c) Drop forging
11. (e) Cross forging
12. (b) Flash
13. (l) Dies
14. (i) Swaging
15. (k) Roll forging

Why a Forged Piece is Stronger

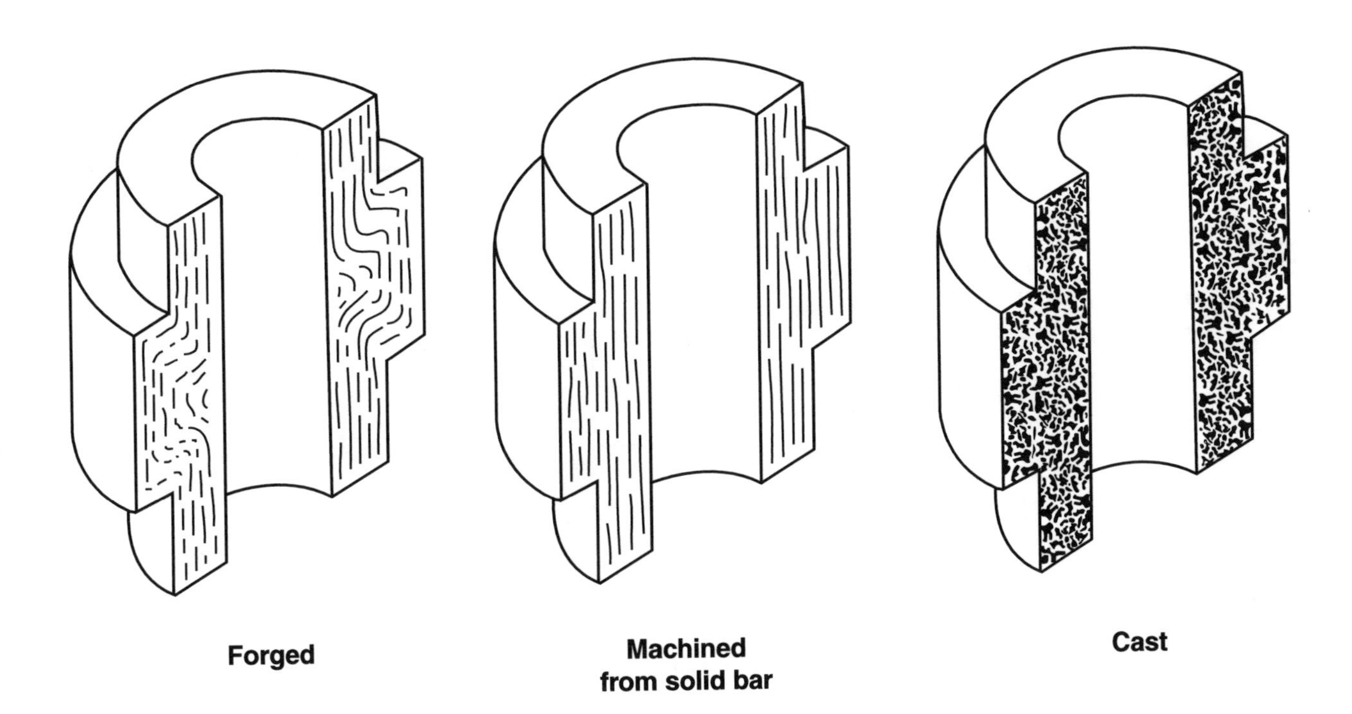

17-2

Drop Forging

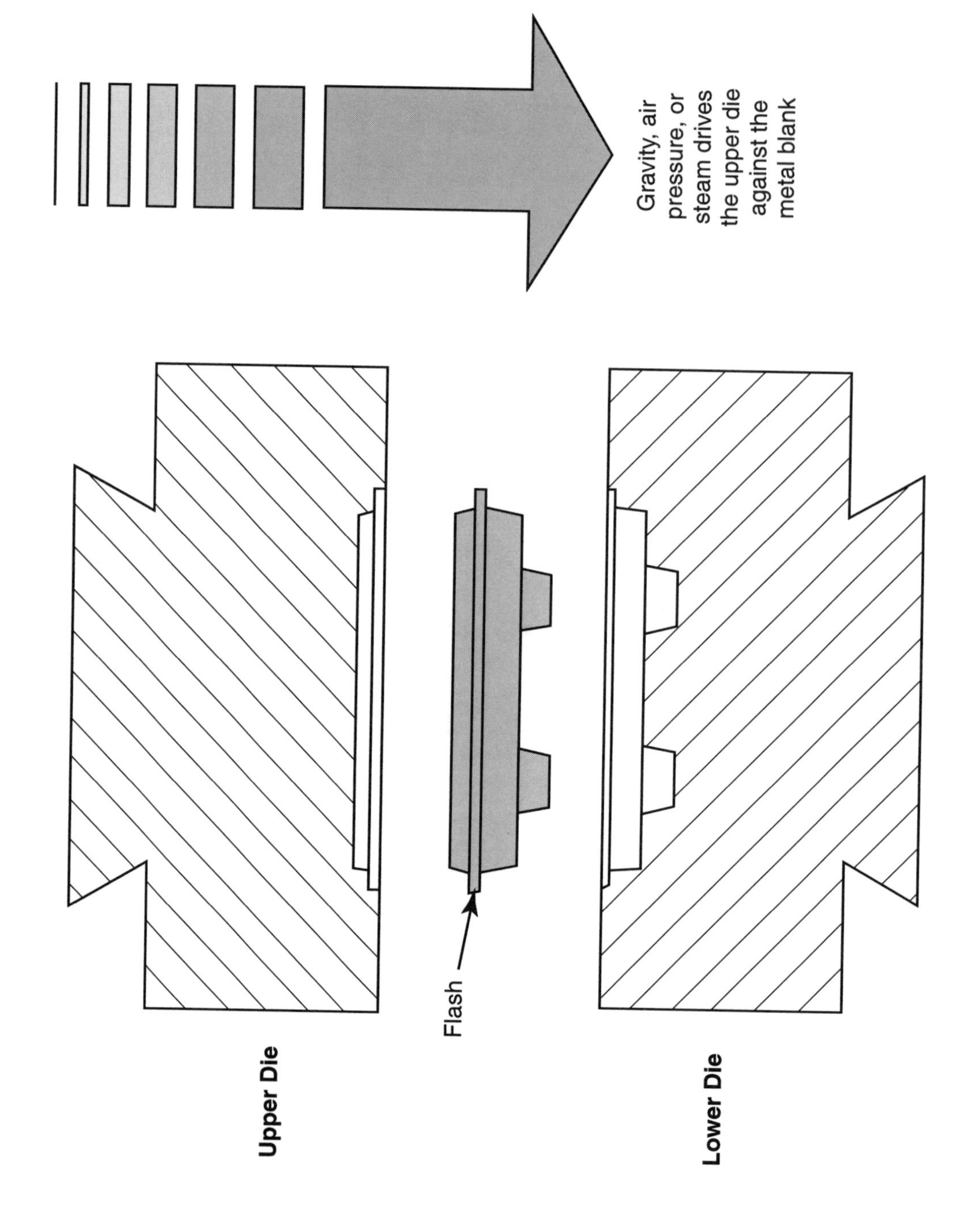

17-3

Press Forging

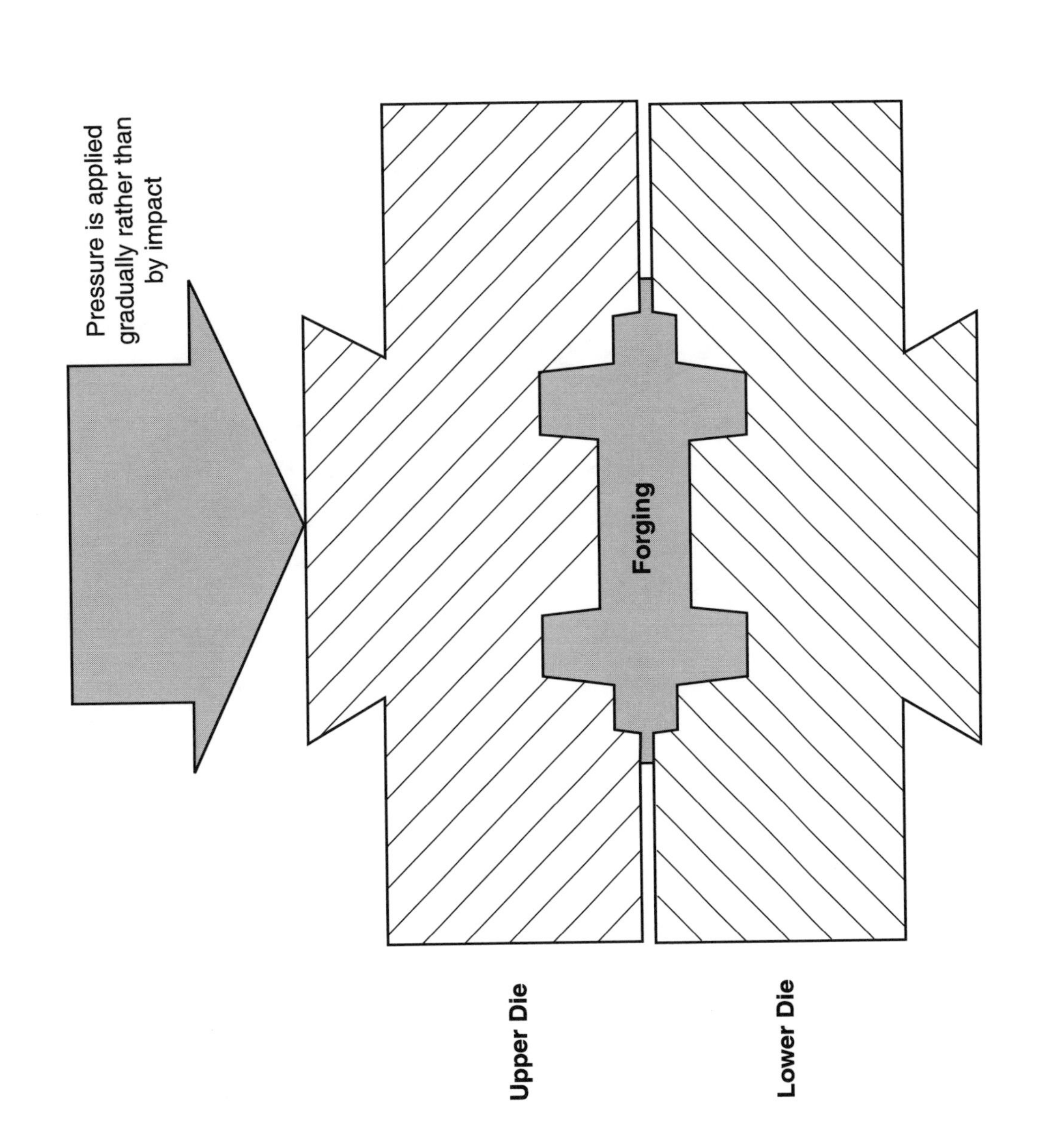

Roll Forging

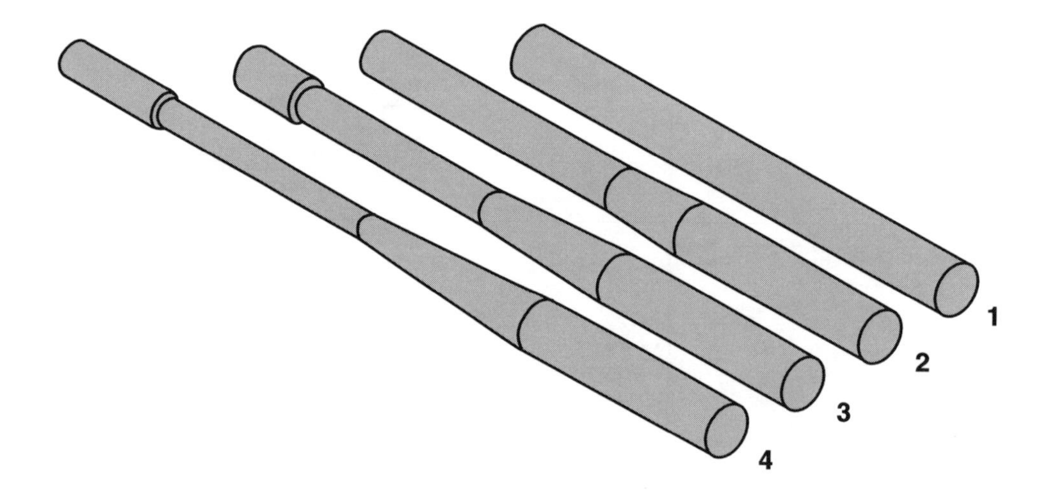

Upset Forging

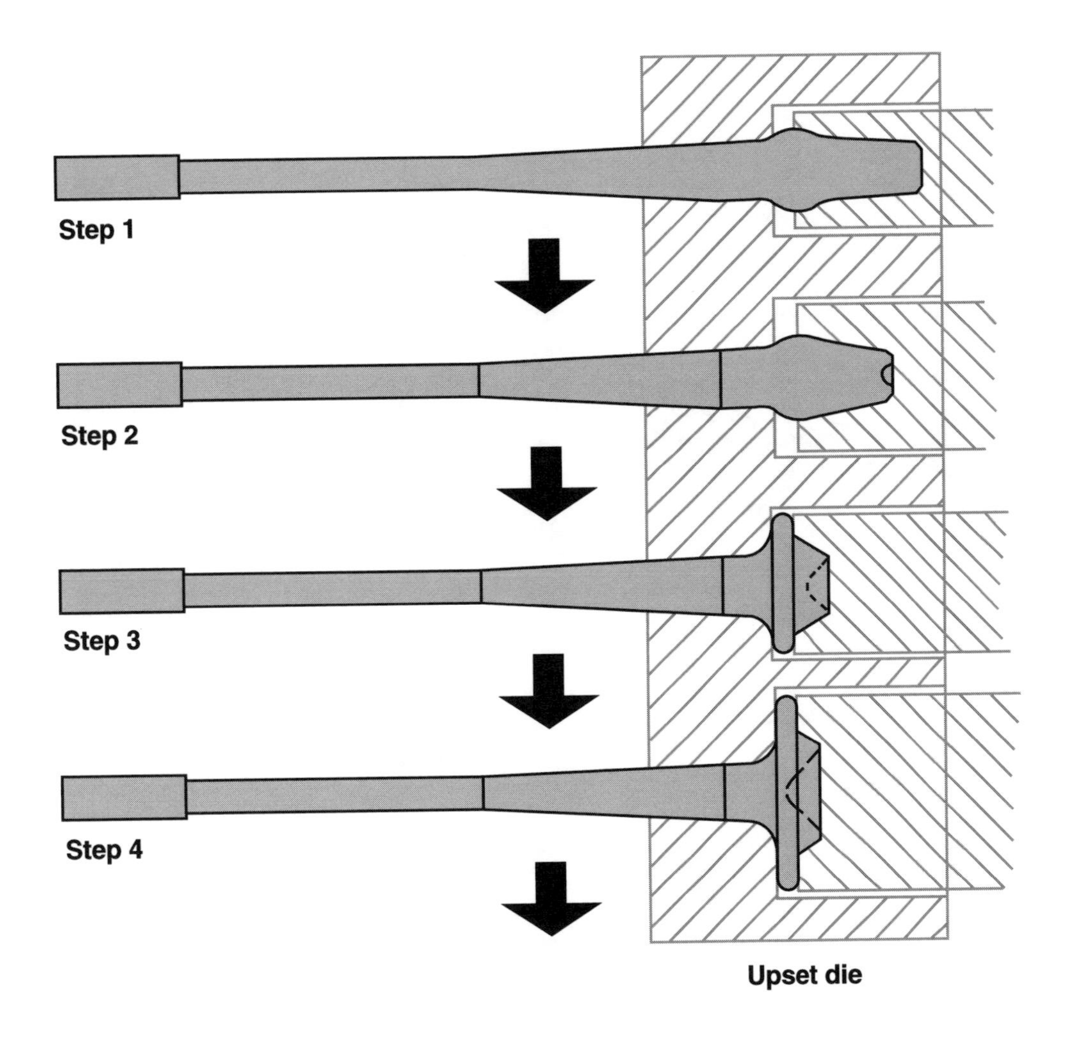

Intermittent Roll Forging

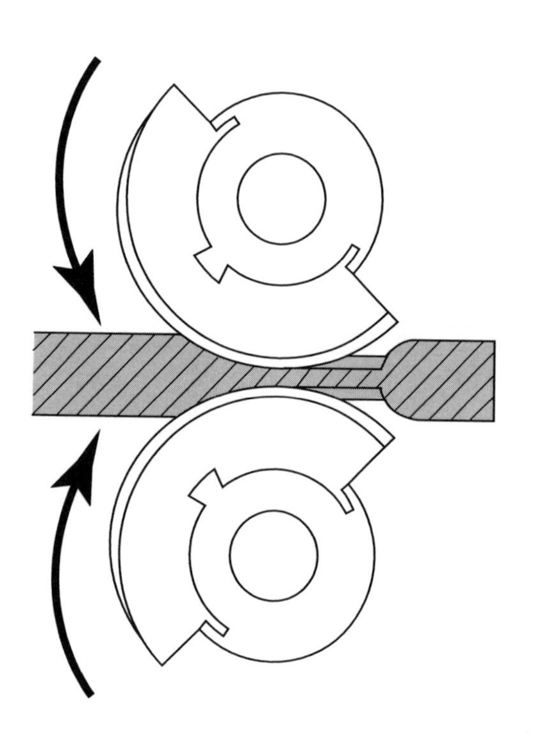

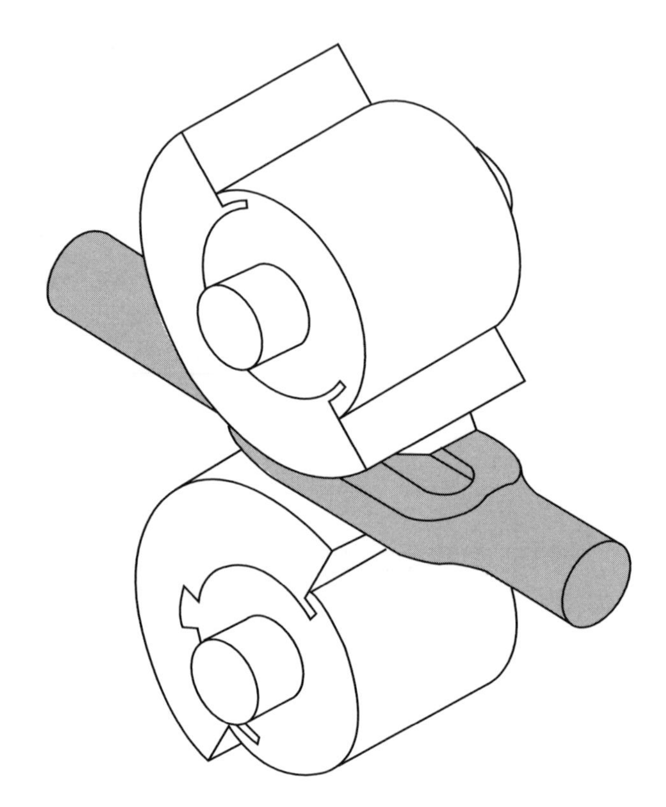

Cross Forging

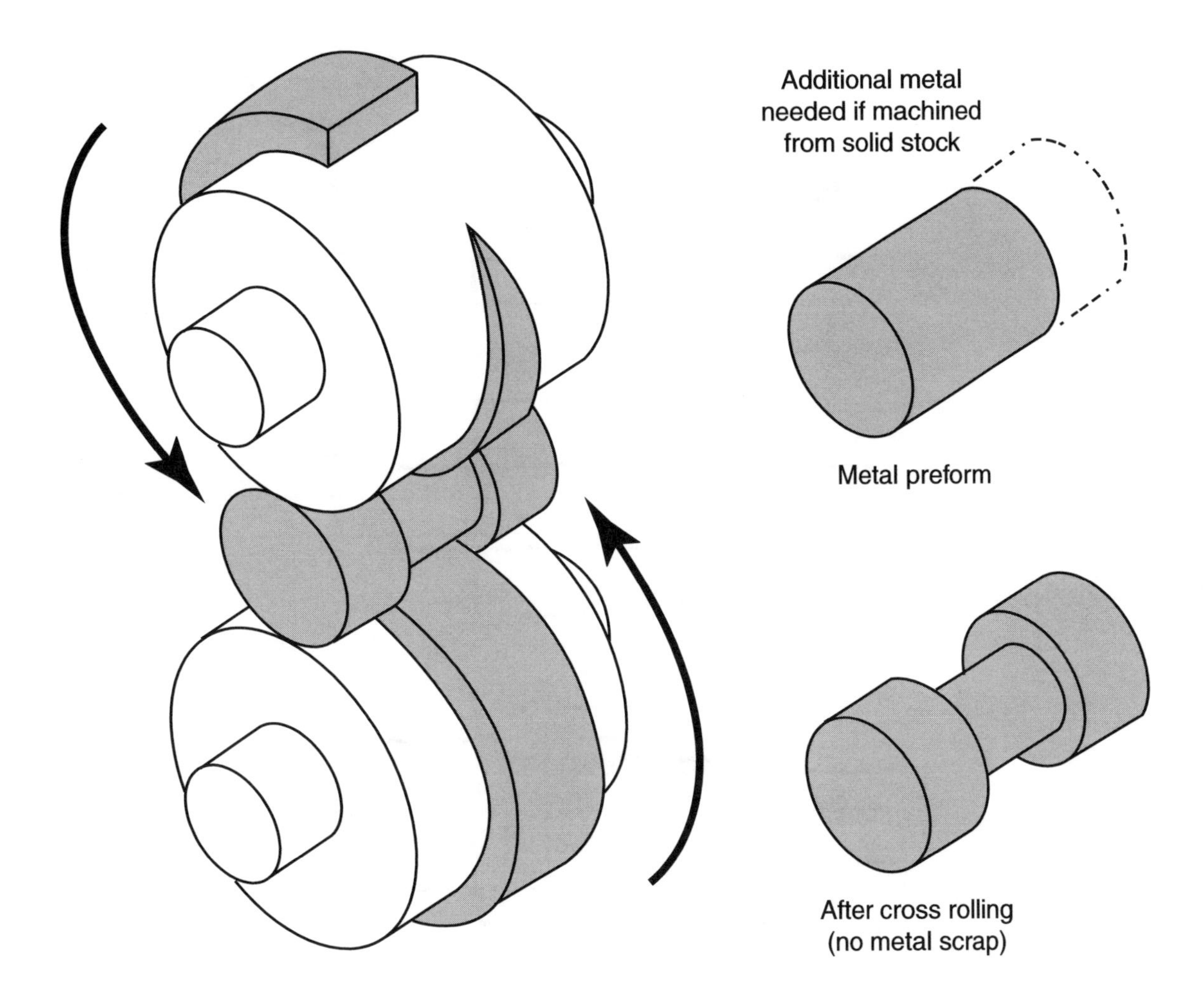

The Basic Cold Forming Processes

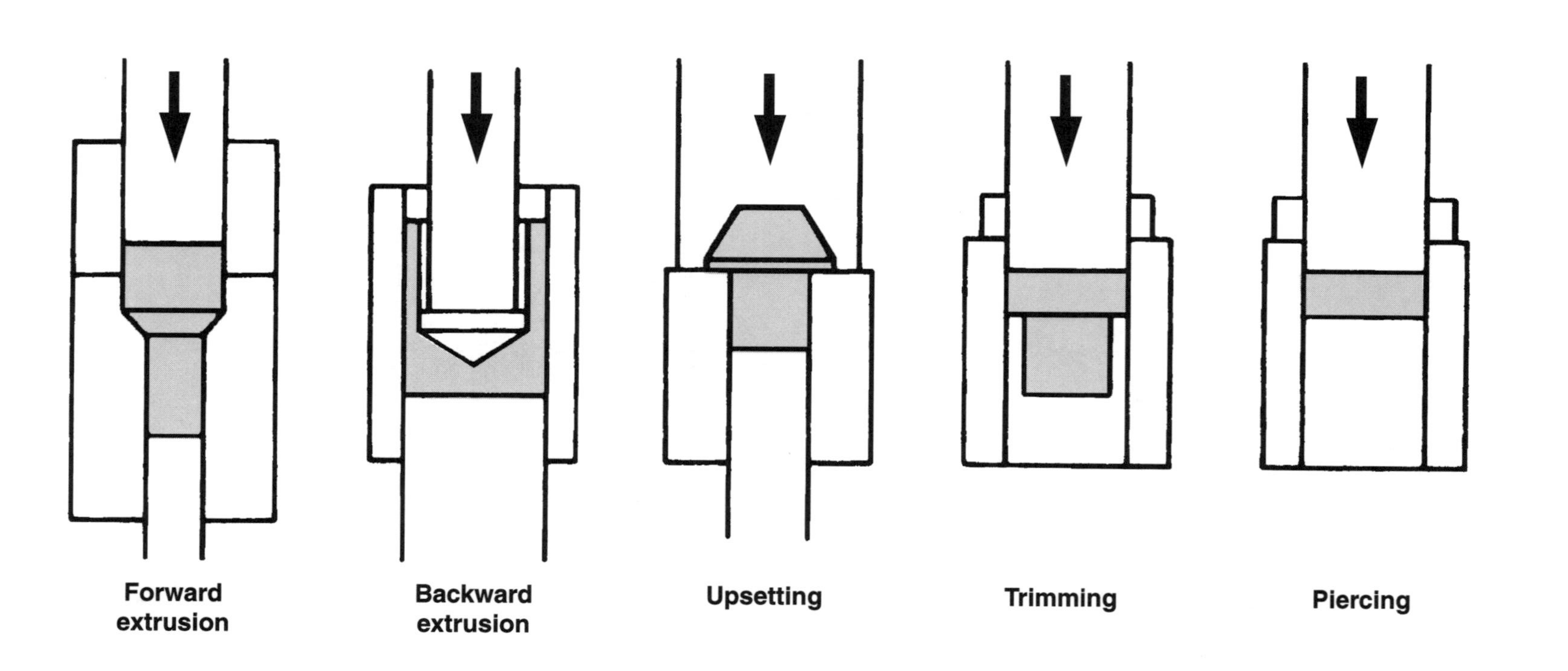

Chapter 17 Quiz

Forging

Name: ______________________ **Date:** ______________ **Period:** ______________

Carefully read each question before attempting to answer it.

1. Forging is the process that _____. 1. ______________
   (a) uses pressure to shape metal
   (b) shapes metal with a hammering action or by squeezing
   (c) requires some metals to be heated before they can be shaped
   (d) All of the above.
   (e) None of the above.

2. A forged piece is _____ than an identical piece that has been cast or machined from a solid bar of metal. 2. ______________
   (a) stronger
   (b) weaker
   (c) no different in strength
   (d) All of the above.
   (e) None of the above.

Match the word or phrase with the following sentences.

________ 3. Used to heat the metal prior to forging.

________ 4. Heated metal is hand forged on this piece of equipment.

________ 5. Tools used to hold metal during hand forging.

________ 6. Available in many shapes and sizes. Used to apply shaping force during hand forging.

________ 7. Fitted to anvil and employed to cut metal.

________ 8. Metal is stretched or lengthened by this forging operation.

________ 9. Forging operation that thickens or bulges metal.

________ 10. Dies replace the flat dies of open die forging.

________ 11. Metal is shaped by gradual application of pressure.

________ 12. Surplus metal that forms at the parting line of dies.

________ 13. Hardened steel blocks with cavities that shape the forging.

________ 14. A form of forging in which metal is shaped by a series of rapid hammerlike blows.

________ 15. Rolls of the desired shape reduce short, thick metal sections into long slender sections.

(a) Tongs
(b) Flash
(c) Drop forging
(d) Open die forging
(e) Cross forging
(f) Forge
(g) Hardies
(h) Anvil
(i) Swaging
(j) Hammers
(k) Roll forging
(l) Dies
(m) Upsetting
(n) Drawing out

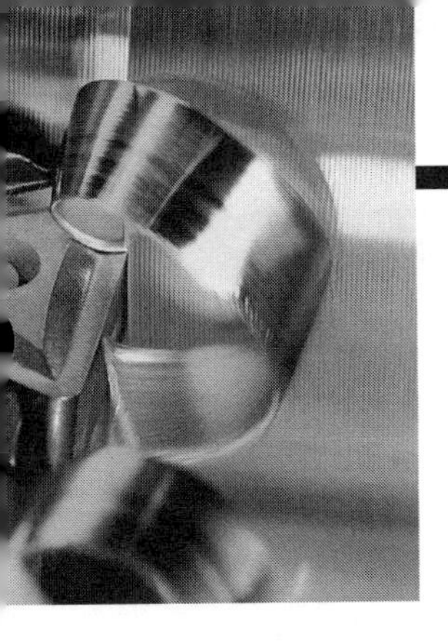

18 Heat Treatment of Metals

Objectives

After studying this chapter, students should be able to:

- Explain the reasons for heat treating metals.
- Summarize the principles of several heat treating processes.
- Demonstrate basic heat treating processes.
- Explain how some hardness testers operate.
- Apply heat treating safety precautions.

Chapter Resources

Text, pages 275–288
  Test Your Knowledge, page 287
  Research and Development, pages 287–288
Workbook, pages 93–94
Instructor's Manual
  Answer keys for:
    Test Your Knowledge Questions
    Workbook
    Chapter Quiz
  Reproducible Masters:
    18-1 Punch Set
    18-2 Small Hammer
    18-3 Chisel
    18-4 Center Punch
    18-5 Chapter Quiz

Guide for Lesson Planning

Prepare a number of center punches in various stages of heat treatment. The punch in the annealed state should show the point blunted because it is not hard enough. A properly heat treated punch will show no sign of wear after use. The fractured pieces of the glass hard punch will illustrate what happens when the steel is hardened but not tempered.

Class Discussion and Demonstration

Have the students read and study the chapter. Review the assignment with them asking appropriate questions such as:

- Name items they believe must be heat treated to do what they were designed to do.
- Heat treating terms.
- Why metals are heat treated.
- How heat treating affects the metal.
- The processes that are included under heat treating.
- Equipment used to determine the hardness of metals.
- Demonstrate annealing, hardening and tempering, case hardening techniques, and furnace safety.
- Emphasize the safety precautions that must be observed when heat treating metals.
- When demonstrating heat treating techniques, be sure there is adequate ventilation and students are wearing appropriate safety equipment.
- Projects should be planned to permit students to develop heat treating skills.

Assignments

1. Assign the Test Your Knowledge questions at the end of the chapter.
2. Assign Chapter 18 in the *Modern Metalworking Workbook.*
3. Assign the chapter quiz. Copy and distribute Reproducible Master 18-5.
4. Permit students to volunteer for the Research and Development activities at the end of the chapter.

Test Your Knowledge

1. heating, cooling
2. Any order: water, brine, oil, liquid nitrogen, blasts of cold air.
3. (a) makes it less brittle
4. annealing
5. Metal is heated to above its normal hardening temperature and allowed to cool slowly in some type of insulating material.
6. A process in which metal parts are heated to slightly above their critical temperature range, then cooled slowly in still air at room temperature. This relieves stresses that may have developed during machining, welding, or forming operations.
7. A heat-treating process that creates a thin, hardened, wear-resistant layer on the outer surface of a material while maintaining a soft and ductile inner core.
8. Evaluate individually. Refer to text Section 18.1.7.
9. pyrometer
10. Any order: Brinell, Rockwell, Shore Scleroscope, Webster.

Workbook

1. controlled heating and cooling of metal for the purpose of obtaining certain desirable changes in its physical characteristics such as hardness, toughness, and resistance to shock
2. Low-carbon, case hardening
3. Any two of the following: safe to operate, quiet, require no elaborate venting system, reach temperature quickly, can be controlled with accuracy.
4. Annealing
5. Quenching
6. Stress relieving
7. Case harden (surface harden)
8. Hardening
9. Tempering
10. Pyrometer
11. Hardness testing
12. Brinell
13. Rockwell
14. Shore scleroscope
15. Evaluate individually. Refer to Section 18.8.
16. Evaluate student work individually.
17. Evaluate student work individually.

Chapter Quiz

1. (h) Case hardening
2. (g) Annealing
3. (f) Stress relieving
4. (j) Surface hardening
5. (b) Hardening
6. (c) Tempering
7. (i) Quenching
8. (d) Pyrometer
9. (k) Hardness testing
10. (a) Brinell Hardness tester
11. (e) Rockwell hardness tester
12. (l) Shore Scleroscope

18-1

PUNCH SET

Material: ∅Drill rod

Heat treat after machining.

A pin punch set can be made by machining one each of the following sizes: ∅2.0, ∅4.0, ∅5.0, ∅7.5.

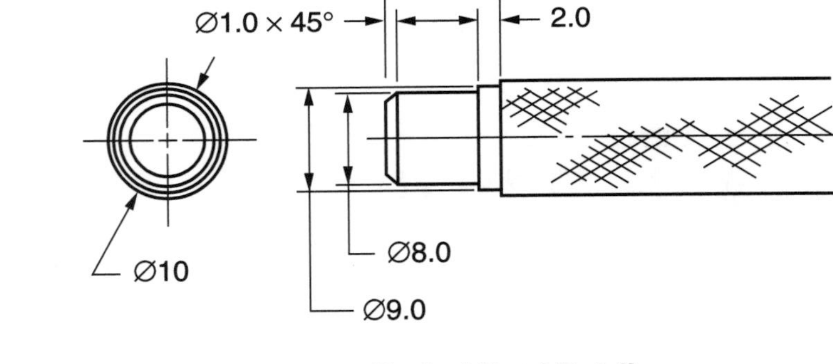

Typical Head Details

Medium knurl

60°

∅3.0

Prick Punch

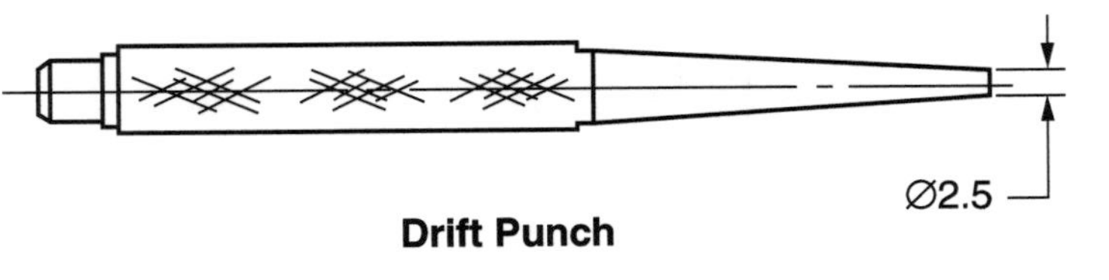

Drift Punch

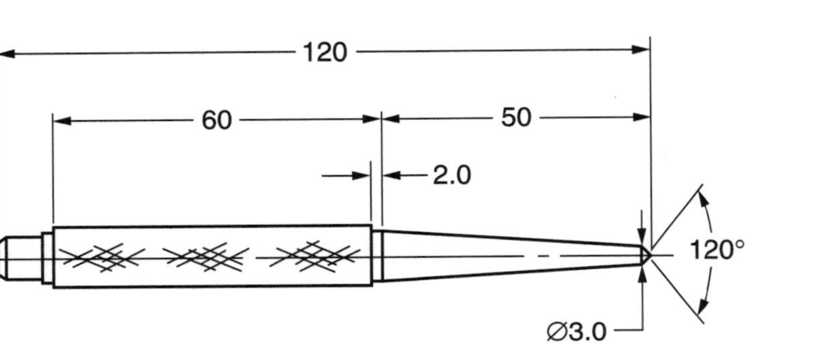

Center Punch

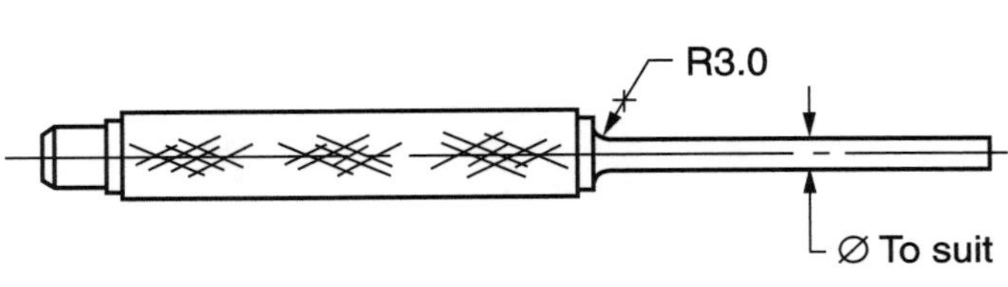

Pin Punch

Metric

Dimensions are in millimeters.

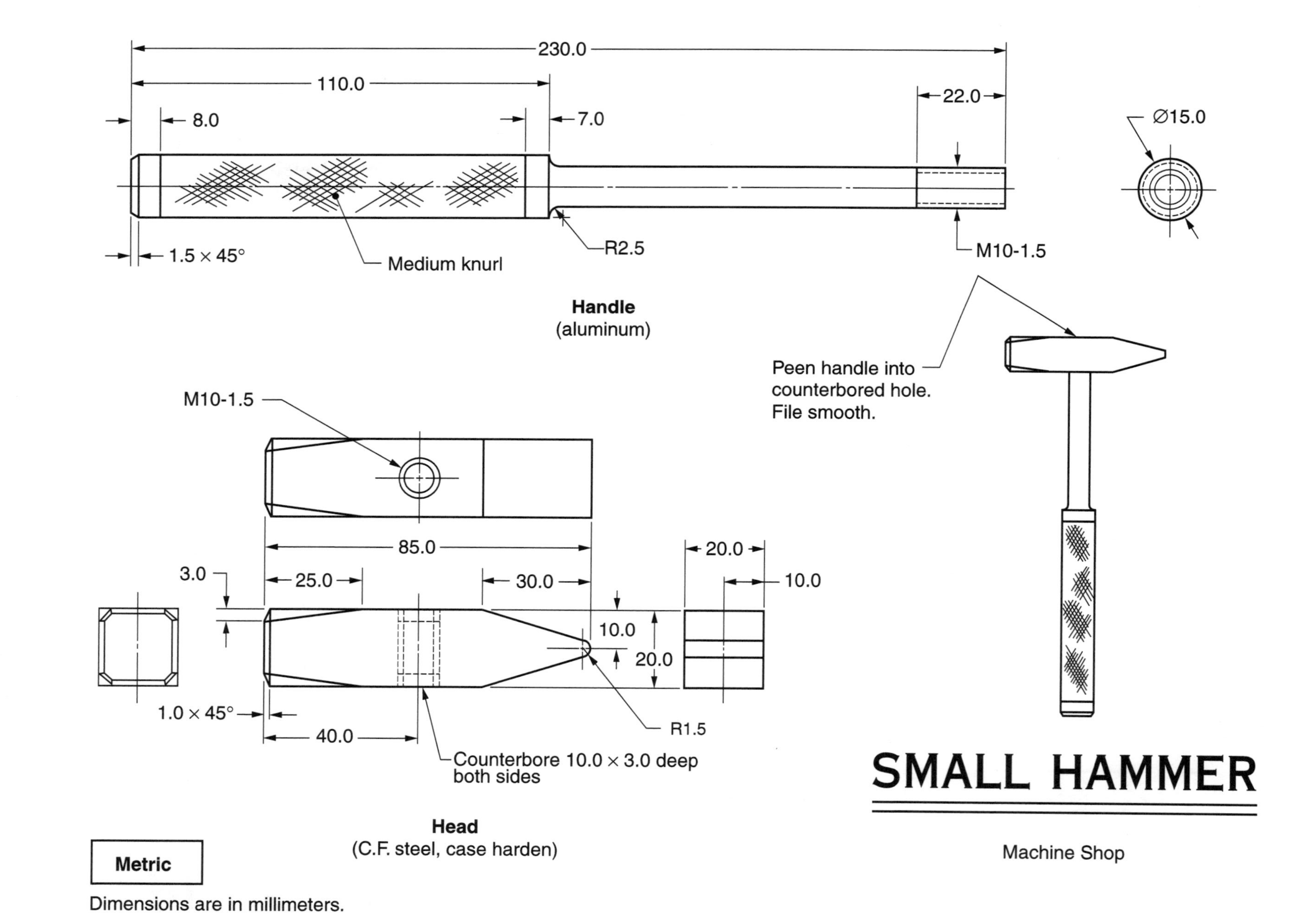
230.0
110.0
22.0
8.0
7.0
Ø15.0
1.5 × 45°
Medium knurl
R2.5
M10-1.5
Handle
(aluminum)
Peen handle into
counterbored hole.
File smooth.
M10-1.5
85.0
20.0
3.0
25.0
30.0
10.0
10.0
20.0
1.0 × 45°
40.0
R1.5
Counterbore 10.0 × 3.0 deep
both sides
SMALL HAMMER
Head
(C.F. steel, case harden)
Metric
Machine Shop
Dimensions are in millimeters.

18-2

CHISEL

1. Material: 7/16 hexagonal or octagonal tool steel.
2. Heat and forge to shape.
3. File smooth and grind chamfer on chisel head.
4. Heat treat and sharpen.

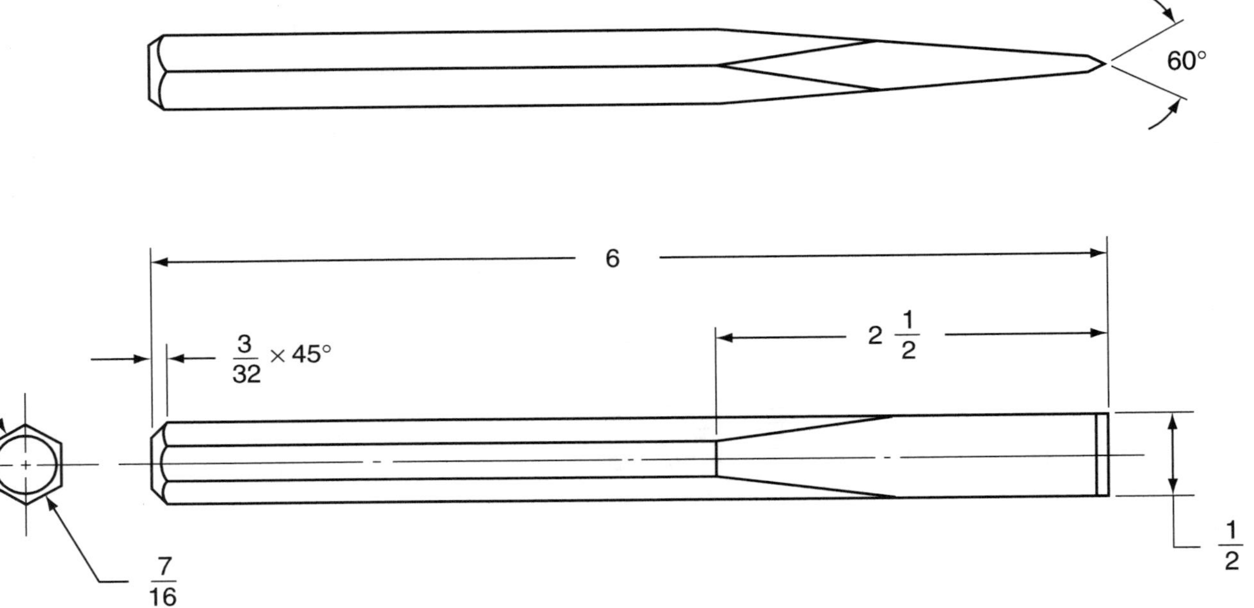

18-3

CENTER PUNCH

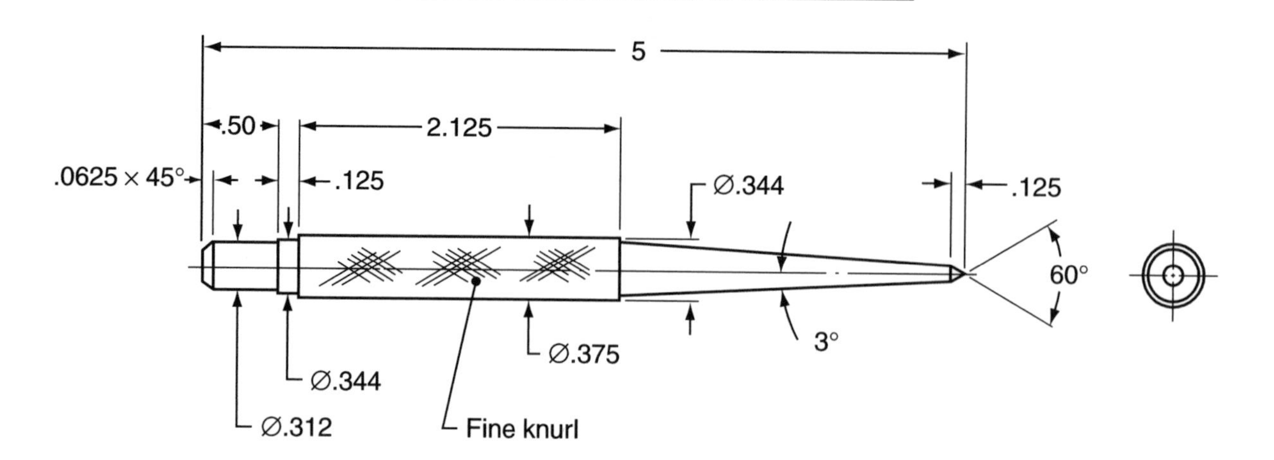

Machining Sequence:

1. Cut a piece of drill rod 5.50″ long.
   A. Face both ends.
   B. Center drill one end.
2. Machine center drilled end as follows:

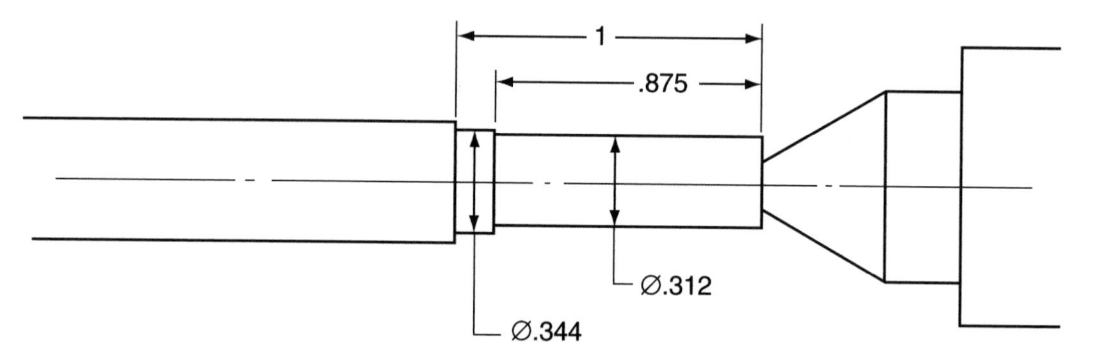

3. Knurl

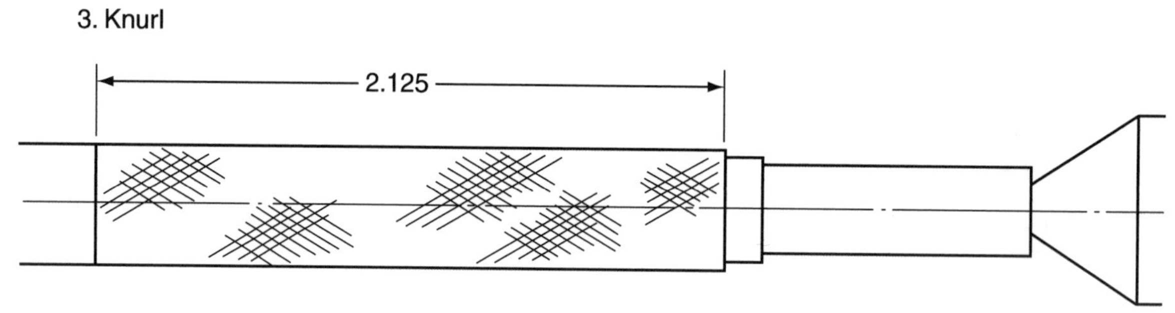

Sheet 1 of 2

4. Reverse work in chuck.
5. Machine to diameter.

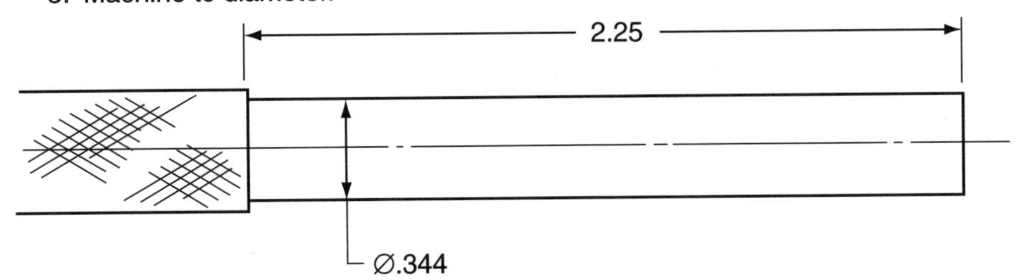

6. Set compound to cut 3° taper.

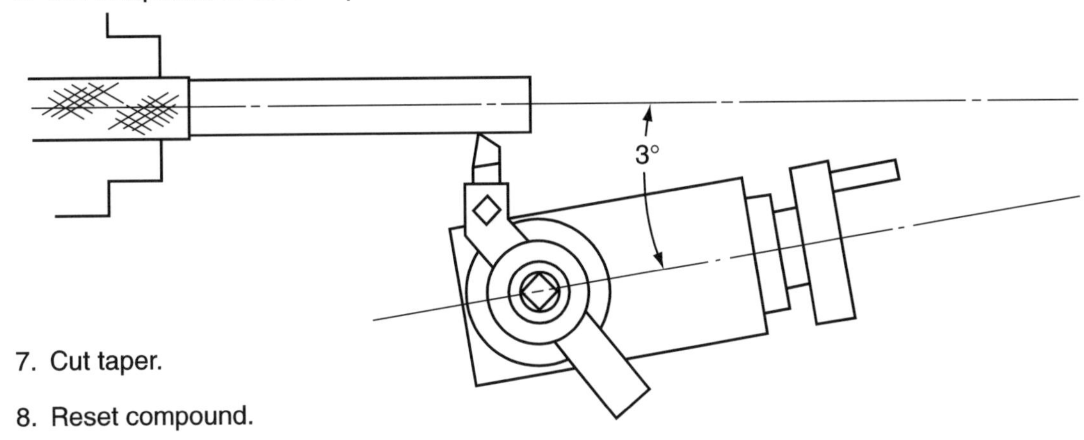

7. Cut taper.
8. Reset compound.
9. Machine point.

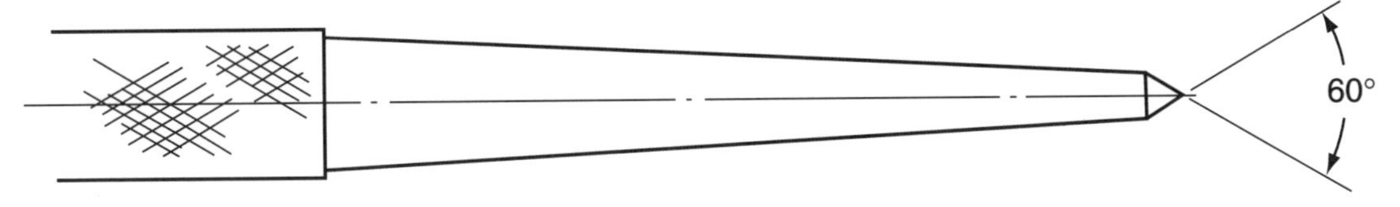

10. Reverse in chuck and finish machine head to size.

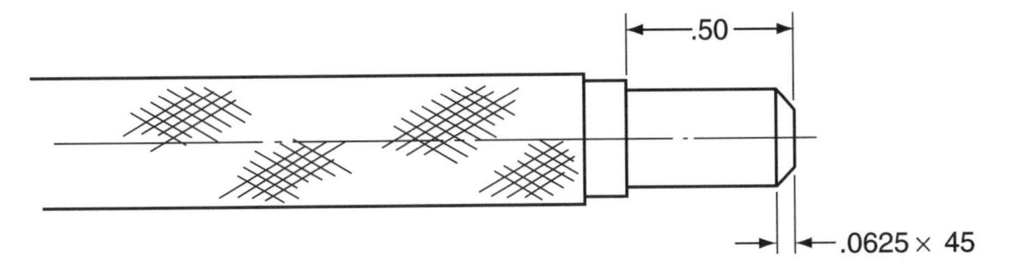

11. Heat treat completed punch.

Sheet 2 of 2

Chapter 18 Quiz

Heat Treatment of Metals

Name: ________________________________ **Date:** ________________________________

Match each word or phrase with the sentence that best describes it.

_________ 1. Produces a hard exterior surface on ferrous metals.

_________ 2. Process used to soften metals to make them easier to machine.

_________ 3. Used to remove internal stresses that have developed in parts that have been cold worked, machined, or welded.

_________ 4. Permits the surface of high carbon and alloy steels to be hardened without affecting the internal structure of the metal.

_________ 5. Technique normally used to obtain optimum physical qualities in steel.

_________ 6. Operation that reduces the brittleness and hardness of the above treated steel.

_________ 7. Cooling heated metal rapidly in water, brine, liquid nitrogen, or with blasts of cold air.

_________ 8. Used to monitor furnace temperature.

_________ 9. Technique of checking whether metal is in required condition of heat treatment.

_________10. Device that checks hardness of metal by forcing a steel ball into the material.

_________11. Device that uses a diamond penetrator to measure metal hardness.

_________12. Portable device that does not mar the surface of the metal being tested.

(a) Brinell Hardness tester
(b) Hardening
(c) Tempering
(d) Pyrometer
(e) Rockwell hardness tester
(f) Stress relieving
(g) Annealing
(h) Case hardening
(i) Quenching
(j) Surface hardening
(k) Hardness testing
(l) Shore Scleroscope

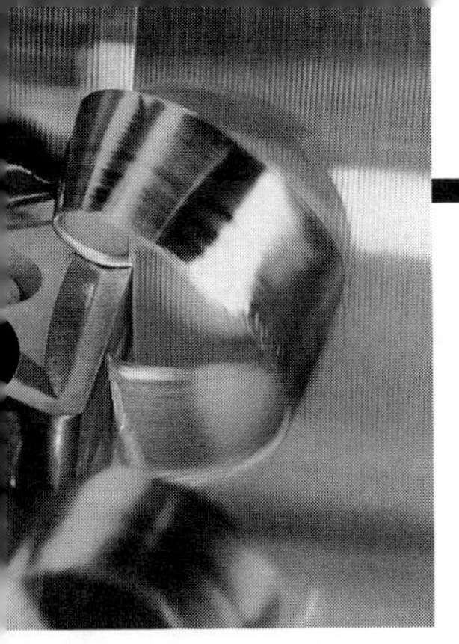

Gas Welding

19

Learning Objectives

After studying this chapter, students should be able to:

- Describe the basic welding processes.
- Identify the parts of an oxyacetylene gas welding outfit.
- Recognize the various weld joints.
- Select the correct protective clothing when gas welding.
- Safely light, adjust, and use a gas torch.
- Select the correct rod and flux for a job.
- Prepare a joint for gas welding.
- Weld using forehand and backhand techniques.
- Explain the difference between welding and brazing.
- Use safety precautions when welding.

Chapter Resources

Text, pages 289–301
  Test Your Knowledge, pages 300–301
  Research and Development, page 301
Workbook, pages 95–98
Instructor's Manual
  Answer keys for:
    Test Your Knowledge Questions
    Workbook
    Chapter Quiz
  Reproducible Masters:
    19-1 Welding Processes
    19-2 Basic Parts of a Welding Torch
    19-3 Basic Weld Joints
    19-4 Chapter Quiz

Guide for Lesson Planning

Have sufficient safety equipment on hand (goggles, shields, etc.) when demonstrating welding techniques. Ventilation must be adequate to remove all fumes. Equipment should be inspected to ensure that it is in safe operating condition. Arrange students so all will be able to see and hear and are clear of any splattering hot weld material.

Gas welding equipment and examples of the five basic joints should be available for the students to examine. Emphasize the safety precautions that must be observed when gas welding.

Class Discussion

Have students read and study the chapter. Review the material using Reproducible Masters 19-1 through 19-3 as overhead transparencies or as handouts. Discuss the assignment by explaining, questioning, and, where necessary, demonstrating the following:

- How welding, brazing, and soldering differ.
- Gas welding equipment, its care and use.
- Gas welding rods.
- Fluxes and why they are used.
- Safety equipment to be worn when gas welding.
- Preparing and adjusting gas welding equipment.
- Lighting the torch.
- Adjusting the flame.
- Five basic types of joints.
- Joint preparation.
- Proper way to gas weld.
- Brazing and how it is done.

Assignments

1. Assign the Test Your Knowledge questions.
2. Assign Chapter 19 in the *Modern Metalworking Workbook.*
3. Assign the chapter quiz. Copy and distribute Reproducible Master 19-4.
4. Permit students to volunteer for the Research and Development activities at the end of the chapter.

Test Your Knowledge

1. melt, fuse together
2. gases, melt, fuse
3. filler metal
4. (a) cylinders
   (b) regulators
   (c) hoses
   (d) torch
5. red
6. green
7. Welding hoses use different types of fittings to prevent improper use or interchanging of hoses.
8. spark igniter, match
9. rusting
10. brass, bronze
11. If allowances are not made for metal contraction as the weld cools.
12. tip cleaner
13. Flux dissolves and facilitates the removal of oxides and other undesirable substances.
14. Evaluate individually. Refer to text Section 19.5.1.
15. Any order: butt, T, lap, edge, corner. Evaluate sketches individually.
16. forehand, backhand
17. The molten spot produced by the torch flame.
18. Brazing
19. braze
20. Evaluate individually. Refer to text Section 19.11.

Workbook

1. Welding is the process of joining metals by heating them to a suitable temperature where they will melt and fuse together. This may be done with or without the application of pressure, and with or without the use of filler metal of a similar composition and melting point as the base metal.
2. make use of burning gases, such as acetylene or hydrogen, to produce the required heat needed for the metal to melt and fuse
3. Evaluate student descriptions individually. Refer to Section 19.2.
4. oxygen, acetylene
5. Spark igniter
6. metal joint, base metal
7. A bronze welding rod is preferred because it has superior strength.
8. Any order: flux cleans the metal, prevents oxidation from forming, promotes a better weld.
9. carburizing
10. oxygen, oxidizing
11. Students answers should include drawings of the butt joint, T-joint, lap joint, edge joint, and corner joint. Refer to Section 19.6.
12. forehand; The flame is in the direction the weld is progressing.
13. backhand; The flame is directed back over the welded portion.
14. Evaluate individually. Refer to text Section 19.11.
15. Evaluate student welds individually.

Chapter Quiz

1. (d) All of the above.
2. cylinders
3. regulators
4. green, red
5. torch
6. (d) All of the above.
7. match, spark lighter
8. Butt joint
9. T-joint
10. Lap joint

Welding Processes

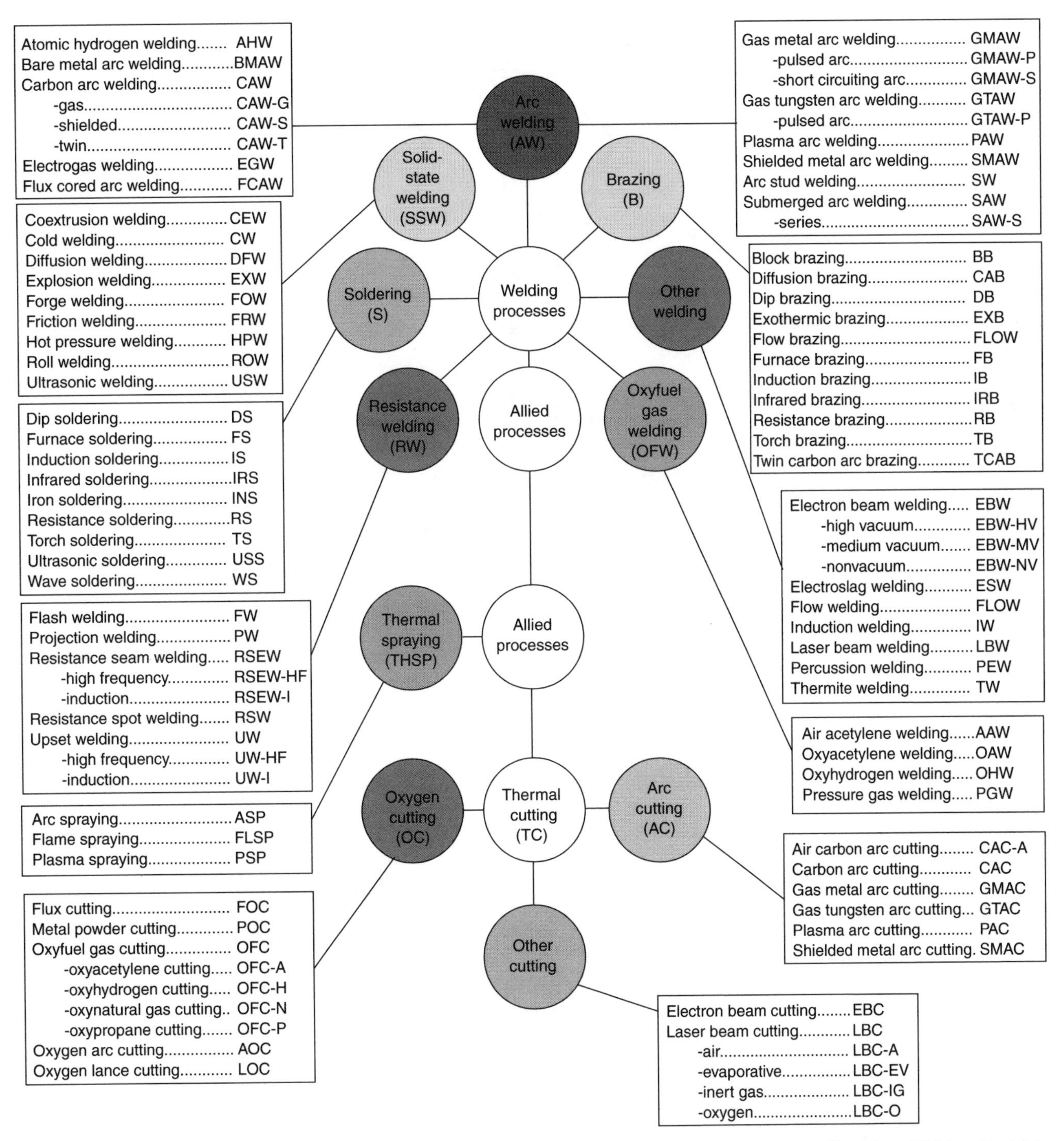

(American Welding Society)

Basic Parts of a Welding Torch

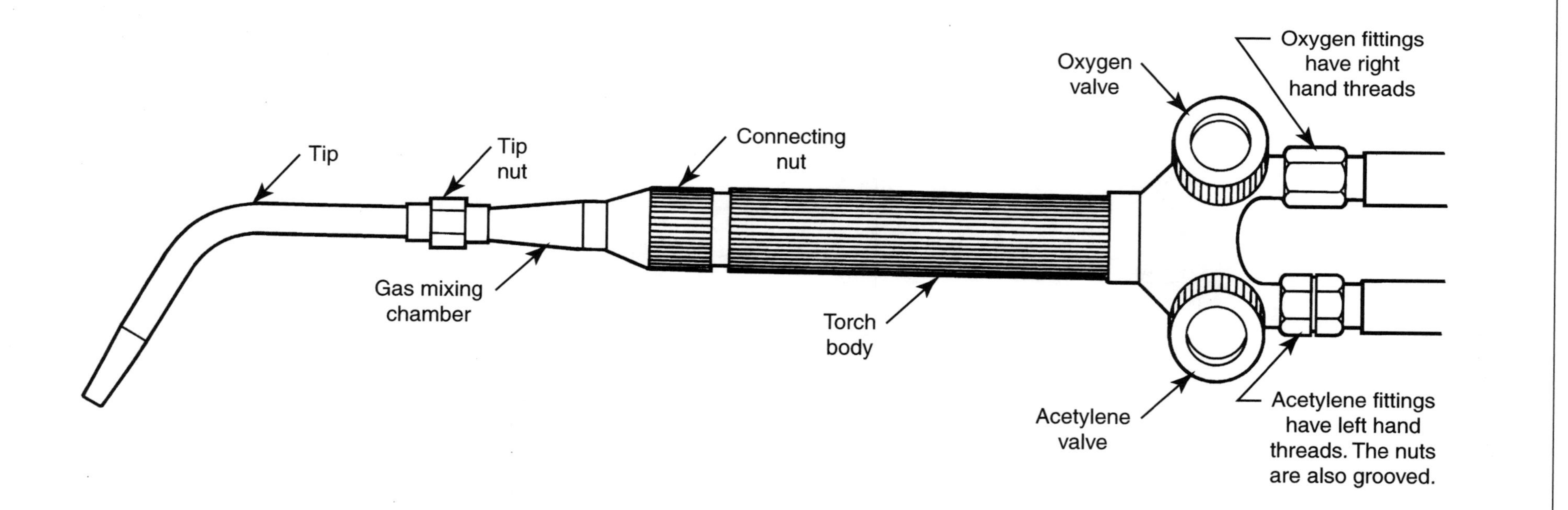

19-2

Basic Weld Joints

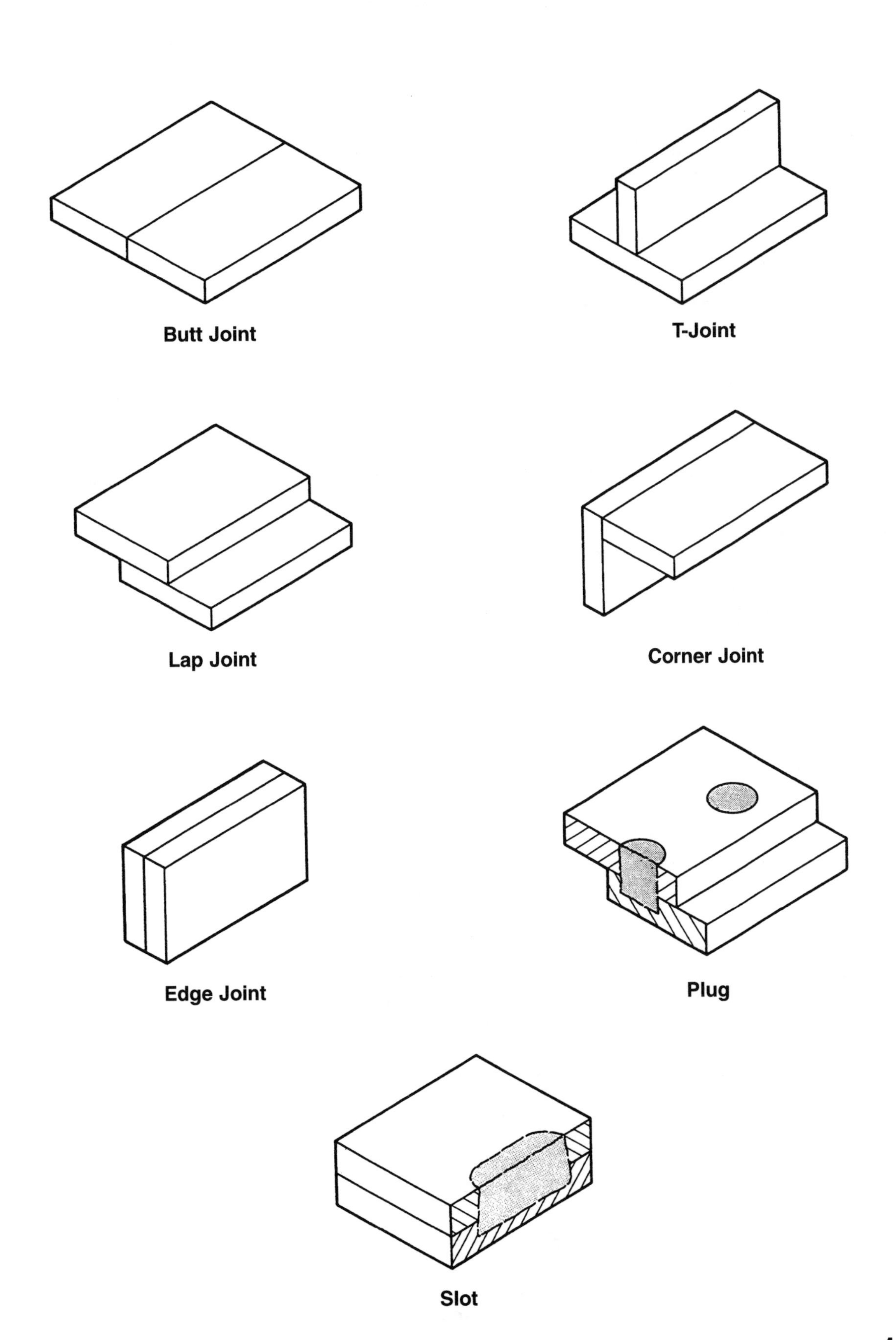

Chapter 19 Quiz
Gas Welding

Name: ______________________ **Date:** ______________________

1. Welding is the process that _____. 1.______________
   (a) makes a permanent joint
   (b) heats the metal until they melt and fuse together
   (c) is a very important facet of modern fabricating techniques
   (d) All of the above.
   (e) None of the above.
2. Gases used in oxyacetylene welding are stored in gas _____. 2.______________
3. Gas pressure is reduced and controlled by _____. 3.______________
4. The hose that carries oxygen to the point of use is colored _____. The acetylene hose is colored _____. 4.______________
   ____
5. The _____ mixes and controls the flow of gases while welding. 5.______________
6. Flux is used on most ferrous metals because _____. 6.______________
   (a) it cleans the metal
   (b) prevents oxidation
   (c) promotes a solid weld
   (d) All of the above.
   (e) None of the above.
7. *Never* light a torch with a(n) _____. *Always* use a(n) _____. 7.______________
   ____

Identify the following weld joints.

8. ______________

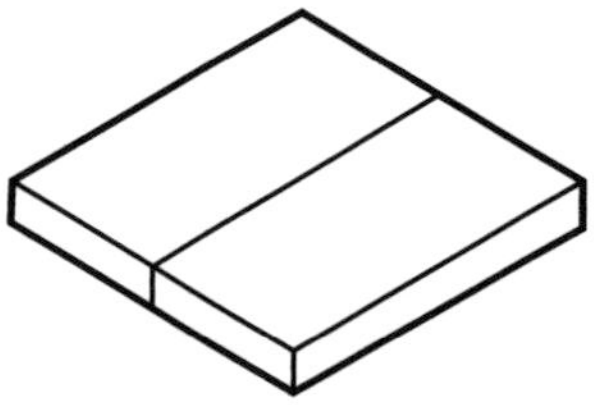

9. ______________

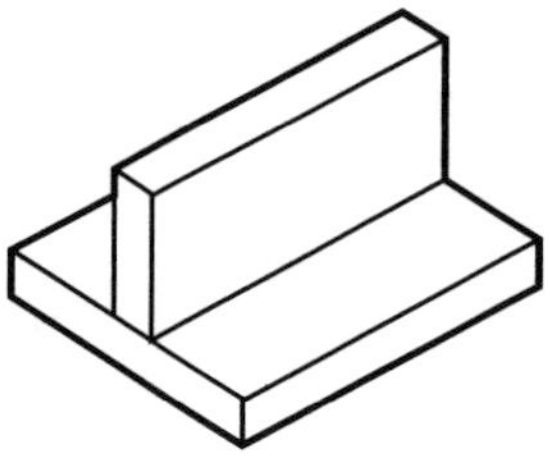

10. ______________

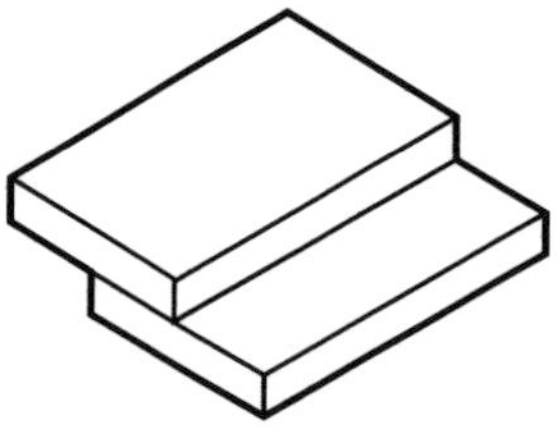

Shielded Metal Arc Welding 20

Learning Objectives

After studying this chapter, students should be able to:

- Explain the shielded metal arc welding process.
- List the various components of a shielded metal arc welding system.
- Prepare equipment and work for welding.
- Select the proper rod for a job.
- Interpret basic welding symbols.
- Dress properly for welding.
- Perform basic arc welding operations.
- Practice welding safety precautions.

Chapter Resources

Text, pages 303–314
  Test Your Knowledge, page 313
  Research and Development, page 313
Workbook, pages 99–102
Instructor's Manual
  Answer keys for:
    Test Your Knowledge Questions
    Workbook
    Chapter Quiz
  Reproducible Masters:
    20-1 Close-up of Arc Welding Procedure
    20-2 Direct Current Electrode Positive Circuit
    20-3 Direct Current Electrode Negative Circuit
    20-4 Electrodes for General Welding
    20-5 Typical Welding Symbols
    20-6 Chapter Quiz

Guide for Lesson Planning

Provide arc welding equipment, safety equipment, and examples of the various arc welded joints available for student examination. Emphasize the safety precautions that must be observed when arc welding.

Class Discussion and Demonstration

Have student read and study the chapter. Review the material using Reproducible Masters 20-1 through 20-5 as overhead transparencies or handouts. Discuss the assignment by explaining, questioning and, if sufficient safety equipment is available, demonstrating the following:

- Arc welding process.
- Safety equipment that must be worn when arc welding and its purpose.
- Arc welding electrodes and their care.
- Basic types of arc welding machines.
- Arc welding equipment.
- Preparing to weld.
- Joint preparation.
- Striking an arc.
- Recommended electrode travel and determining welding speed.
- Weld characteristics.
- Welding symbols, why they are used, and their importance.

Students must wear approved eye protection and safety equipment when observing arc welding demonstrations. There must also be adequate ventilation to remove fumes.

Assignments

1. Assign the Test Your Knowledge questions at the end of the chapter.
2. Assign Chapter 20 in the *Modern Metalworking Workbook.*
3. Assign the chapter quiz. Copy and distribute Reproducible Master 20-6.
4. Permit students to volunteer for Research and Development activities at the end of the chapter.

Test Your Knowledge

1. electric current
2. Any order: power source, cables (leads), electrode holder, electrodes, cleaning accessories, tools to handle hot metals, protective clothing.
3. (d) All of the above.
4. To protect the eyes and face from the harmful rays of the electric arc and spatter of molten metal.
5. To provide the welder with specific welding instructions.
6. Evaluate individually. Refer to Figure 20-15.
7. It must be cleaned and carefully fitted.
8. Evaluate individually. Refer to text Section 20.4.1.
9. Evaluate individually. Refer to Figure 20-23.
10. Serious burns will result from stepping on hot weld spatter. Trousers with cuffs are dangerous since molten weld spatter can fall into them and start a fire.

Workbook

1. A joining technique that makes use of an electric current to produce the heat needed to cause the metal to melt and fuse together.
2. Direct current (dc) is electric current that flows only in one direction. Alternating current (ac) is electricity that reverses its direction of flow in regular intervals.
3. According to their current output.
4. Electrodes are metal rods covered with a baked flux coating.
5. They support the welding arc and provide filler metal to the joint.
6. The flux melts and cleans the oxides from the base metal. It also acts as insulation, slowing down the cooling rate of the weldment. This helps relieve internal strains that develop from rapid changes in temperature.
7. It provides the means to make a solid, electrically-sound ground connection with the work or welding table.
8. When the positive lead from the power source is connected to the electrode, and the negative lead is connected to the work.
9. When there is a negative lead from the power source to the electrode, and a positive lead connected to the work.
10. Evaluate individually. Refer to Section 20.2.
11. Evaluate individually. Refer to Section 20.2.
12. Evaluate individually. Refer to Section 20.3.
13. Evaluate individually. Refer to Section 20.4.
14. Evaluate student beads individually.
15. Evaluate student welds individually.

Chapter Quiz

1. produce the heat needed to cause the metals to melt and fuse together
2. Welding goggles do not provide adequate protection from welding rays of arc welding for the face and eyes.
3. Any three of the following: gauntlet type gloves, leather apron, leather sleeves, fire resistant coveralls, heavy leather safety shoes.
4. welding arc, filler metal
5. scratch
6. To prevent them from pulling apart as they are welded.
7. The weld will be weaker than the base metal.
8. Weld symbols
9. Evaluate individually. Refer to Figure 20-15.
10. Evaluate individually. Refer to Figure 20-15.

20-1

Close-up of Arc Welding Procedure

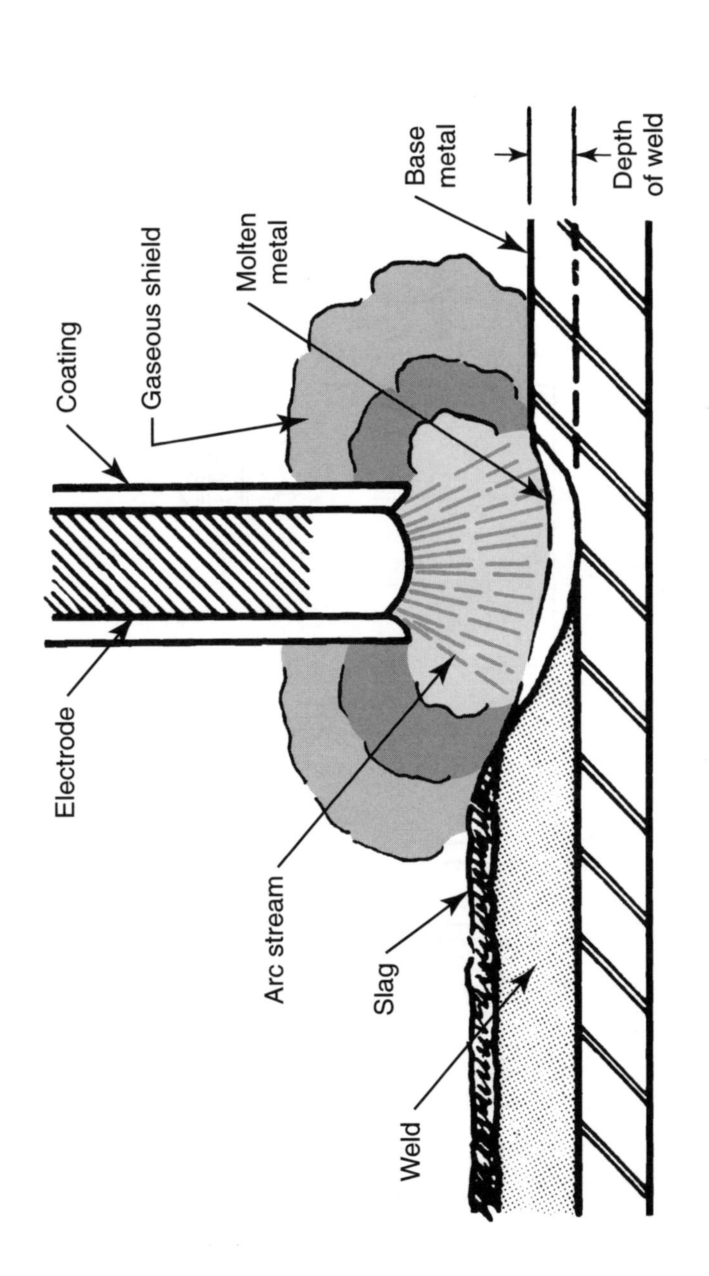

Direct Current Electrode Positive Circuit

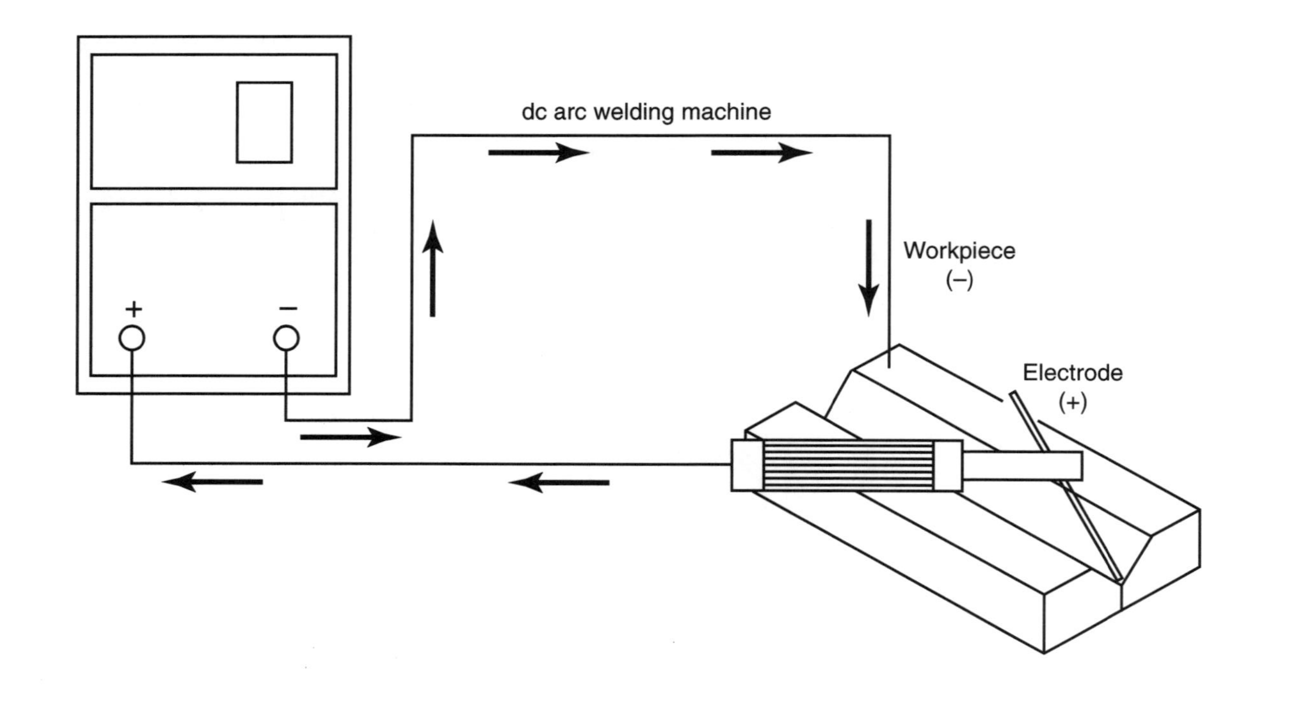

Direct Current Electrode Negative Circuit

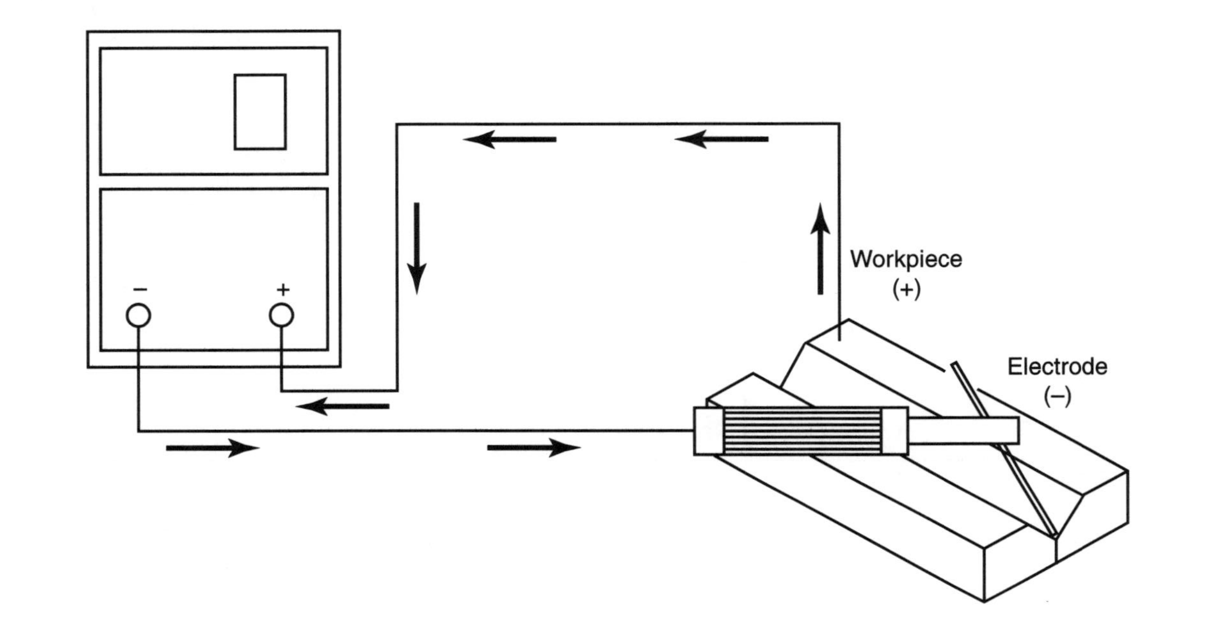

20-3

Electrodes for General Welding

AWS No.	Welding Current	Color Marking End (E) Secondary (S)	Welding Position	Penetration and Characteristics	Application
E-6010	dc, reverse polarilty	None	All	Deep pentration; thin slag, easy to remove; forceful arc.	Building construction, pipe lines, pressure tanks, bridges, ship building, storage tanks, machinery frames.
E-6011	ac or dc, reverse polarity	Blue (S)	All	Deep penetration; thin slag, easy to remove; forceful arc.	Same as E-6010. For use on ac welding current, not all are usable with limited-input welders.
E-6012	dc straight polarity or ac	White (S)	All	Medium penetration; heavy slag; soft arc.	Work with poor fit-up, high-speed welding, light gauge welding, build-up.
E-6013	ac or dc, straight polarity	Brown (S)	All	Medium to shallow penetration; light slag, easy to remove; soft arc.	Mild steel repair, sheet metal, auto bodies, general farm welding and repair wtih limited-input welders.
E-6014	ac or dc	Brown (S)	All	Iron powder coating; medium to shallow penetration; drag technique; exceptionally easy to operate; little spatter; heavy slag almost self-removing.	High-speed welding, for ease of operation in all general welding. This is an E-6013 electrode with iron powder added to the coating.
E-6016	dc reverse polarity or ac	Orange (S)	All	Medium to deep penetration; forceful arc; medium to heavy slag.	Higher carbon steels, alloy steels, armor plate, auto bumpers. Not usable with limited-input welders.

Typical Welding Symbols

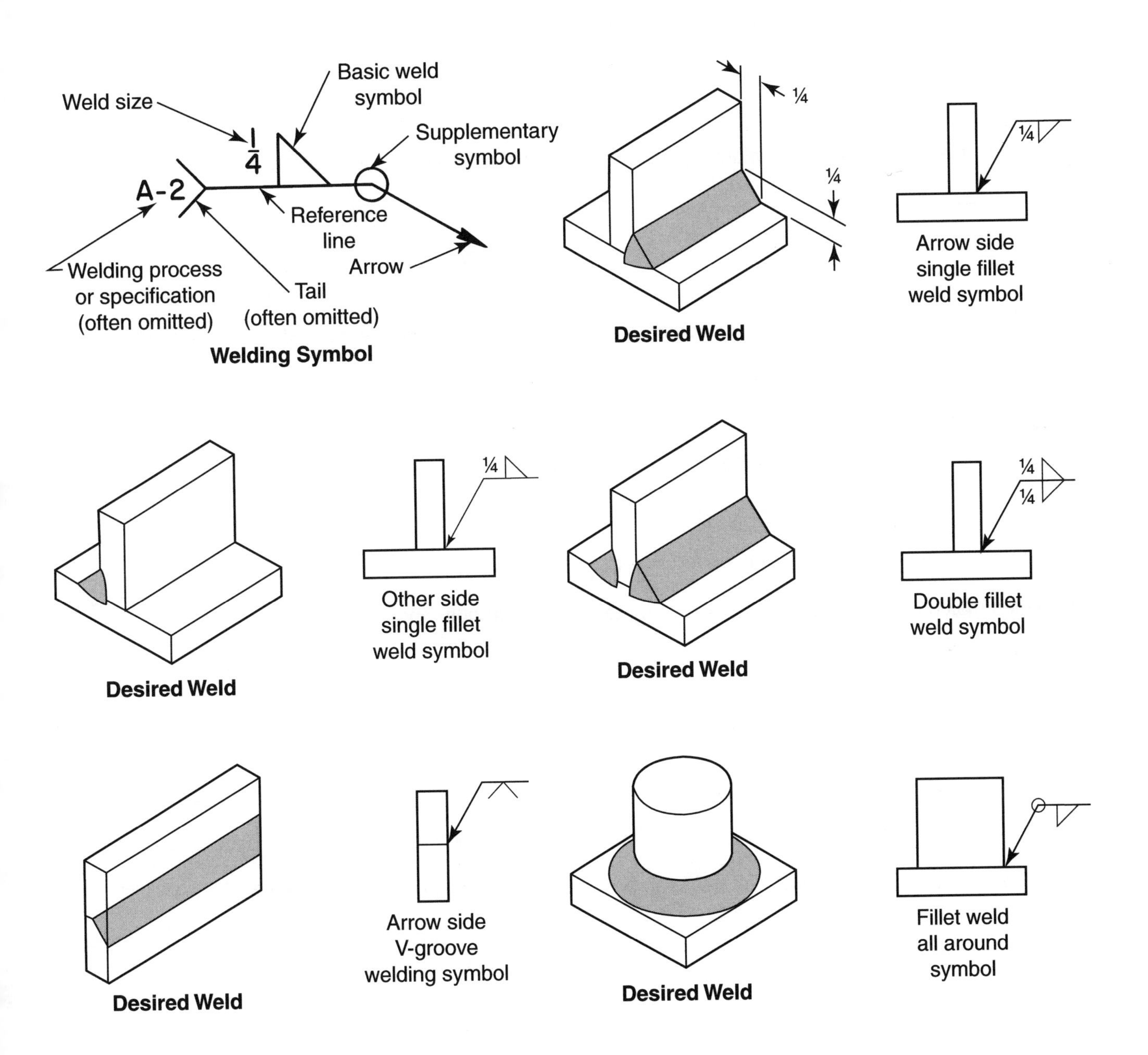

Chapter 20 Quiz

Shielded Metal Arc Welding

Name: ______________________________ **Date:** ______________________________

1. Arc welding makes use of an electric current to ______________________________
____
____.

2. Why is it dangerous to wear welding goggles instead of a welding helmet when arc welding?
____
____
____

3. In addition to the welding helmet, what other three protective items should be worn when arc welding?
____
____

4. Electrodes are used to support the _____, and supply _____. 4.______________
____

5. The _____ method is the recommended way to strike the arc. 5.______________

6. When making butt welds, why is it necessary to tack weld at both ends of the joint?
____
____

7. Why must welds penetrate 100%? ______________________________
____

8. _____ are used on drawings to give the welder specific welding instructions. 8.______________

9. Make a sketch of a T-joint with fillet welds on both sides of the joint. Add the weld symbol for a 1/4″ fillet weld.

10. Draw the weld symbol for a 5/16″ all around fillet weld.

21 Other Welding Processes

Objectives

After studying this chapter, students should be able to:

- Identify various welding processes currently used by industry.
- Describe the operation of various welding processes.
- Explain when the various welding processes might be employed.

Chapter Resources

Text, pages 315–328
  Test Your Knowledge, pages 326–327
  Research and Development, page 327
Workbook, pages 103–104
Instructor's Manual
  Answer keys for:
    Test Your Knowledge Questions
    Workbook
    Chapter Quiz
  Reproducible Masters:
    21-1 GTAW Welding Set-up
    21-2 GMAW Welding Set-up
    21-3 Flux-cored Arc Welding Process
    21-4 Submerged Arc Welding Technique
    21-5 Stud Welding Technique
    21-6 Spot Welding Sequence
    21-7 Seam Welding Joint
    21-8 Projection Welding Technique
    21-9 Ultrasonic Welding Technique
    21-10 Cold Welding Technique
    21-11 Chapter Quiz

Guide for Lesson Planning

Have students review the chapter paying particular attention to the illustrations. Review the material using Reproducible Masters 21-1 through 21-10 as overhead transparencies or handouts. Discuss the assignment by explaining and questioning student understanding of the following:

- GTAW and GMAW welding.
- Flux-cored arc welding.
- Submerged arc welding.
- Resistance welding.
- Electron beam welding.
- Friction welding
- Ultrasonic welding.
- Cold welding.
- Laser welding.
- Micro welding.
- Flame spraying.

Assignments

1. Assign the Test Your Knowledge questions at the end of the chapter.
2. Assign Chapter 21 in the *Modern Metalworking Workbook.*
3. Assign the chapter quiz. Copy and distribute Reproducible Master 21-11.
4. Permit students to volunteer for one or more of the Research and Development activities at the end of the chapter.

Test Your Knowledge

1. (m) Ultrasonic welding
2. (b) Resistance welding
3. (c) Electron beam welding
4. (d) MIG
5. (a) GMAW and GTAW
6. (g) Submerged arc welding
7. (i) Micro welding
8. (j) Projection welding
9. (k) Flash welding
10. (e) TIG
11. (l) Upset welding

12. (o) Laser welding
13. (h) Arc stud welding
14. (p) Thermal spraying
15. (n) Cold welding

Workbook

1. Tungsten inert gas (TIG)
2. Metal inert gas (MIG)
3. Submerged arc (SAW)
4. Arc stud welding (SW)
5. Resistance welding
6. Resistance spot welding (RSW)
7. Electron beam welding (EBW)
8. Friction welding (FRW)
9. Ultrasonic welding (USW)
10. Micro welding
11. Cold welding
12. Laser beam welding (LBW)
13. Flame spraying
14. Thermo spraying
15. Plasma spraying

Chapter Quiz

1. (d) Metal inert gas welding
2. (g) Tungsten inert gas welding
3. (h) Submerged arc welding
4. (j) Arc stud welding
5. (b) Resistance welding
6. (a) Electron beam welding
7. (e) Friction welding
8. (c) Micro welding
9. (f) Flame spraying
10. (i) Laser welding

GTAW Welding Set-up

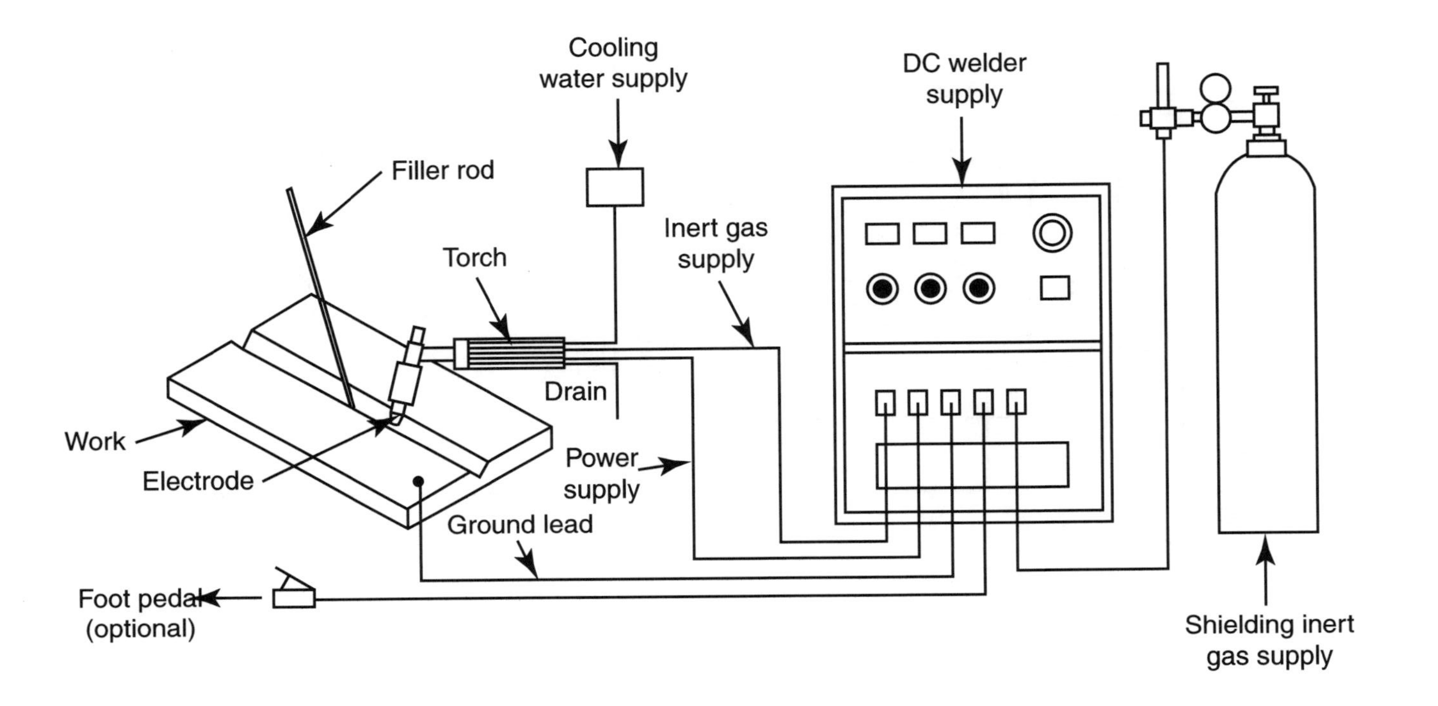

21-1

GMAW Welding Set-up

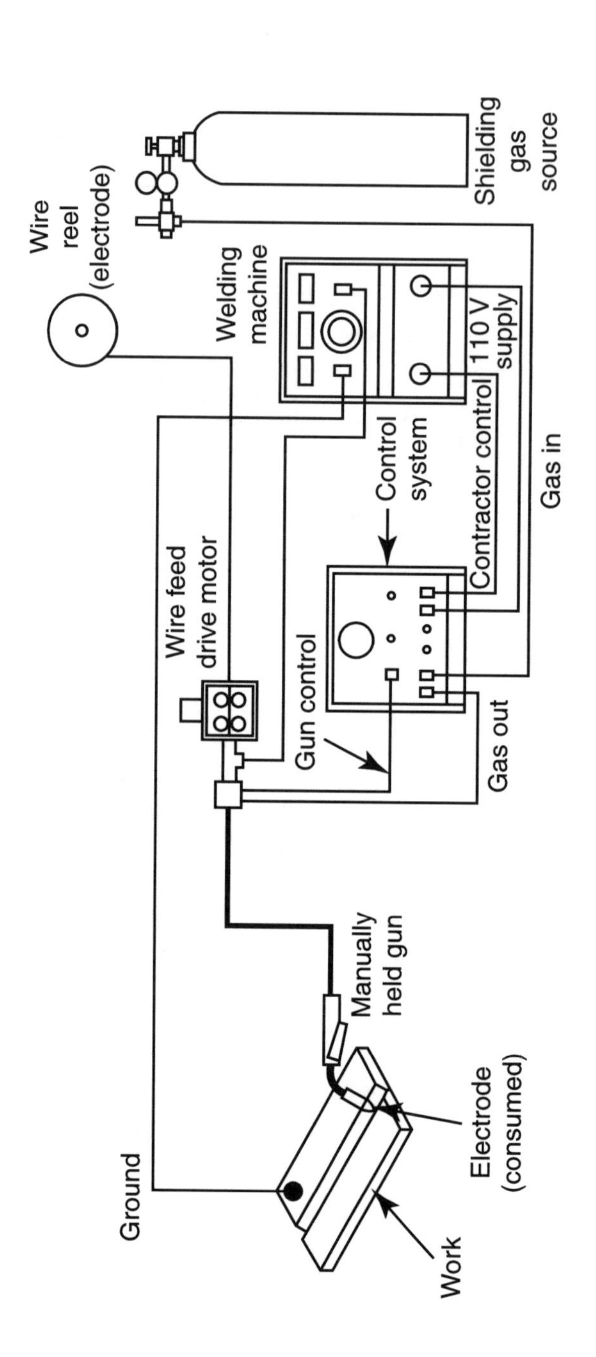

Flux-cored Arc Welding Process

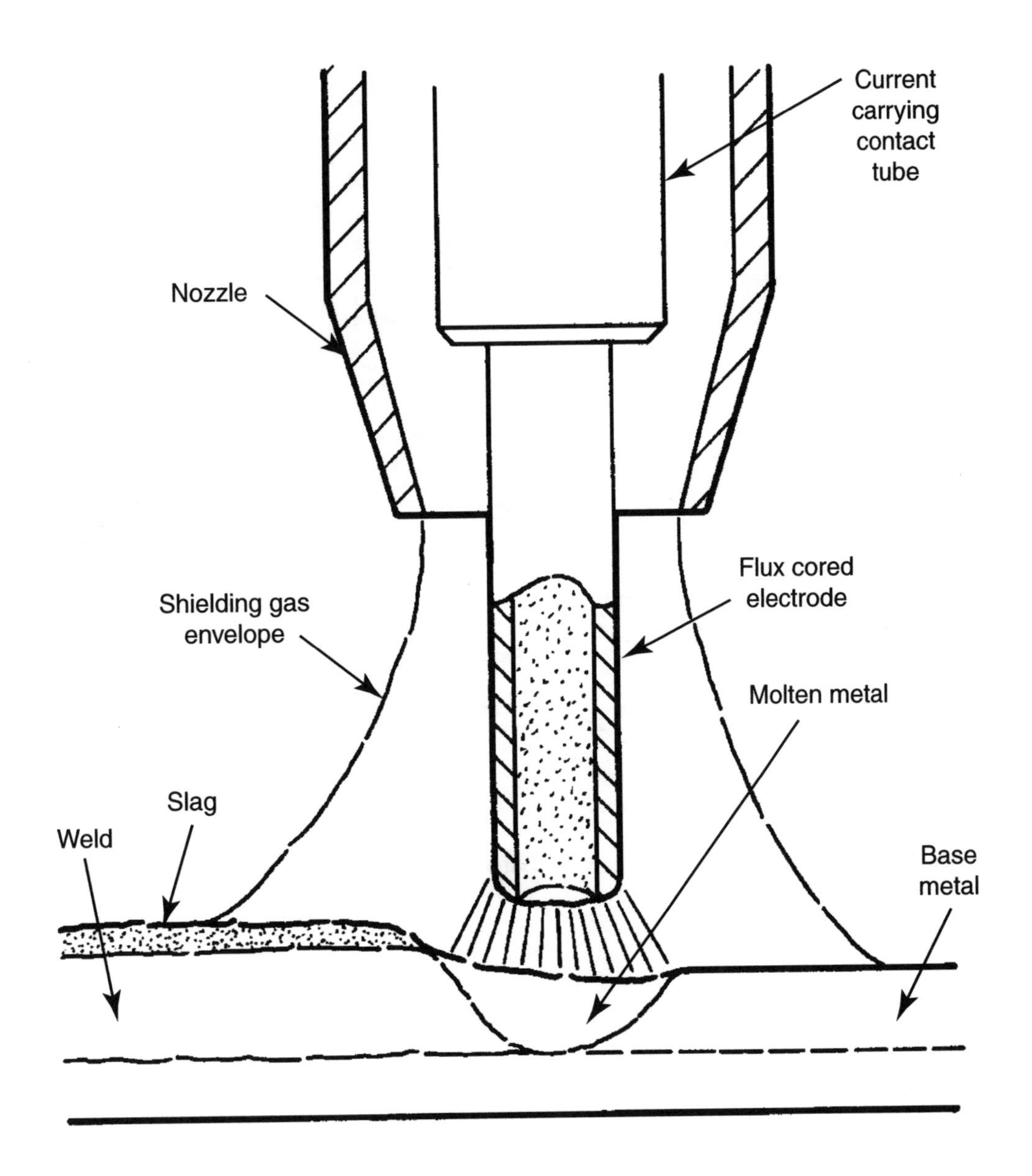

Submerged Arc Welding Technique

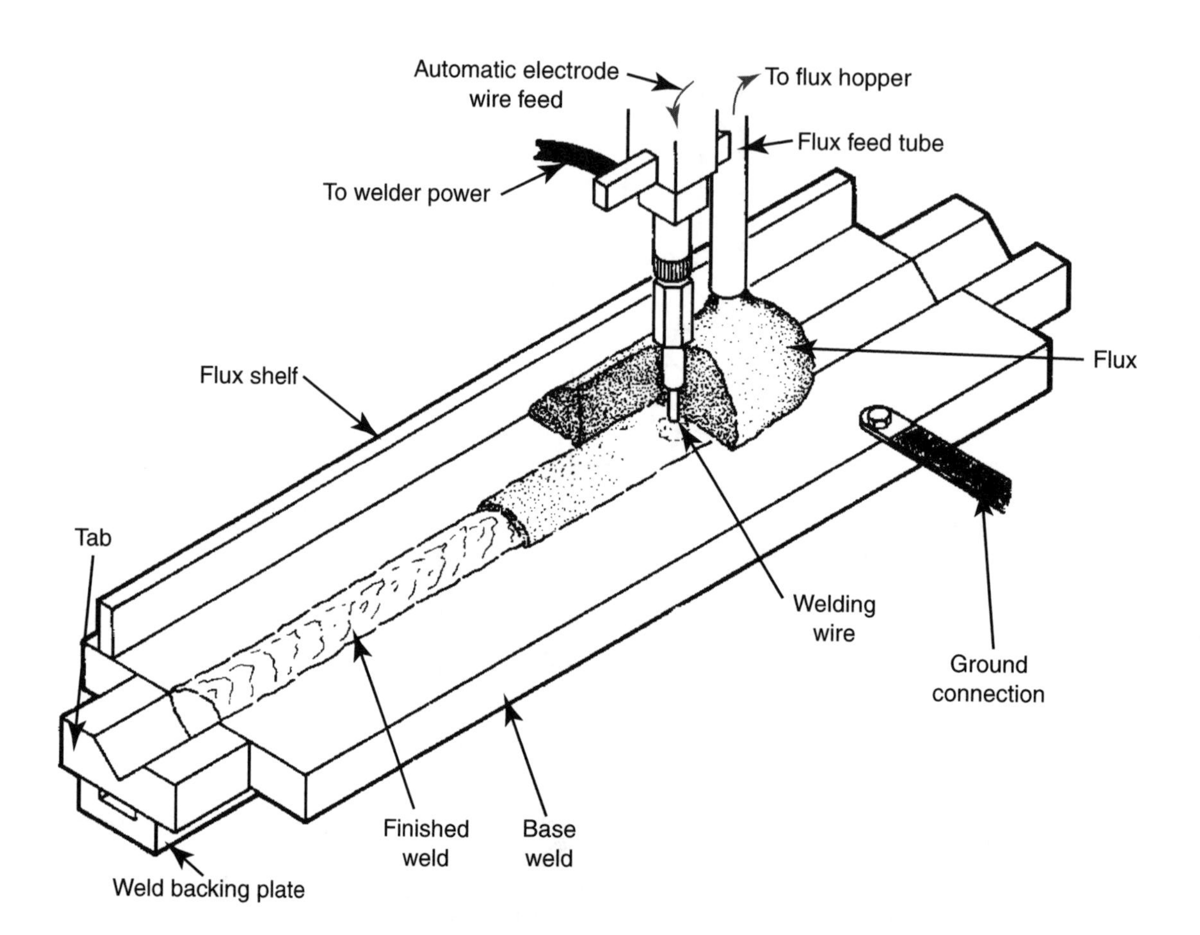

Stud Welding Technique

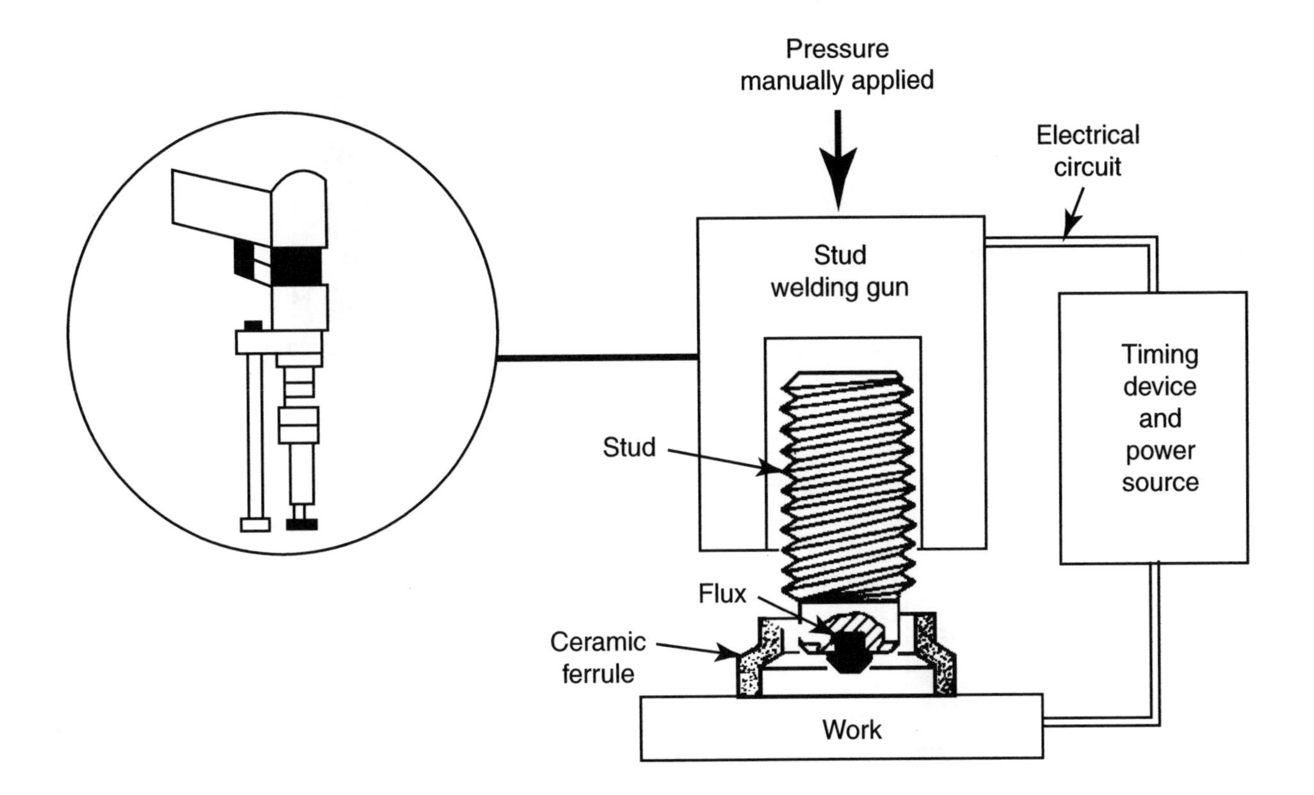

Spot Welding Sequence

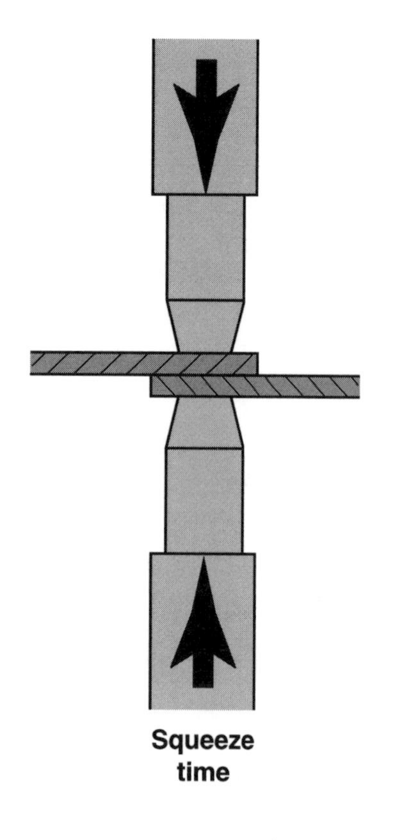

Squeeze time

Weld time

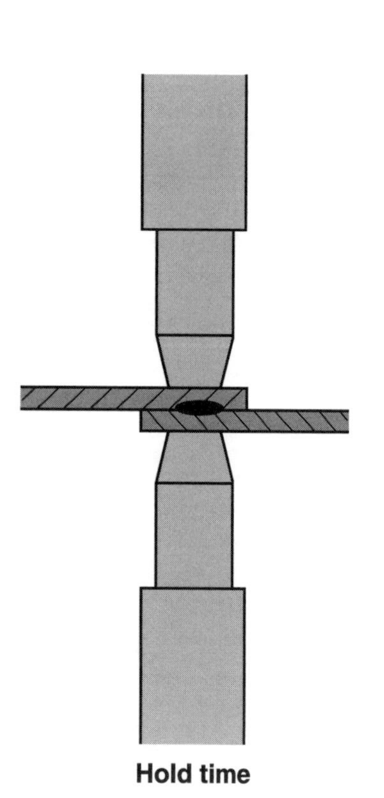

Hold time

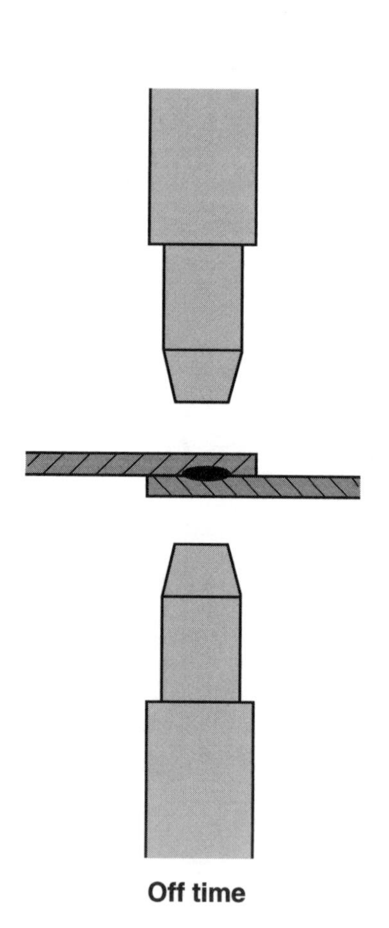

Off time

Seam Welding Joint

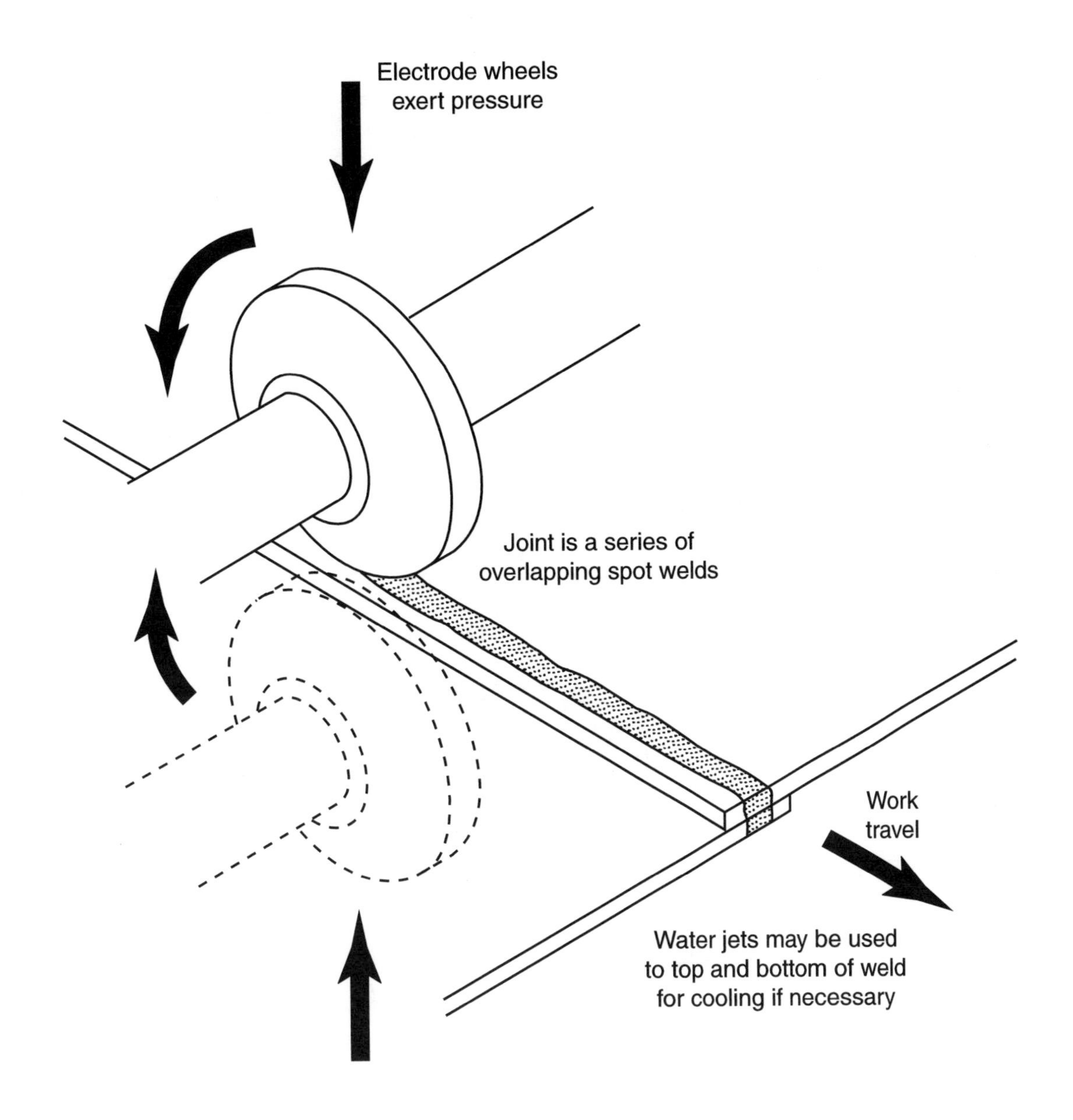

Projection Welding Technique

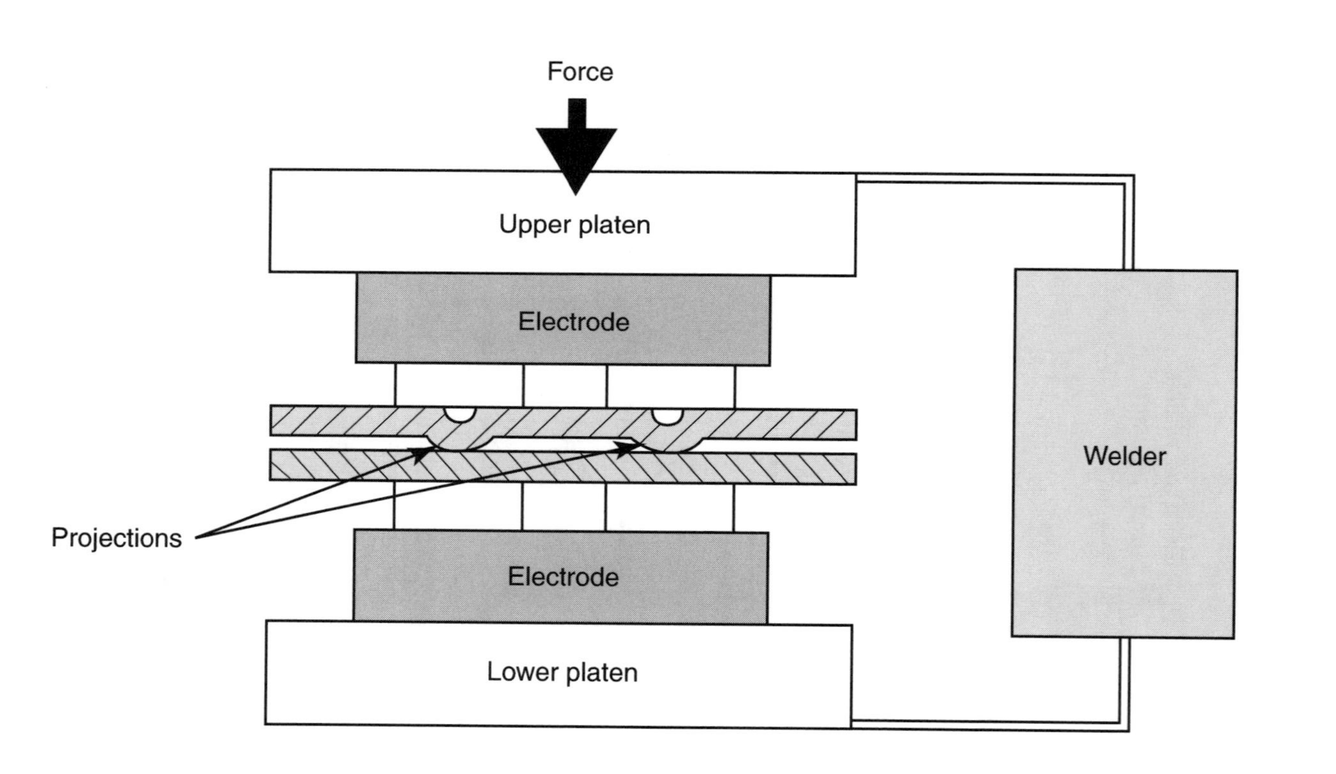

21-8

Ultrasonic Welding Technique

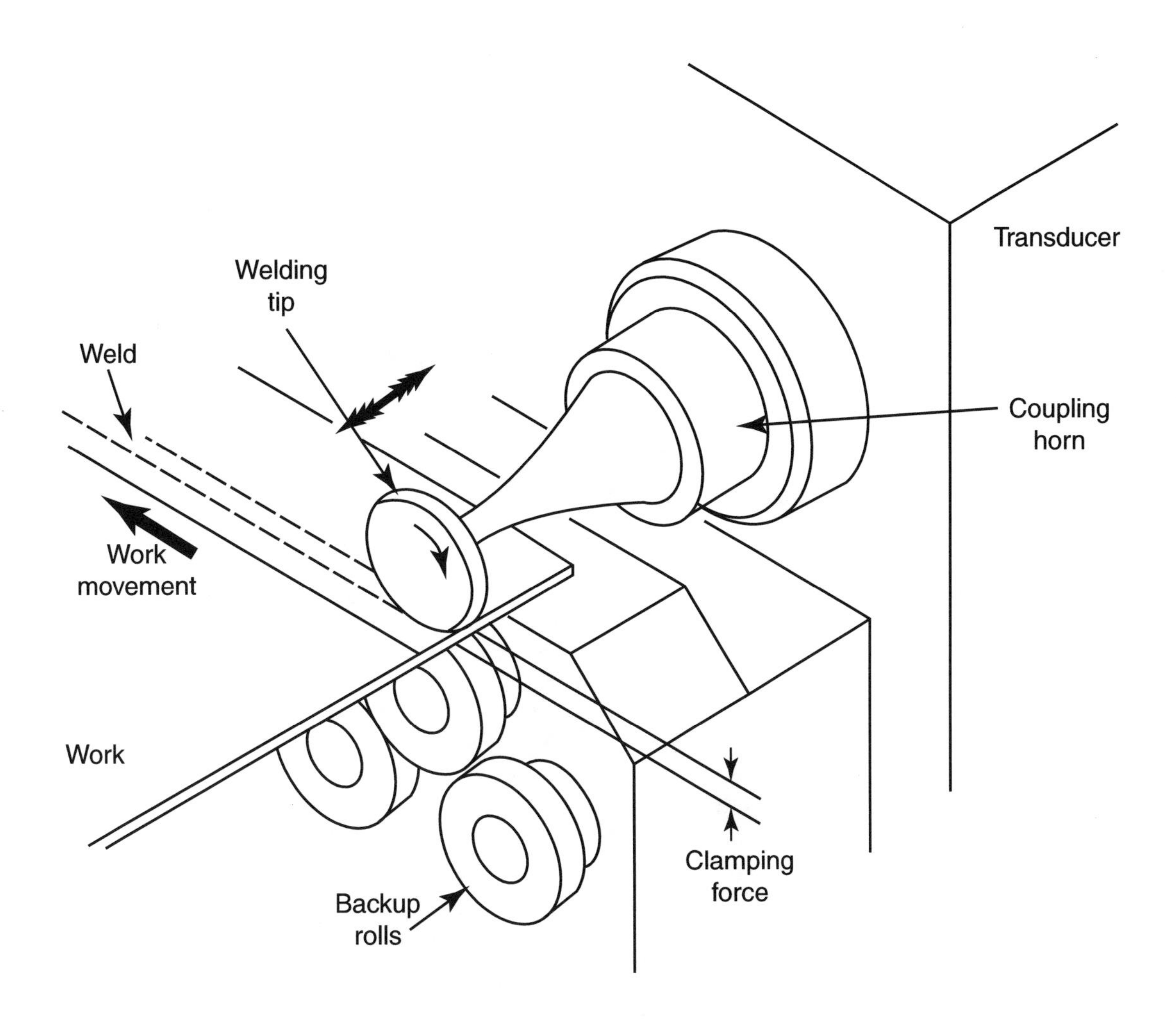

Cold Welding Technique

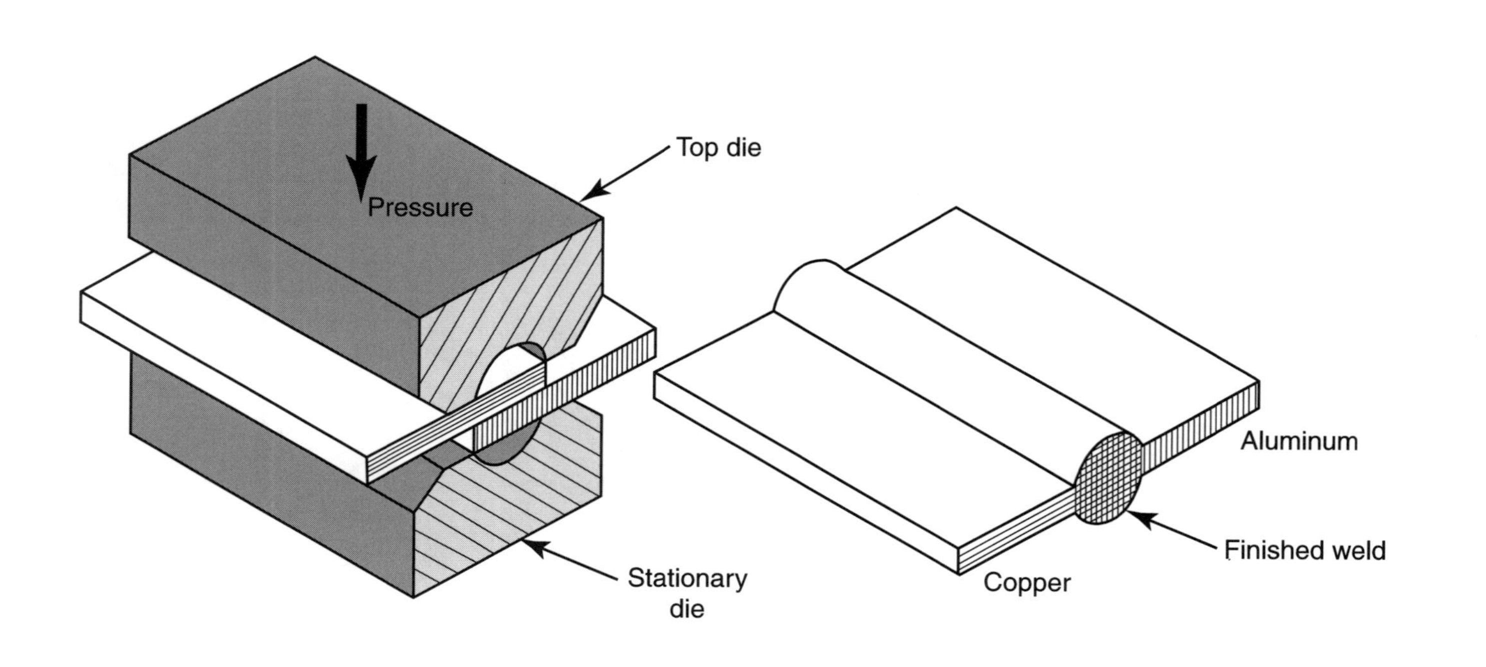

21-10

Chapter 21 Quiz

Other Welding Processes

Name: ______________________________ **Date:** ______________________________

Match each word or phrase with the sentence that best describes it.

_______ 1. A gas shielded welding technique in which the electrode melts and contributes filler metal to the joint.

_______ 2. A gas shielded welding technique that uses a permanent electrode (the electrode is not consumed).

_______ 3. A type of arc welding where the consumable metal electrode is shielded by a blanket of flux that covers the weld area.

_______ 4. Fusion is produced by an electric arc between the metal stud, or similar part, and the work surface.

_______ 5. A group of welding techniques that use pressure, electric current, electrical resistance of the work, and the resulting heat to join metal sections.

_______ 6. Makes use of fast moving electrons to supply the energy to melt and fuse the parts being joined. Welds must be made in a vacuum.

_______ 7. Uses frictional heat and pressure to produce full strength welds in a matter of seconds.

_______ 8. Welding technique used to attach leads to microcircuits.

_______ 9. Metal is brought to its melting temperature and sprayed onto a surface to produce a metal coating.

_______ 10. Uses a narrow and intense beam of light that can be onto an area only a few microns in diameter.

(a) Electron beam welding
(b) Resistance welding
(c) Micro welding
(d) Metal inert gas welding
(e) Friction welding
(f) Flame spraying
(g) Tungsten inert gas welding
(h) Submerged arc welding
(i) Laser welding
(j) Arc stud welding

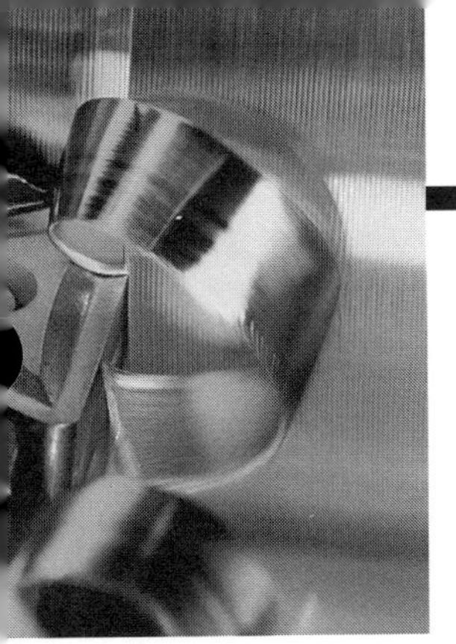

Metal Finishes 22

Learning Objectives

After studying this chapter, students should be able to:

- Explain why finishes are applied to metals.
- Describe several finishing techniques.
- Apply several types of finishes.
- Determine the abrasive(s) best suited for a job.
- Demonstrate how to polish some metals by buffing.
- Follow metal finishing safety rules.

Chapter Resources

Text, pages 329–342
  Test Your Knowledge, pages 340–341
  Research and Development, page 341
Workbook, pages 105–108
Instructor's Manual
  Answer keys for:
    Test Your Knowledge Questions
    Workbook
    Chapter Quiz
  Reproducible Master:
    22-1 Chapter Quiz

Guide for Lesson Planning

Have a selection of metal products, with various types of finishes described in the text, available for students to examine.

Class Discussion and Demonstration

Have students read all or part of the chapter. Review and discuss the assignment with them asking questions such as:

- Reasons finishes are used on metals. Have students give examples of each type.
- Name the many types of finishes used on metals.
- Polishing with abrasives.
- Types of abrasives.
- Mechanical finishes.
- Metal coatings.
- Methods employed to apply organic coatings.
- Inorganic coatings.
- Anodizing.
- Film coatings.
- Buffing and polishing.
- Safety procedures to be observed when applying metal finishes.
- There should be adequate ventilation in the area where organic finishes are applied.
- Appropriate safety equipment must be worn when anodizing and electroplating metals.
- Emphasize that hands must be washed thoroughly after hand polishing with abrasives, applying organic coatings, buffing, anodizing, and electroplating metals.

Assignments

1. Assign the Test Your Knowledge questions at the end of the chapter.
2. Assign Chapter 22 in the *Modern Metalworking Workbook.*
3. Assign the chapter quiz. Copy and distribute Reproducible Master 22-1.
4. Permit students to volunteer for one or more of the Research and Development activities at the end of the chapter.

Test Your Knowledge

1. Evaluate individually. Refer to Section 22.1.
2. (c) hide scratches and small dents
3. (c) thoroughly cleaned
4. (a) a hard substance used to wear away another material
5. Any order: aluminum oxide, silicon carbide, crocus.
6. supported

7. smooth, bright
8. satin
9. molten zinc
10. Electroplating
11. Sputtering
12. Any three of the following: paints, varnishes, lacquers, enamels, epoxies, plastic base.
13. Any order: brushing, spraying, roller coating, dipping, flow coating.
14. Anodizing
15. glass
16. chemical blackening
17. Evaluate individually. Refer to text Section 22.8.1.
18. tool marks
19. Melted lubricant causes the abrasive to adhere to the buffing wheel.
20. Evaluate individually. Refer to Section 22.9.

Workbook

1. Any hard substance that can be used to wear away another material.
2. the substance, grain size, backing material, and the manner the abrasive is bonded to the backing material
3. Any order: appearance, protection, identification, cost reduction.
4. Emery
5. Aluminum oxide
6. Crocus
7. Evaluate individually. Refer to text Section 22.3.5.
8. Sandblasting
9. Wire brushing
10. hot dipping
11. Electroplating
12. melting point
13. Through the evaporation of their solvents.
14. hardener or catalyst
15. Any order: brushing, spraying, roller coating, dipping, flow coating.
16. Anodizing
17. Electrobrightening
18. Chemical blackening
19. Evaluate individually. Refer to Section 22.9.
20. Review student abrasive samples individually.

Chapter Quiz

1. Any order: appearance, protection, identification, cost reduction.
2. anodizing
3. To prevent surface defects from showing.
4. (b) any hard, sharp substance that can be used to wear away another material
5. buffing
6. Wire brushing
7. zinc
8. emery
9. Silicon carbide
10. Any order: brushing, spraying, roller coating, dipping, flow coating.

Chapter 22 Quiz

Metal Finishes

Name: ______________________________ **Date:**______________ **Period:**______________

Carefully read each question before attempting to answer it.

1. Finishes are applied to metals for many reasons. List the four most important.

____

____

____

____

2. Through a process called _____, the finish is made part of the surface of aluminum. 2.______________

3. Why is the surface of some metal sheet embossed or textured?

____

____

____

____

4. An abrasive is commonly thought of as _____. 4.______________
   (a) a tool used to polish metal
   (b) any hard, sharp substance that can be used to wear away another material
   (c) natural or manufactured
   (d) All of the above.
   (e) None of the above.

5. The procedure that produces a bright mechanical finish is known as _____. 5.______________

6. _____ is a process that produces a smooth, satin sheen on metal. 6.______________

7. Galvanized sheet is produced by dipping steel in molten _____. 7.______________

8. _____ is a natural abrasive. It is black in color and cuts slowly. 8.______________

9. _____ is the hardest and sharpest of the manufactured abrasives. 9.______________

10. List five methods of applying organic coatings.

____

____

____

____

____

GRINDING

23

LEARNING OBJECTIVES

After studying this chapter, students should be able to:

- Explain the operation of typical grinding machines.
- Adjust and prepare a grinding machine for operation.
- Describe the operation of several types of precision grinding machines.
- Observe grinding safety rules.

CHAPTER RESOURCES

Text, pages 343–362
  Test Your Knowledge, page 360
  Research and Development, pages 360–361
Workbook, pages 109–112
Instructor's Manual
  Answer keys for:
    Test Your Knowledge Questions
    Workbook
    Chapter Quiz
  Reproducible Masters:
    23-1 Grinding Machine Operation
    23-2 A Properly Spaced Tool Rest
    23-3 Plunge Grinding
    23-4 Centerless Grinding
    23-5 Traverse Grinding
    23-6 Surface Grinding—Planer Type
    23-7 Surface Grinding—Rotary Type
    23-8 Chapter Quiz

GUIDE FOR LESSON PLANNING

Grinding can best be taught by dividing the chapter into two segments—offhand grinding and precision grinding.

PART I—OFFHAND GRINDING

The first segment should be concerned with offhand grinding equipment normally available in the school shop/lab. When demonstrating grinding techniques, be sure that:

- Equipment is properly adjusted with all guards and safety devices in place.
- Students are wearing approved eye protection.
- Grinding wheels are solid, dressed, and run true.
- Students understand what metals may be ground or may not be ground on shop/lab grinders.

Have students read and study the assignment. Review and discuss it with them asking questions such as:

- Definition of grinding.
- Types of bench and pedestal grinders.
- Adjusting the tool rest. Why is this important?
- Dressing a grinding wheel.
- Safety procedures to be followed when operating grinders.

PART II—PRECISION GRINDING

Have students read the remainder of the chapter. Review the assignment with them. Discuss the following:

- Abrasive belt grinding.
- Precision grinding machines.
- Work holding devices.
- Grinding wheels and how they are classified.
- Review grinding safety.

ASSIGNMENTS

1. Assign the Test Your Knowledge questions.
2. Assign Chapter 23 in the *Modern Metalworking Workbook.*
3. Assign the chapter quiz. Copy and distribute Reproducible Master 23-8.

4. Permit students to volunteer for one or more of the Research and Development activities at the end of the chapter.

Test Your Knowledge

1. wheel, belt
2. Evaluate individually. Refer to Sections 23.1 and 23.2.
3. offhand
4. (c) the work is manipulated by hand until the desired shape is obtained
5. Any order: dry type, wet type.
6. The tool rest should be about 1/16" (1.5 mm) away from the grinding. This prevents the work from being wedged between the wheel and the tool rest.
7. (a) as the abrasive particles become dull
8. diamond dressing tool
9. Loading and glazing.
10. manufactured
11. By supporting it with a wire and striking it with a light metal piece.
12. clear ringing sound
13. Any order: abrasive type, grain size, grade, structure, bond.
14. Evaluate individually. Refer to Figure 23-13.
15. Any order: surface, cylindrical, internal, tool and cutter, form.
16. flat
17. Evaluate individually. Refer to Figure 23-20.
18. magnetic chuck
19. Cylindrical
20. centerless

Workbook

1. By rotating an abrasive wheel or belt against the work.
2. (d) All of the above.
3. (c) to within 1/16" (1.5 mm) of the wheel
4. Suspend grinding wheel on string or wire and tap it with a light metal rod. A solid wheel will give off a ringing sound.
5. (b) true the wheel and remove the glaze
6. Evaluate individually. Refer to text Section 23.3.2.
7. remove excess material and to finish hardened steel parts to very accurate sizes with extremely fine surface finishes
8. (e) Form grinding
9. (a) Cylindrical grinding
10. (b) Centerless grinding
11. (f) Surface grinding
12. (c) Internal grinding
13. planer-type
14. To neutralize the magnetic field created when the work was mounted on a magnetic chuck.
15. Traverse
16. the bonding material wears away slowly enough to get maximum use from the individual abrasive grains, yet it would also wear away rapidly enough to permit the dulled particles to drop off and expose sharp new grains
17. Loading, glazing
18. (a) The abrasive
    (b) Grit size
    (c) Grade
    (d) Structure
    (e) Bond type
19. Evaluate individually. Refer to Section 23.7.
20. Evaluate student safety posters individually.

Chapter Quiz

1. (d) All of the above.
2. 1/16"
3. suspending it on a string or wire and tapping the side of the wheel lightly with a metal rod or screwdriver handle
4. true, remove
5. (g) Offhand grinding
6. (c) Flexible shaft grinder
7. (a) Cylindrical grinding
8. (h) Centerless grinding
9. (b) Internal grinding
10. (e) Tool and cutter grinding
11. (d) Surface grinding
12. (i) Form grinding
13. (f) Magnetic chuck
14. Evaluate individually. Refer to Section 23.7.

Grinding Machine Operation

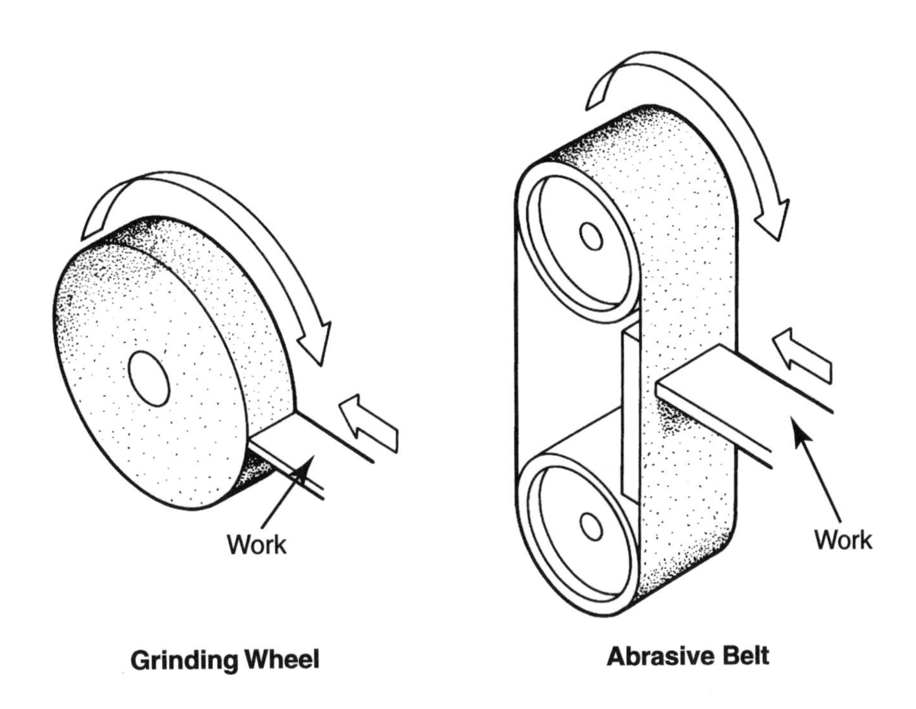

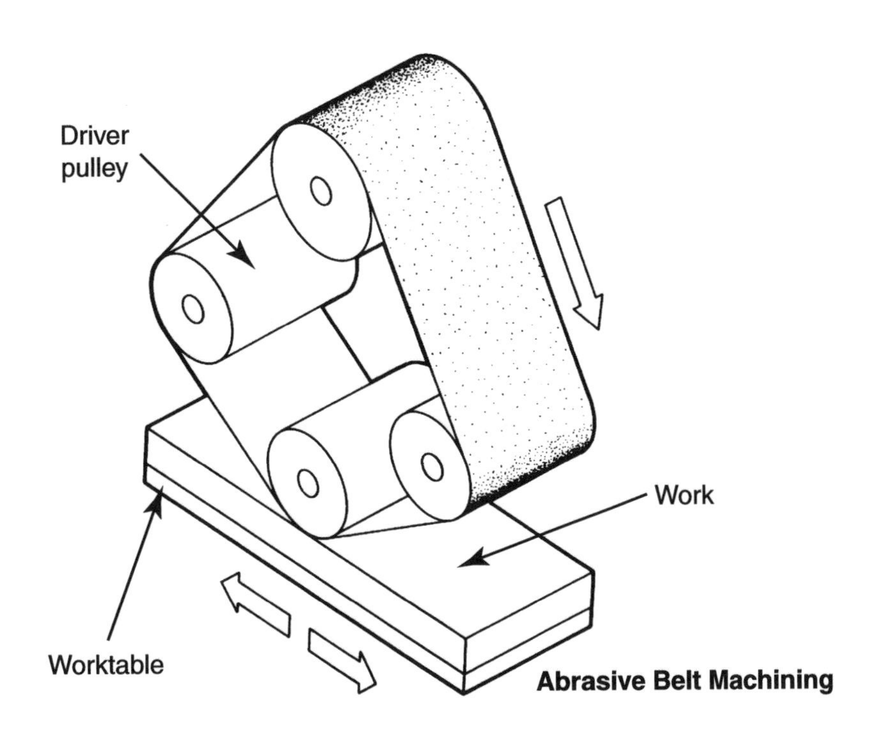

A Properly Spaced Tool Rest

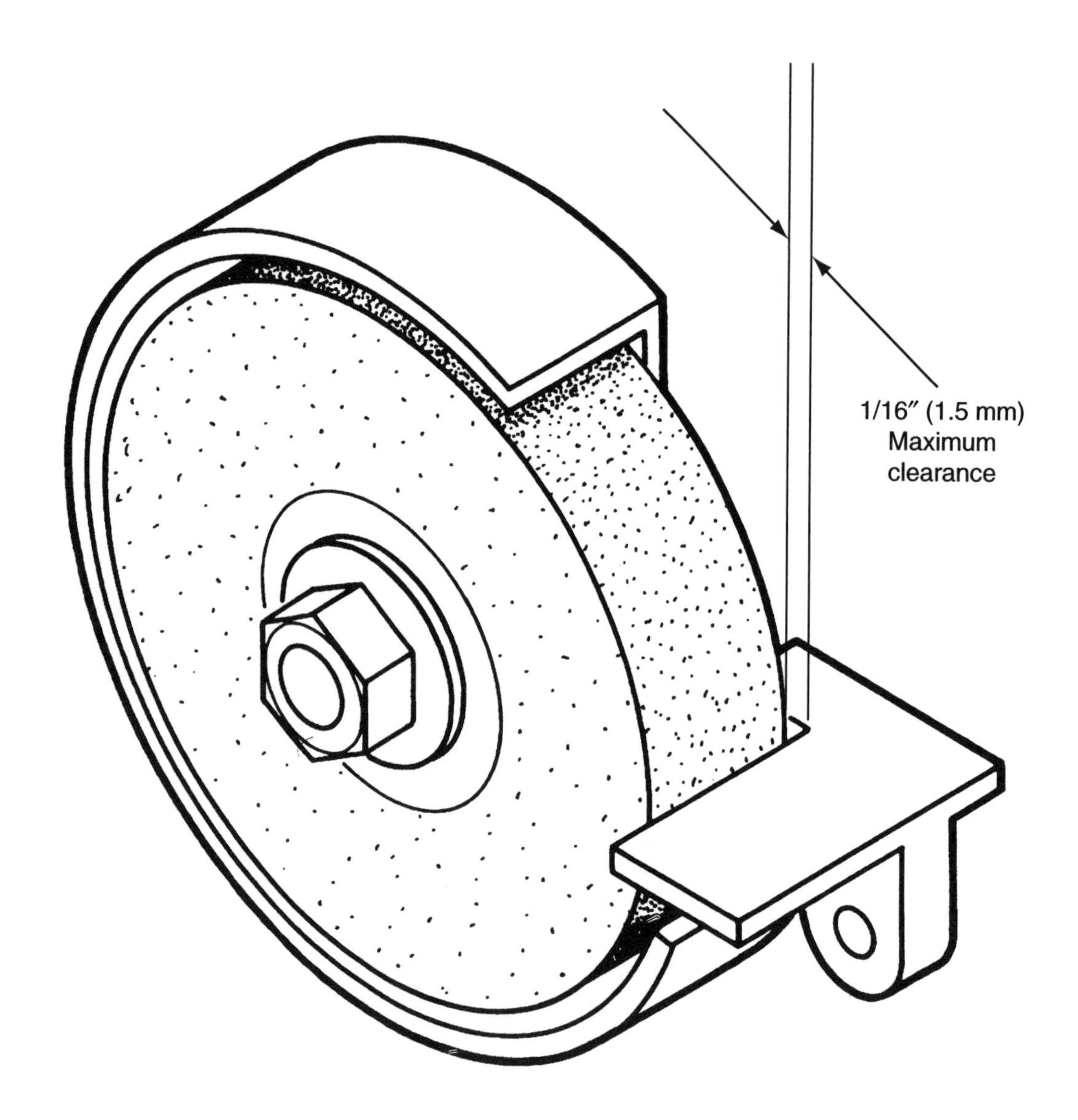

Plunge Grinding

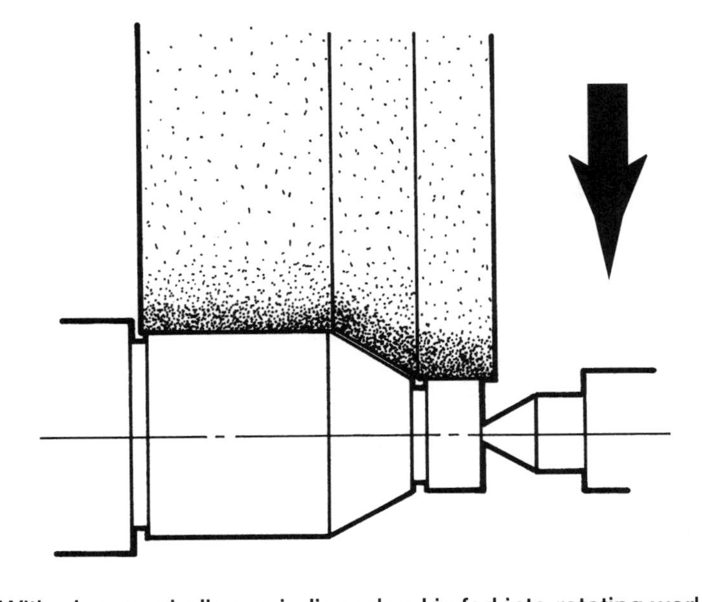

With plunge grinding, grinding wheel is fed into rotating work. Since work is no wider than grinding wheel, reciprocating motion is not needed.

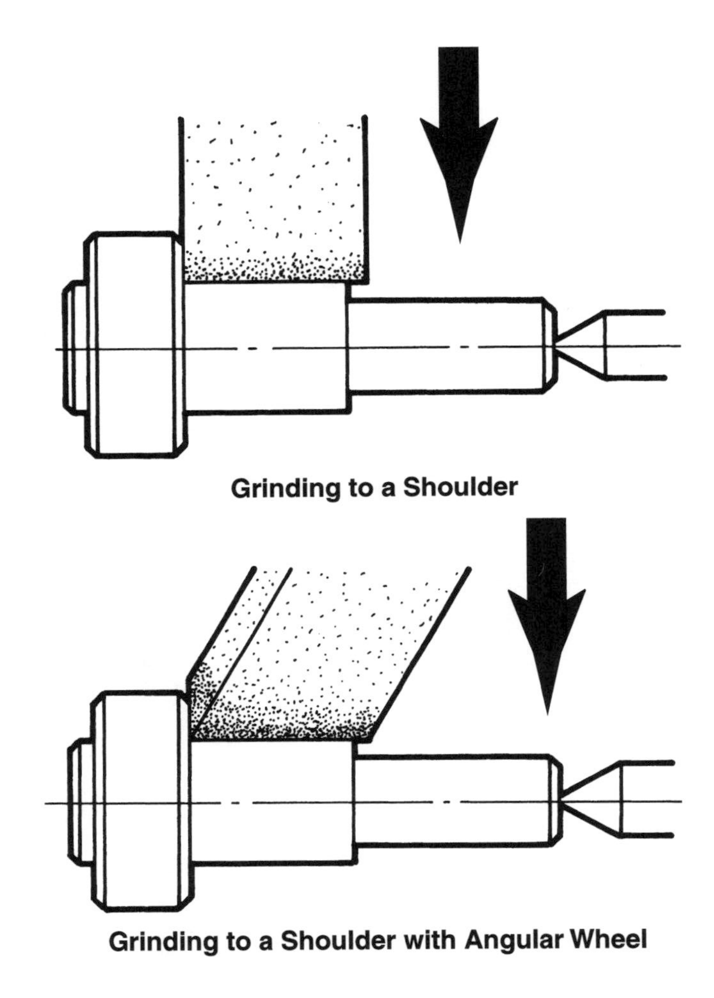

Grinding to a Shoulder

Grinding to a Shoulder with Angular Wheel

23-3

Centerless Grinding

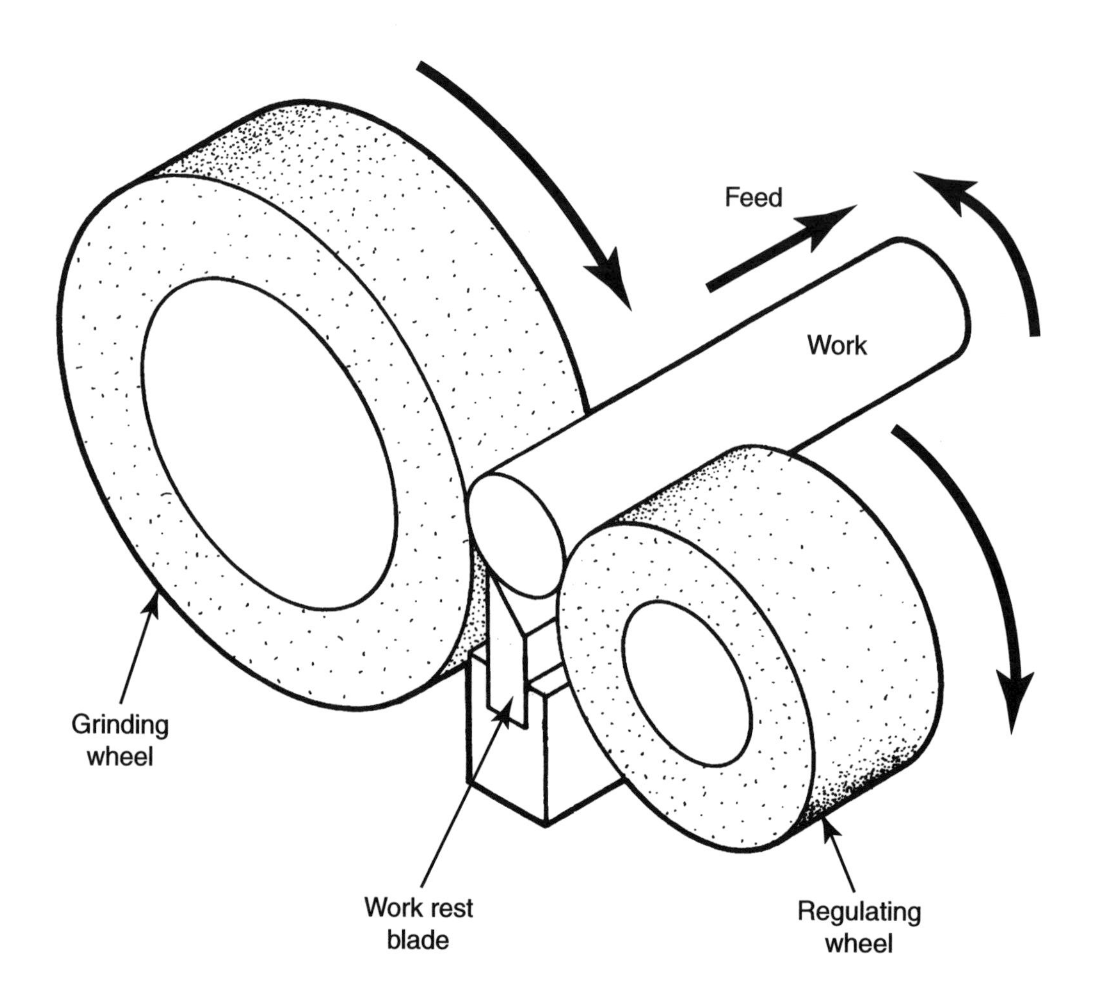

Traverse Grinding

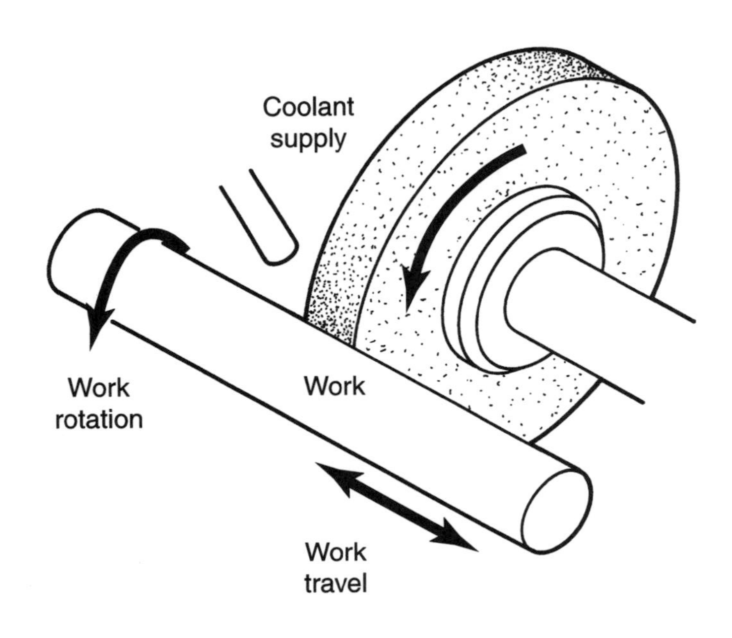

The rotating work moves past the rotating grinding wheel

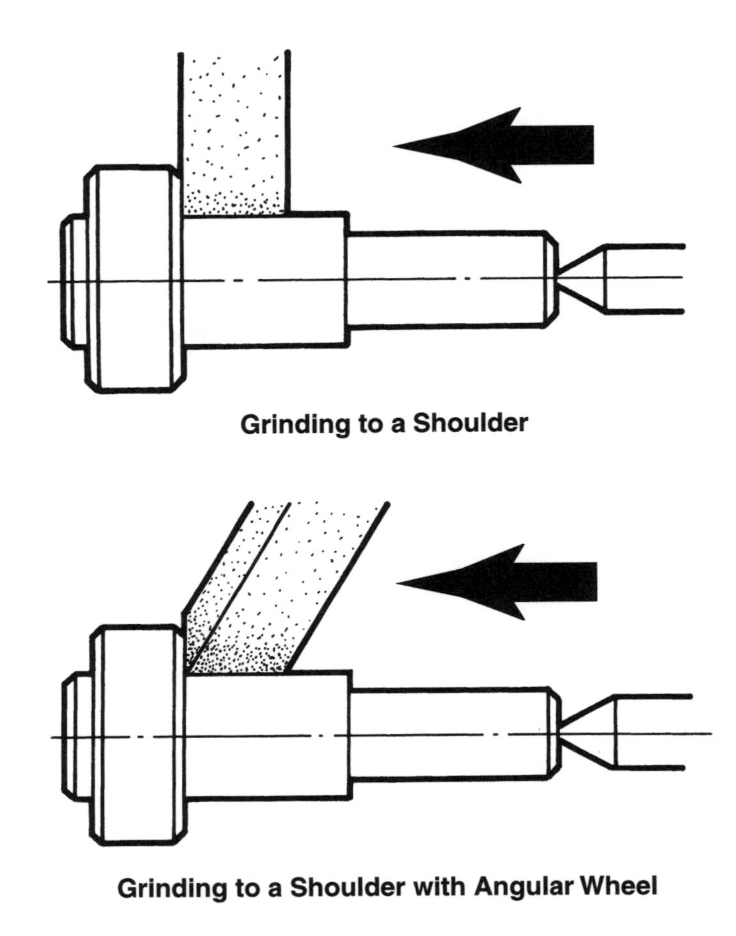

Grinding to a Shoulder

Grinding to a Shoulder with Angular Wheel

Surface Grinding—Planer Type

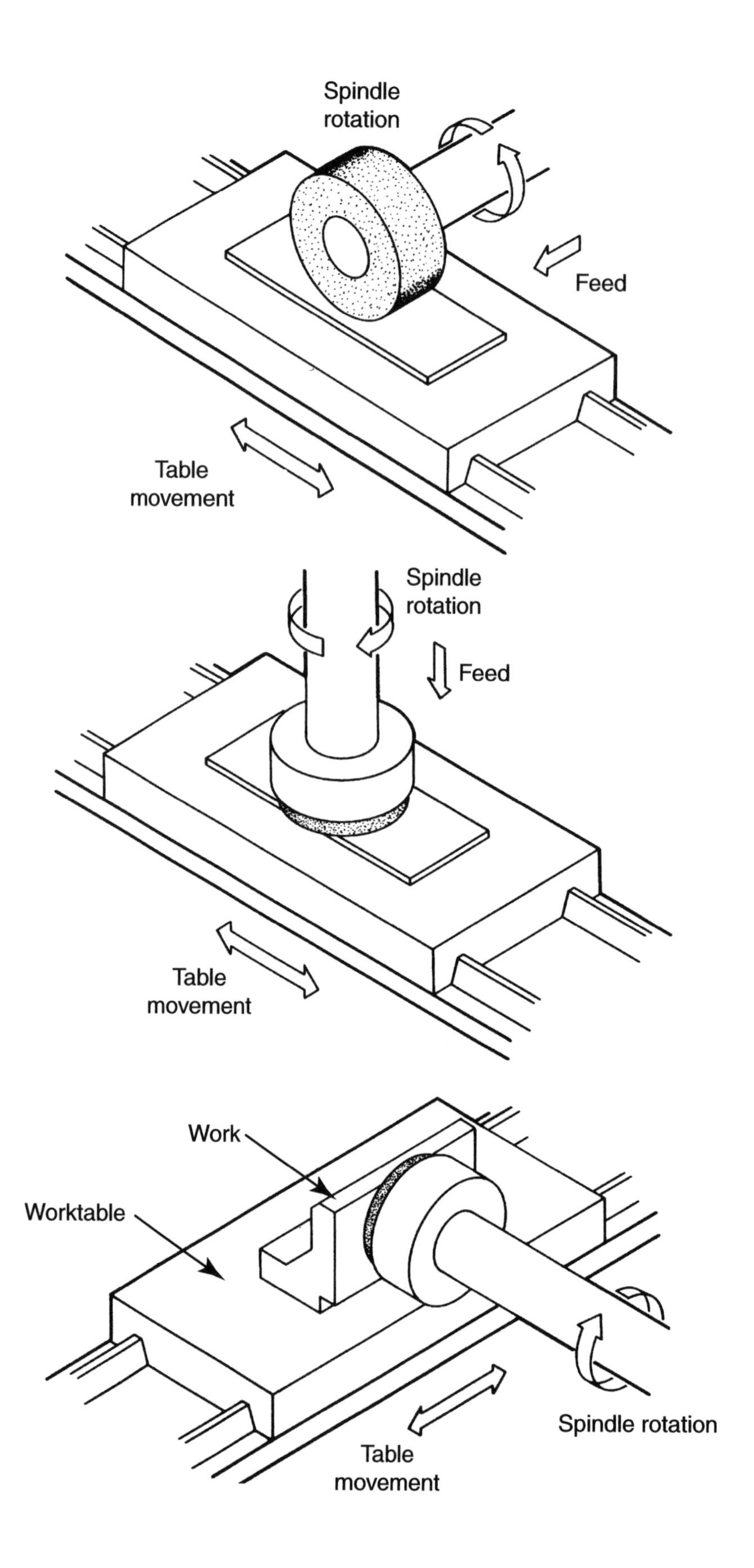

Surface Grinding—Rotary Type

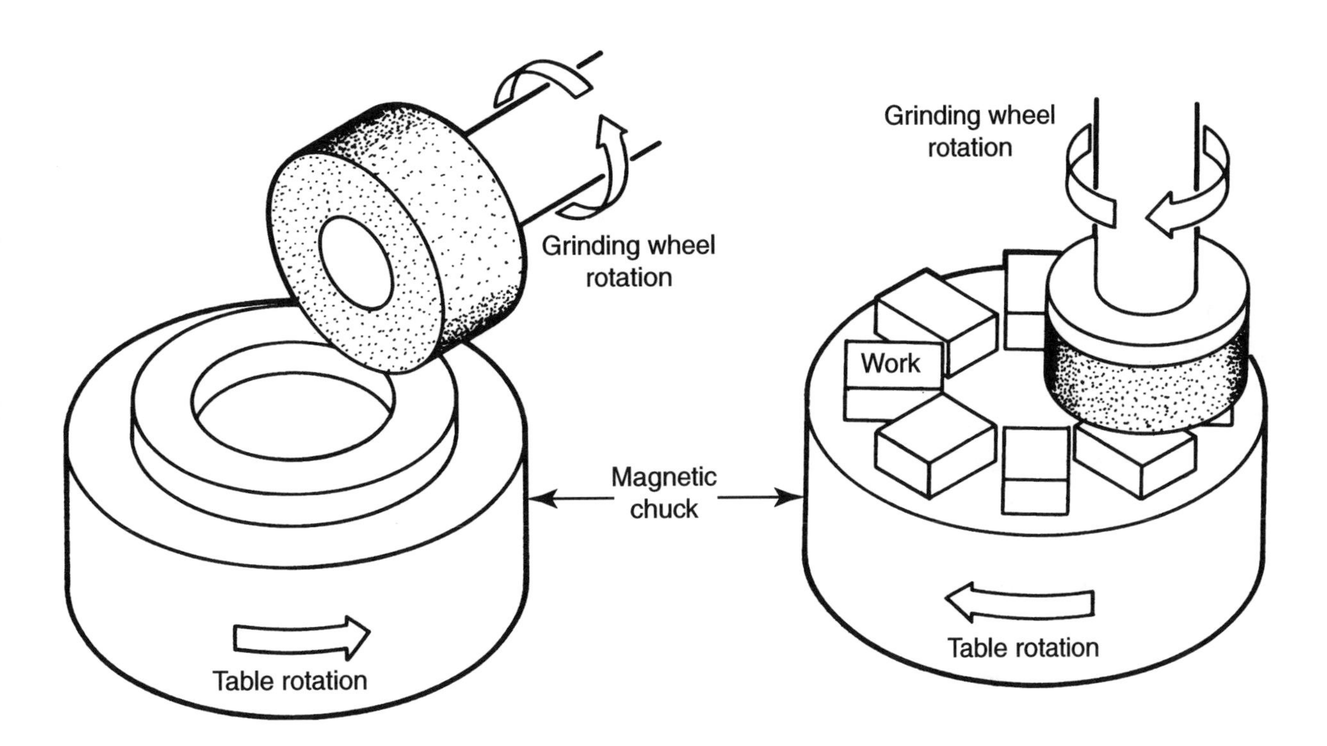

Chapter 23 Quiz

Grinding

Name: ______________________ **Date:** ____________ **Period:** ____________

1. Grinding is the operation that _____. 1.__________
   (a) produces fine surface finishes and close tolerances
   (b) is employed to sharpen tools
   (c) removes material by rotating an abrasive wheel or belt against the material
   (d) All of the above.
   (e) None of the above.
2. To prevent work being ground on a bench or pedestal grinder from being wedged between the wheel and the tool rest, it is recommended that the rest be adjusted to within _____ inch of the wheel. 2.__________
3. A grinding wheel can be checked for soundness by ______________________________
   ____
   ____.
4. A wheel dresser is used on a grinding wheel to _____ and/or to _____ any glaze. 4.__________
   ____

Match the sentence with the correct word or phrase.

_______ 5. The work does not require great accuracy in size or shape, and the work is manipulated by hand.
_______ 6. Used extensively on jobs such as polishing dies.
_______ 7. Work is mounted between centers and rotates while in contact with the grinding wheel.
_______ 8. Work is supported on a blade and is fed automatically between a regulating or feed wheel and the grinding wheel.
_______ 9. Used to secure a fine surface finish and accuracy on internal diameters.
_______10. Technique to sharpen milling cutters, reamers, and taps.
_______11. Method for grinding flat surfaces.
_______12. The grinding wheel is shaped to produce the required shape of the work.
_______13. Device employed to hold work for grinding.

(a) Cylindrical grinding
(b) Internal grinding
(c) Flexible shaft hand grinder
(d) Surface grinding
(e) Tool and cutter grinding
(f) Magnetic chuck
(g) Offhand grinding
(h) Centerless grinding
(i) Form grinding

14. List five safety precautions that should be observed when grinding.
   1)______________________________
   2)______________________________
   3)______________________________
   4)______________________________
   5)______________________________

Drills and Drilling Machines

Learning Objectives

After studying this chapter, students should be able to:

- Select and safely use the correct drill(s) and drilling machine for a given job.
- Make safe setups on a drill press.
- Understand drill size classifications.
- Calculate cutting speeds.
- Select the proper coolant for material being drilled.
- Align work for drilling.
- Identify various types of drilling machines.
- Observe drilling safety rules.

Chapter Resources

Text, pages 363–392
  Test Your Knowledge, page 391
  Research and Development, page 392
Workbook, pages 113–118
Instructor's Manual
  Answer keys for:
    Test Your Knowledge Questions
    Workbook
    Chapter Quiz
  Reproducible Masters:
    24-1 Types of Drills
    24-2 Parts of a Twist Drill
    24-3 Centering Round Stock in a V-Block
    24-4 Countersinking, Counterboring, and Spotfacing
    24-5 Chapter Quiz

Guide for Lesson Planning

Have available an assortment of drilling equipment—drills, drill gage, center drill, sleeve, socket, drift, center finder, vises, parallels, etc., for student examination. Have students read and study all or part of the chapter.

Using the appropriate Reproducible Masters, review the assignment with them discussing the following:

- Definition of a machine tool.
- Drill press can be used for many machining operations.
- How drill press size is determined.
- How a drill cuts. (The spiral flutes do *not* pull the drill into the material.)
- Why tool is called a twist drill.
- Types of drills.
- Drill sizes.
- Ways of determining drill size.
- Parts of a drill.
- How drills are mounted in a drill press.
- Work-holding devices and set ups.
- Cutting speeds and feeds and their importance.
- Using a center finder to position drill. (Demonstrate)
- Proper way to drill a hole. (Demonstrate)
- Cutting compounds. When and when not to use them.
- Reason for pilot hole.
- Holding and centering round stock for drilling.
- Reamers and reaming.
- Countersinking, counterboring, and spotfacing.
- Industrial applications of the drill press.
- Emphasize drilling safety, especially the importance of mounting the work solidly to the work table to prevent the dangerous merry-go-round.

Before demonstrating drill press operations, be sure tools and equipment are in safe operating condition. All students should be able to observe the demonstration and wearing approved eye protection.

Assignments

1. Assign the Test Your Knowledge questions.
2. Assign Chapter 24 in the *Modern Metalworking Workbook.*

3. Assign the chapter quiz. Copy and distribute Reproducible Master 24-5.
4. Permit students to volunteer for one or more of the Research and Development activities at the end of the chapter.

Test Your Knowledge

1. (c) diameter work piece that can be drilled on center
2. high-speed steel (HSS)
3. Straight
4. Taper
5. oil hole
6. Flutes. Any order: help form the cutting edge of the drill point; curl the chips for easier removal; form channels through which the chips can escape from the hole; allow the lubricant and coolant to get down to the cutting edge.
7. Any order: numbers (#80 to #1), letters (A to Z), inches and fractions (1/64″ to 3 1/2″), metric (3.0 mm to 76.0 mm).
8. Any order: micrometer, drill gage.
9. sleeve
10. A drill socket makes it possible to use drills with taper shanks larger than the opening in the drill press spindle.
11. drift
12. Without constant pressure, the drill will chatter. This will cause it to dull rapidly, chip the cutting edges, or break.
13. (d) All of the above.
14. No
15. burning, breaking
16. Any order: lip clearance, length and angle of lips, proper location of dead center.
17. (a) larger than
18. The drill may break, permanent damage to the drilling machine can result, and the hole produced will be oversized and often out-of-round.
19. drill point gage
20. 118
21. center finder (wiggler)
22. pilot hole
23. partially
24. depth gage
25. Countersinking
26. counterbore
27. Spotfacing
28. A reamed hole is extremely accurate in size and has a fine surface finish.
29. jobber's
30. smaller

Workbook

1. By the largest diameter of a circular piece that can be drilled on center.
2. twist drills
3. oil hole
4. Microdrills
5. The entire drill does not have to be replaced when the cutting edges are worn—only the inserts are changed.
6. (a) Number
   (b) Letter
   (c) Inches and fractions
   (d) Metric
7. micrometer, drill gage
8. (e) Lip clearance
9. (d) Heel
10. (b) Dead center
11. (l) Chuck
12. (a) Drill point
13. (k) Web
14. (i) Tang
15. (f) Straight shank
16. (c) Lips
17. (g) Taper shank
18. (j) Margin
19. (h) Flutes
20. (a) Sleeve
    (b) Socket
    (c) Drift
21. It enlarges taper shank tools so they will fit in the drill press spindle.
22. Tools with a taper shank larger than the drill press spindle can often be used by fitting a socket.
23. It is used to separate taper shank tools.
24. When improperly clamped it will spring and move, causing drill damage or breakage. Serious injury can result from work becoming loose and spinning about.
25. Any or all of the following: to cool the cutting tool; prevent chips from becoming welded to the drill lips; to reduce friction; improve hole finish; aid in the rapid removal of chips from the hole.
26. Any order: lip clearance, length and angle of lips, proper location of the dead center.
27. countersink
28. counterbore

29. spotfacing
30. (b) has a better finish and is more accurate than
31. smaller
32. Evaluate individually. Refer to Section 24.9.
33. 8000 rpm
34. 1800 rpm
35. 1000 rpm (rounded from 1013 rpm)
36. 650 rpm (rounded from 666 rpm)
37. 4200 rpm (rounded from 4166 rpm)
38. Evaluate student use of center finder or wiggler individually.
39. Evaluate student ability to sharpen drills individually.
40. Evaluate student ability to center holes on round stock individually.

Chapter Quiz

1. the largest diameter that can be drilled on center
2. Any order: inches and fractions, numbers, letters, metric.
3. micrometer, drill gage
4. sleeve
5. drift
6. mounted solidly
7. (b) apply and maintain constant pressure to advance the drill into the work
8. cutting
9. feed
10. Any two of the following: cool drill, improve finish, aid in rapid removal of chips, reduce friction, prevent chips from welding to drill lips.
11. wiggler, center finder
12. (d) All of the above.
13. Reaming
14. countersinking
15. Spotfacing
16. Evaluate individually. Refer to Section 24.9.

Notes

Types of Drills

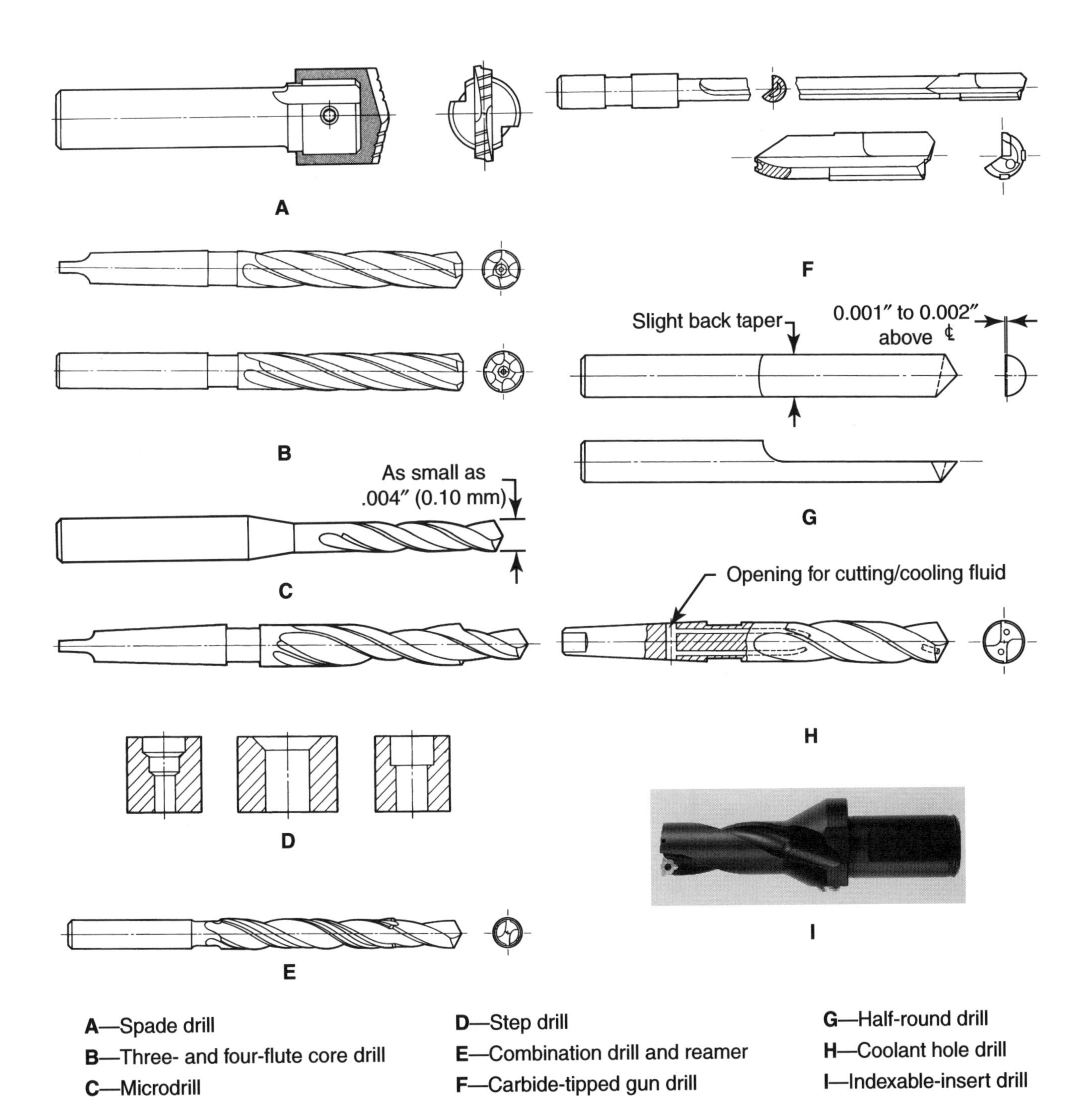

A—Spade drill
B—Three- and four-flute core drill
C—Microdrill
D—Step drill
E—Combination drill and reamer
F—Carbide-tipped gun drill
G—Half-round drill
H—Coolant hole drill
I—Indexable-insert drill

Parts of a Twist Drill

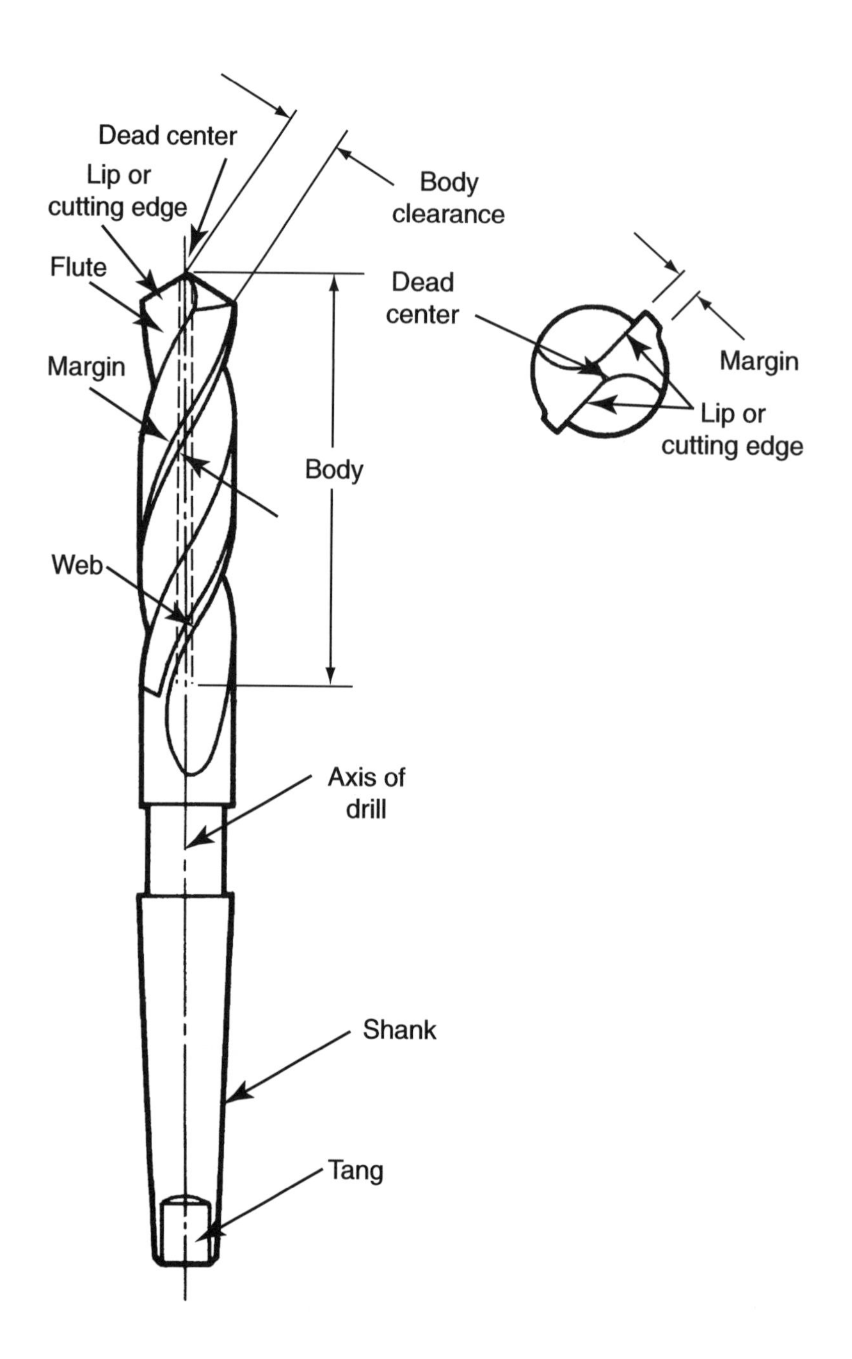

Centering Roundstock in a V-Block

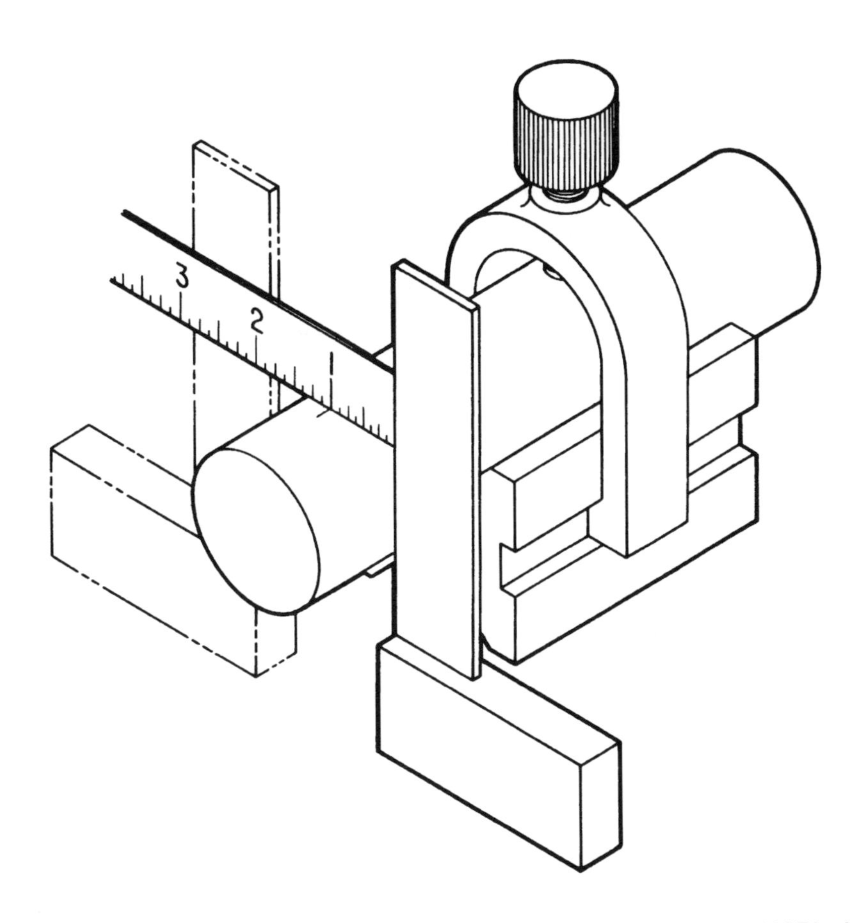

To align the hole for drilling through exact center, place the work and V-Block on the drill press table or on a surface plate. Rotate the punch mark until it is upright. Place a steel square on the flat surface with the blade against the round stock as shown above. Measure from the square blade to the punch mark, and rotate the stock until the measurement is the same when taken from both sides of the stock.

Countersinking, Counterboring, and Spotfacing

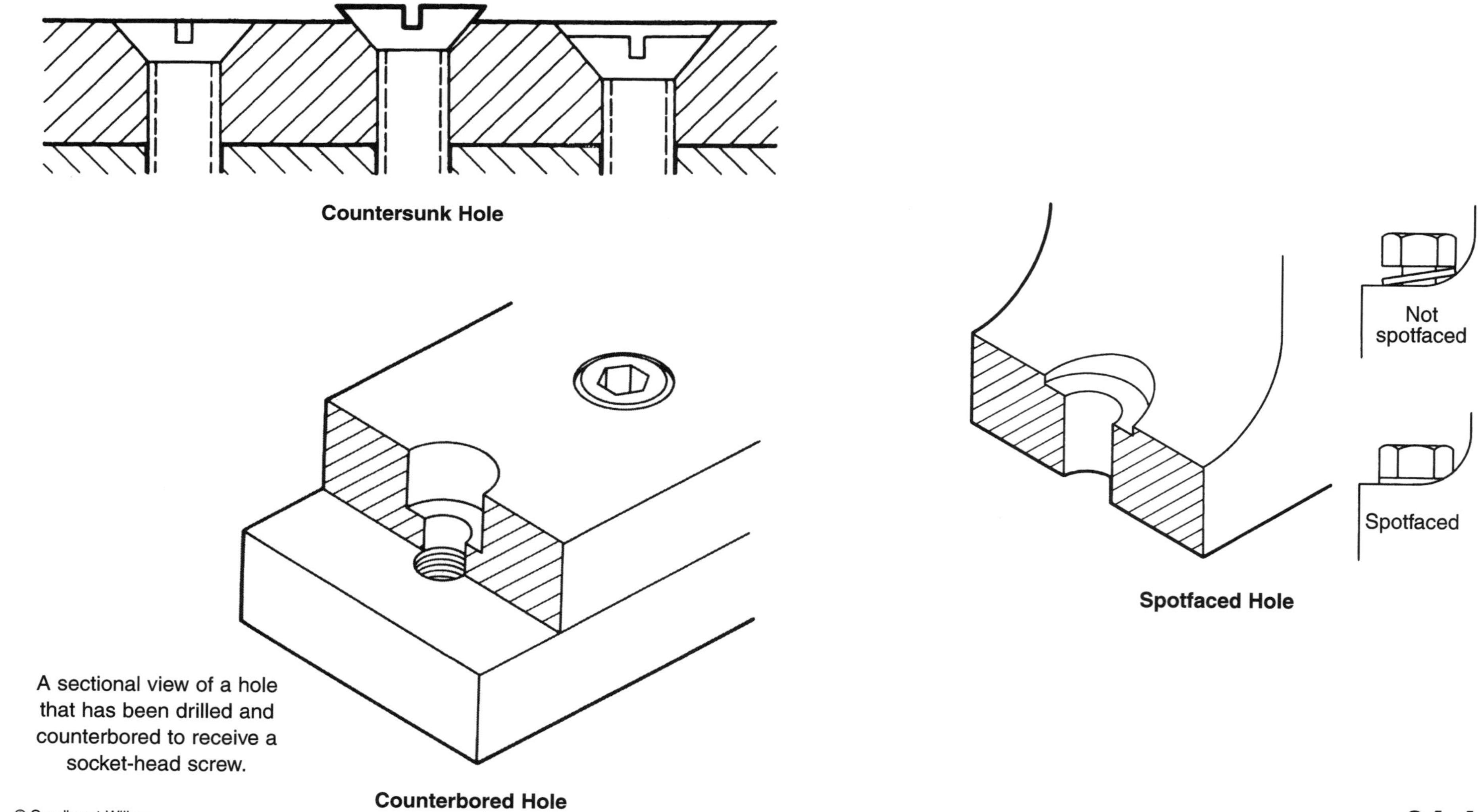

Countersunk Hole

Counterbored Hole

A sectional view of a hole that has been drilled and counterbored to receive a socket-head screw.

Spotfaced Hole

Chapter 24 Quiz

Drills and Drilling Machines

Name: ______________________ Date: ____________ Period: ____________

1. Drill press size is determined by ______________________________________

2. List the four drill size categories.

____

____

3. ______________________
____

3. Drill diameter can be determined by checking it with a(n) _____ or a(n) _____.

4. ______________________

4. A drill with a taper shank smaller than the drill press spindle taper must have its shank enlarged by fitting it with a(n) _____.

5. ______________________

5. Taper shank drills are separated from the above device with a(n) _____, *never* with a file tang.

6. ______________________

6. For safety purposes, when using a vise to hold your work for drilling, the vise must be _____ to the drill press table.

7. ______________________

7. To have a drill cut into the work at a constant rate, _____.
   (a) allow the flutes to pull the drill into the work
   (b) apply and maintain constant pressure to advance the drill into the work
   (c) operate the drill at a constant speed
   (d) All of the above.
   (e) None of the above.

8. ______________________

8. The speed that the drill rotates is called the _____ speed.

9. ______________________

9. The distance the drill advances into the work with each revolution is known as _____.

10. List two reasons cutting fluids are used when drilling.

____

____

11. ______________________
____

11. It is difficult to align a drill with the centerline by eye. To assist with this operation, use a(n) _____ or _____.

12. ______________________

12. By first drilling a pilot or lead hole when using larger diameter drills, _____.
   (a) cutting pressure is greatly reduced
   (b) hole accuracy is improved
   (c) the drill will cut much faster
   (d) All of the above.
   (e) None of the above.

Chapter 24 Quiz *(Continued)*

13. _____ is the operation that produces hole diameters that are extremely accurate and have an excellent surface finish. 13.__________

14. The operation that cuts a chamfer in a hole to permit a flat-headed fastener to be inserted with the head flush with the work surface is called _____. 14.__________

15. _____ is the operation that machines a circular spot on a rough surface to furnish a solid bearing surface for a bolt head or nut. 15.__________

16. List four safety rules that should be observed when operating a drill press.

 1) ____________________

 ____

 2) ____________________

 ____

 3) ____________________

 ____

 4) ____________________

 ____

Sawing and Cutoff Machines

Learning Objectives

After studying this chapter, students should be able to:

- Describe the operation of the three principal metal cutting power saws.
- Select and mount the proper blade for a job.
- Prepare a power saw for operation.
- Mount work properly for sawing.
- Safely operate a power saw.
- Practice power saw safety rules.

Chapter Resources

Text, pages 393–402
  Test Your Knowledge, page 401
  Research and Development, page 401
Workbook, pages 119–122
Instructor's Manual
  Answer keys for:
    Test Your Knowledge Questions
    Workbook
    Chapter Quiz
  Reproducible Masters:
    25-1 Reciprocating, Continuous Band, and Cutoff Saw Operation
    25-2 Tooth Set and Tooth Shape
    25-3 Correct Way to Hold Work
    25-4 Cutting Pressure
    25-5 Chapter Quiz

Guide for Lesson Planning

Have students read and study all or part of the chapter. Review and discuss the assignment with them. Question them on:

- Types of power saws and how they operate.
- Proper way to mount, position, and cut material.
- Blade selection.
- Vertical band saw.
- Proper way to mount and tension band blades on machine. (Demonstrate)
- Circular type metal cutting saws.
- Safety precautions to be observed when power sawing.

When demonstrating power sawing be sure all students can see and observe the operation. Approved eye protection must also be worn.

Assignments

1. Assign the Test Your Knowledge questions at the end of the chapter.
2. Assign Chapter 25 in the *Modern Metalworking Workbook.*
3. Assign the chapter quiz. Copy and distribute Reproducible Master 25-5.
4. Permit students to volunteer for one or more of the Research and Development activities at the end of the chapter.

Test Your Knowledge

1. A back-and-forth (reciprocating) cutting action.
2. (c) at least three teeth are in contact with the work at all times during the cutting sequence
3. feed
4. Flexible-back blades and all-hard blades. Evaluate descriptions individually.
5. Evaluate individually. Refer to text Section 25.2.2.
6. Band saws use a continuous blade that moves in one direction.
7. Any order: more precise, greater speed, little waste.
8. The old cut is too narrow and will pinch and ruin the new blade.
9. Circular metal-cutting saws use a round, flat blade that rotates into the work.
10. Evaluate individually. Refer to Section 25.6.

Workbook

1. Any order: reciprocating blade saw, continuous or band-type saw, circular blade or abrasive wheel cutoff saw. Evaluate descriptions individually.
2. coarse, fine
3. The three-tooth rule means at least three teeth are cutting at all times.
4. Flexible-back
5. All-hard
6. tensioned
7. Band saws use a continuous blade that moves in one direction.
8. It is faster, more precise, and creates little waste.
9. Evaluate individually. Refer to text Section 25.2.2.
10. (a) Raker
    (b) Wavy
11. Evaluate sketches individually. Refer to text Figure 25-12.
12. Evaluate individually. Refer to Section 25.4.
13. Any order: abrasive cut-off saw, cold circular saw, friction saw.
14. The friction saw. If teeth are on the blade, their primary use is to carry oxygen to the cutting area.
15. Evaluate individually. Refer to Section 25.7.

Chapter Quiz

1. (a) Reciprocating saw
   (b) Band or continuous blade saw
   (c) Circular type saw
2. at least three teeth cutting at the same time
3. large pieces, soft material
4. hard material, small work
5. direction of cut
6. Evaluate individually. Refer to Section 25.7.

Reciprocating, Continuous Band, and Cutoff Saw Operation

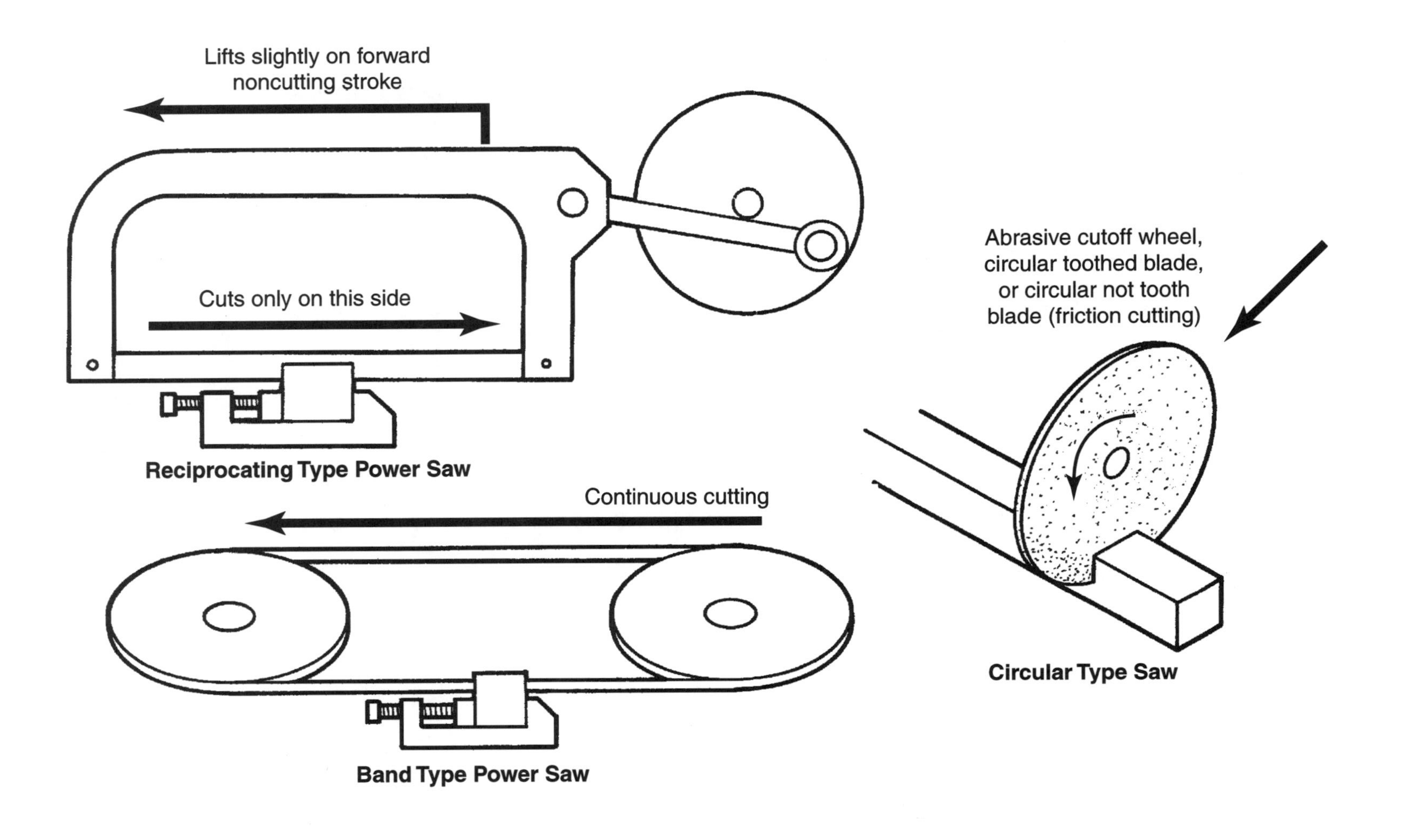

25-1

Tooth Set and Tooth Shape

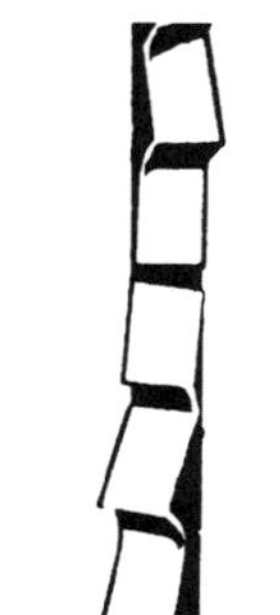

Raker **Wavey**

Saw blades commonly have raker or wavy teeth. Raker teeth are preferred for general use, cutting large solid sections, and cutting thick plate.

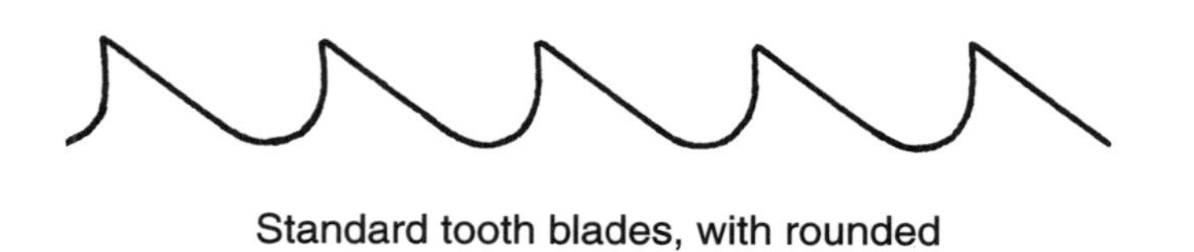

Standard tooth blades, with rounded gullets, are usually best for most ferrous metals, hard bronzes, and hard brasses.

Skip tooth blades provide for more chip clearance without weakening the blade body. They are recommended for cutting aluminum, magnesium, copper, and soft brasses.

Hook tooth blades offer two advantages over skip tooth blades—easier feeding and less "gumming up."

Correct Way to Hold Work

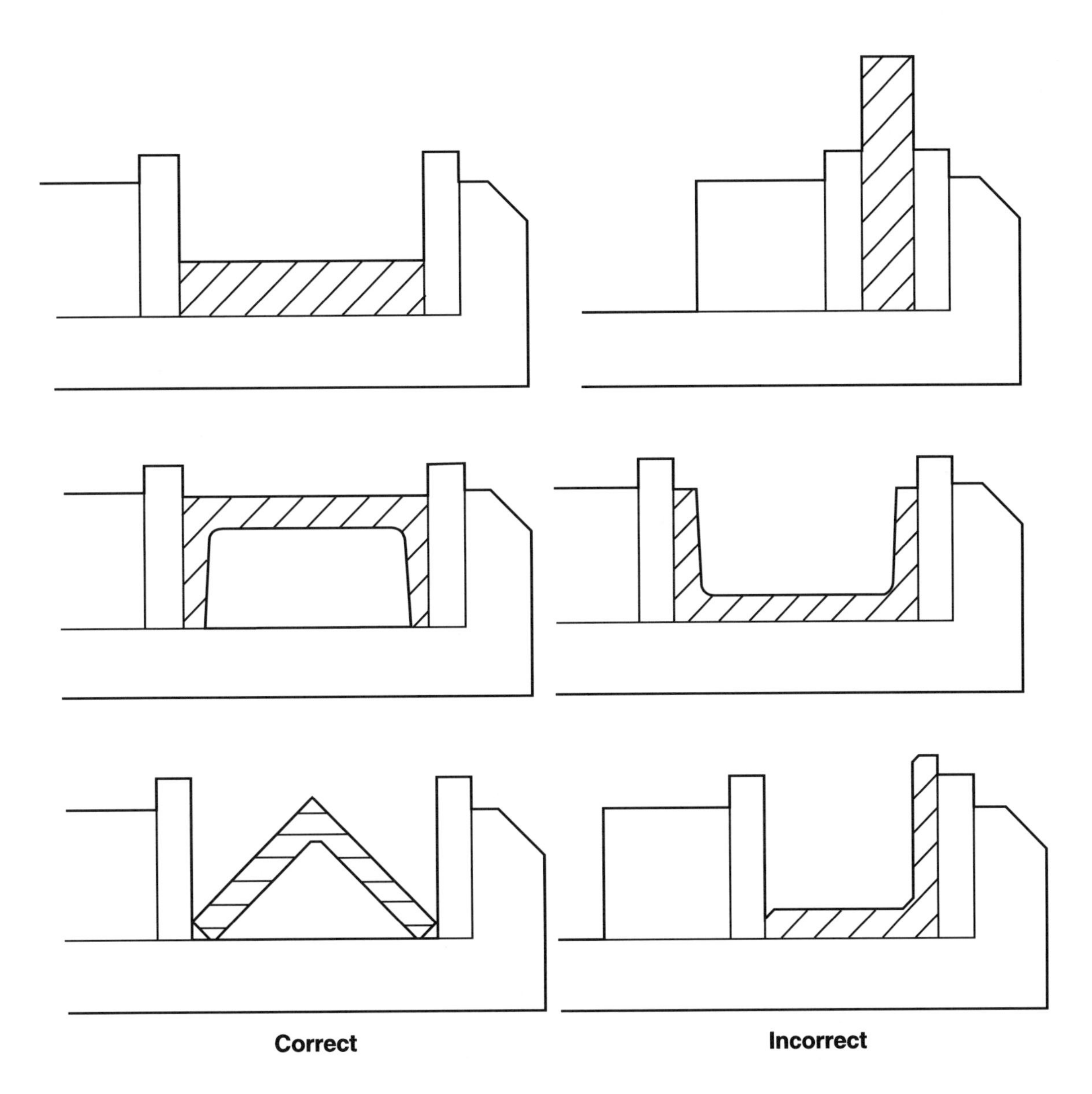

Recommended ways to hold sharp-cornered work for cutting. A carefully planned setup will ensure that at least three teeth will be cutting, greatly extending blade life.

Cutting Pressure

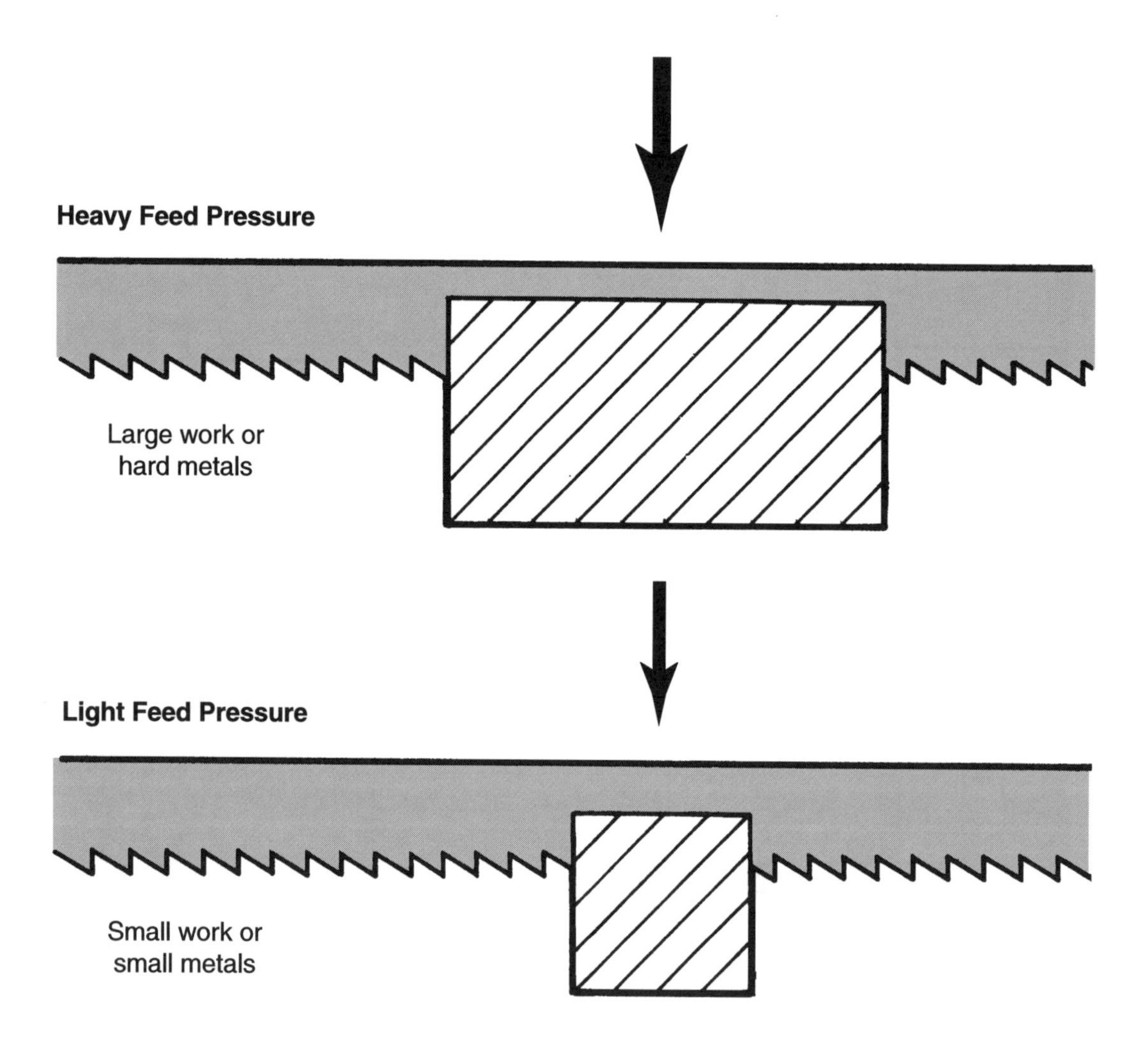

Apply heavy feed pressure on hard metals and large work. Use light pressure on soft metals and work with small cross sections

Chapter 25 Quiz

Sawing and Cutoff Machines

Name: ______________________ **Date:**____________ **Period:**____________

1. Identify the three types of metal cutting power saws described in the following statements. 1.__________
   (a) The blade cuts only on the back stroke.
   (b) The blade is continuous cutting.
   (c) Uses a round blade made of steel or of an abrasive.

2. What does the "three tooth rule" for sawing mean? ______________________________
   ____
   ____

3. A coarse tooth blade with heavy pressure is used to cut _____ and _____ material. 3.__________
   ____

4. A fine tooth blade with light pressure is used to cut _____ or _____ and _____ material. 4.__________
   ____
   ____

5. Blades must be installed so the teeth will cut on the _____. 5.__________

6. List five safety rules that should be observed when using a power saw.
   1) ______________________________
   ____
   2) ______________________________
   ____
   3) ______________________________
   ____
   4) ______________________________
   ____
   5) ______________________________
   ____

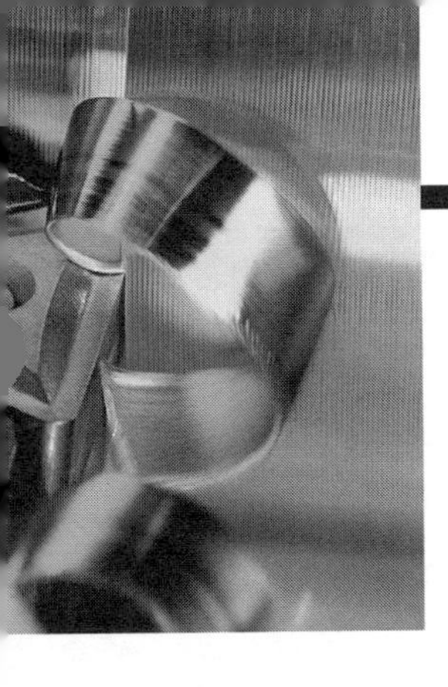

METAL LATHE

LEARNING OBJECTIVES

After studying this chapter, students should be able to:

- Describe how a lathe operates.
- Identify the various parts of a lathe.
- Safely set up and operate a lathe using various work-holding devices.
- Sharpen lathe cutting tools.
- Practice lathe safety rules.

CHAPTER RESOURCES

Text, pages 403–434
  Test Your Knowledge, page 432-433
  Research and Development, page 434
Workbook, pages 123–126
Instructor's Manual
  Answer keys for:
    Test Your Knowledge Questions
    Workbook
    Cutting Speed Problems
    Chapter Quiz
  Reproducible Masters:
    26-1 How a Lathe Operates
    26-2 How a Lathe is Measured
    26-3 Parts of a Lathe
    26-4 Cutting Tool Shapes
    26-5 Calculating Cutting Speeds
    26-6 Cutting Speed Problems
    26-7 Properly Drilled Center Hole
    26-8 Chapter Quiz

GUIDE FOR LESSON PLANNING

Due to the amount of material covered, it would be advisable to divide the chapter into several segments. Teach the segments that best suit your program.

PART I—MAJOR PARTS OF THE LATHE

Have students read and study Sections 26.1 through 26.4. Using the appropriate Reproducible Masters, review and discuss the following:

- How the lathe operates.
- How lathe size is determined.
- Major parts of the lathe.
- How to prepare the lathe for operation. (Demonstrate)
- Proper way to clean a lathe. (Demonstrate)

PART II—LATHE SAFETY

Have students read and study Section 26.5, *Lathe Safety*. Review and discuss the assignment with them.

PART III—LATHE CUTTING TOOLS AND TOOL HOLDERS

Large models of cutting tool shapes made from wood and painted to simulate metal will aid in teaching this portion of the chapter.

Have students read and study Sections 26.6 and 26.7. Review the assignment with them and discuss the following:

- Types of cutting tools.
- Tool holders.
- Cutting tool shapes and how they are used.
- How to sharpen a cutter bit. (Demonstrate)

Students can practice sharpening cutter bits using sections of 1/4″ or 5/16″ square mild steel rod rather than on expensive cutter bits.

PART IV—CUTTING SPEEDS, FEEDS, AND DEPTH OF CUT

Have students read and study Section 26.8. Using the appropriate Reproducible Masters, review the assignment with them. Instruct the students on how to calculate cutting speeds.

Discuss:

- The meaning of the term *cutting speed.*
- Roughing cuts.
- Finish cuts.
- How to determine *depth of cut.* Check your lathe feed dials to determine whether they are direct reading or not direct reading.

Part V—Work-Holding Attachments and Facing, Turning, and Parting Operations

Have an assortment of work-holding attachments available including a 3-jaw universal chuck, 4-jaw independent chuck, Jacobs chuck, face plate and dogs, collets, etc., for students to examine.

Have students read and study Sections 26.9 through 26.14. Review the assignment with them and, asking appropriate questions, discuss the following:

- Turning between centers. (Show lathe set up.)
- Preparing lathe and work for turning between centers.
- Proper way to position cutting tool. (Demonstrate)
- Turning stock between centers. (Demonstrate)
- Types of chucks.
- Advantages of using chucks.
- Why long work mounted in chuck must be supported by tailstock center.
- The correct way to mount and remove chuck from lathe.
- How to center work in a 4-jaw independent chuck.
- Turning work held in a chuck.
- Precautions that must be taken when turning work held in chuck.

Assignments

1. Assign the Test Your Knowledge questions.
2. Assign Chapter 26 in the *Modern Metalworking Workbook.*
3. Assign the chapter quiz. Copy and distribute Reproducible Master 26-8.
4. Permit students to volunteer for one or more of the Research and Development activities at the end of the chapter.

Test Your Knowledge

1. (a) move faster or slower if the carriage is engaged to the lead screw
2. (b) Changes spindle speed.
3. (d) None of the above.
4. (d) None of the above.
5. (c) headstock
6. (d) compound rest
7. (a) tool post
8. (a) saddle
9. (b) Engages the half-nuts for thread cutting.
10. (b) Engages the clutch for automatic power feed.
11. (a) Moves the carriage right and left on the ways.
12. (c) work against the edge of a cutting tool; the tool travels or feeds into or across it
13. (d) None of the above.
14. (b) carefully check over and lubricate it
15. (a) independent
16. (b) universal
17. (a) the automatic power feed
18. (a) lead
19. (d) Cross-slide
20. ways
21. Any order: chuck, between centers, collets, mounted on a faceplate.
22. Any order: independent, universal, Jacobs.
23. spindle nose
24. (b) Tool post
25. (c) Tool post wedge
26. (d) Lead screw
27. (h) Dead center
28. (g) 3-jaw chuck
29. (a) Ways
30. (f) 4-jaw chuck
31. (e) Apron
32. (a) 700 rpm—roughing cut, 1150 rpm—finishing cut
    (b) 400 rpm—roughing cut, 500 rpm—finishing cut
    (c) 80 rpm—roughing cut, 100 rpm—finishing cut

Workbook

1. (e) Tailstock
2. (b) Headstock
3. (c) Spindle
4. (a) Lathe bed

5. (l) Half-nut
6. (j) Feed change levers
7. (k) Index plate
8. (f) Carriage
9. (g) Lead screw
10. (h) Quick-change gearbox
11. (d) Back gears
12. (i) Feed mechanism
13. Roughing
14. Facing
15. diameter, ways
16. Any order: driving the lathe, holding and rotating the work, holding and moving the cutting tool.
17. knockout bar

Cutting Speed Problems

1. 2000
2. 200
3. 150
4. 200

Chapter Quiz

1. (m) Roughing cut
2. (b) Tool holder
3. (e) Universal chuck
4. (i) Chucks
5. (k) Finish turning
6. (o) Cutter bit
7. (j) Independent chuck
8. (f) Facing
9. (g) Bed
10. (d) Live center
11. (c) Chatter
12. (h) Dog
13. (n) Face plate
14. (l) Half-nuts
15. (a) Parting

Notes

How a Lathe Operates

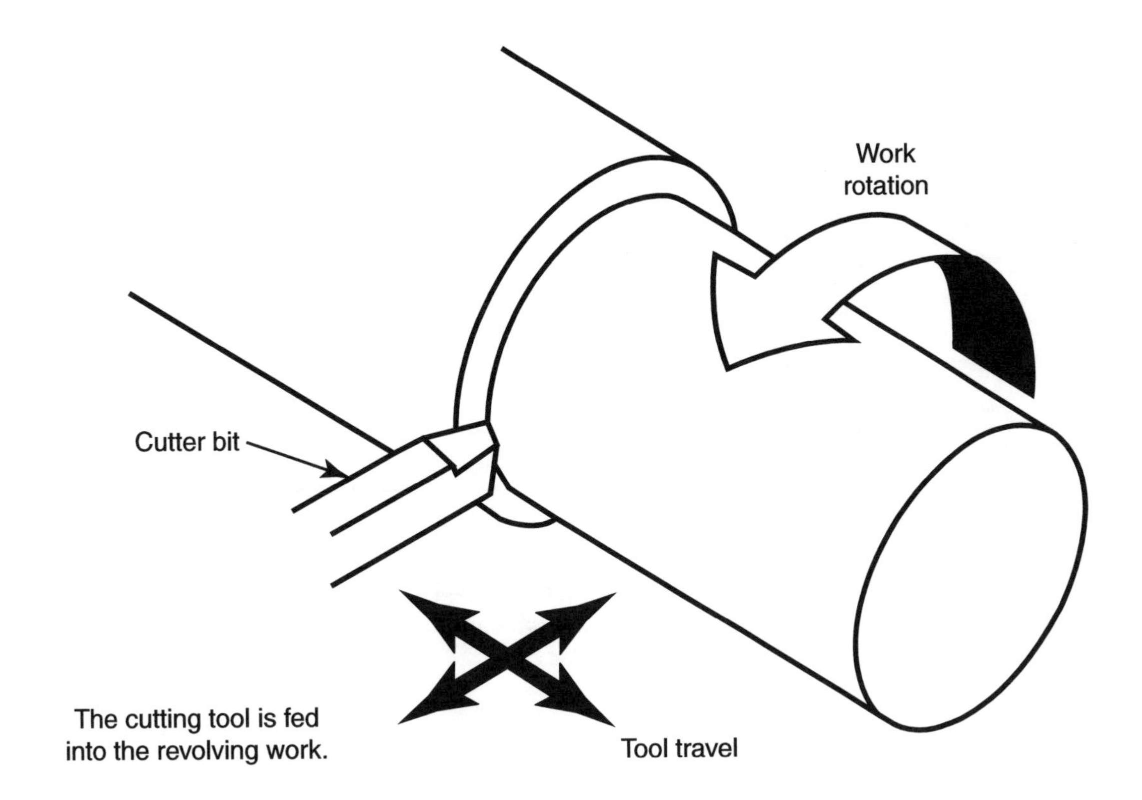

How a Lathe is Measured

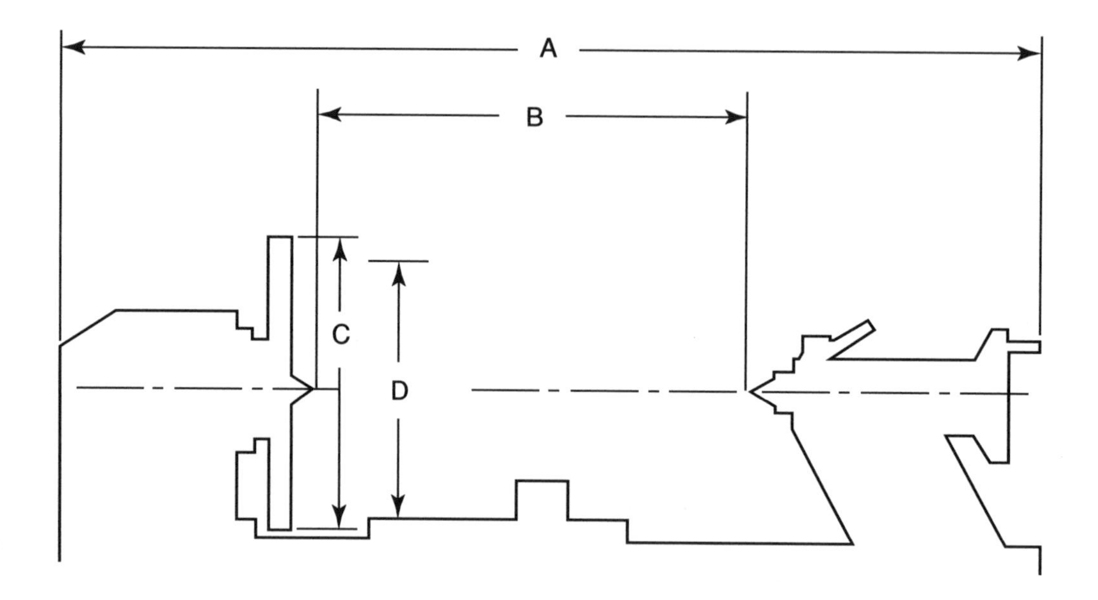

A—Length of bed. B—Distance between centers. C—Diameter of work that can be turned over the ways. D—Diameter of work that can be turned over the cross-slide.

Parts of a Lathe

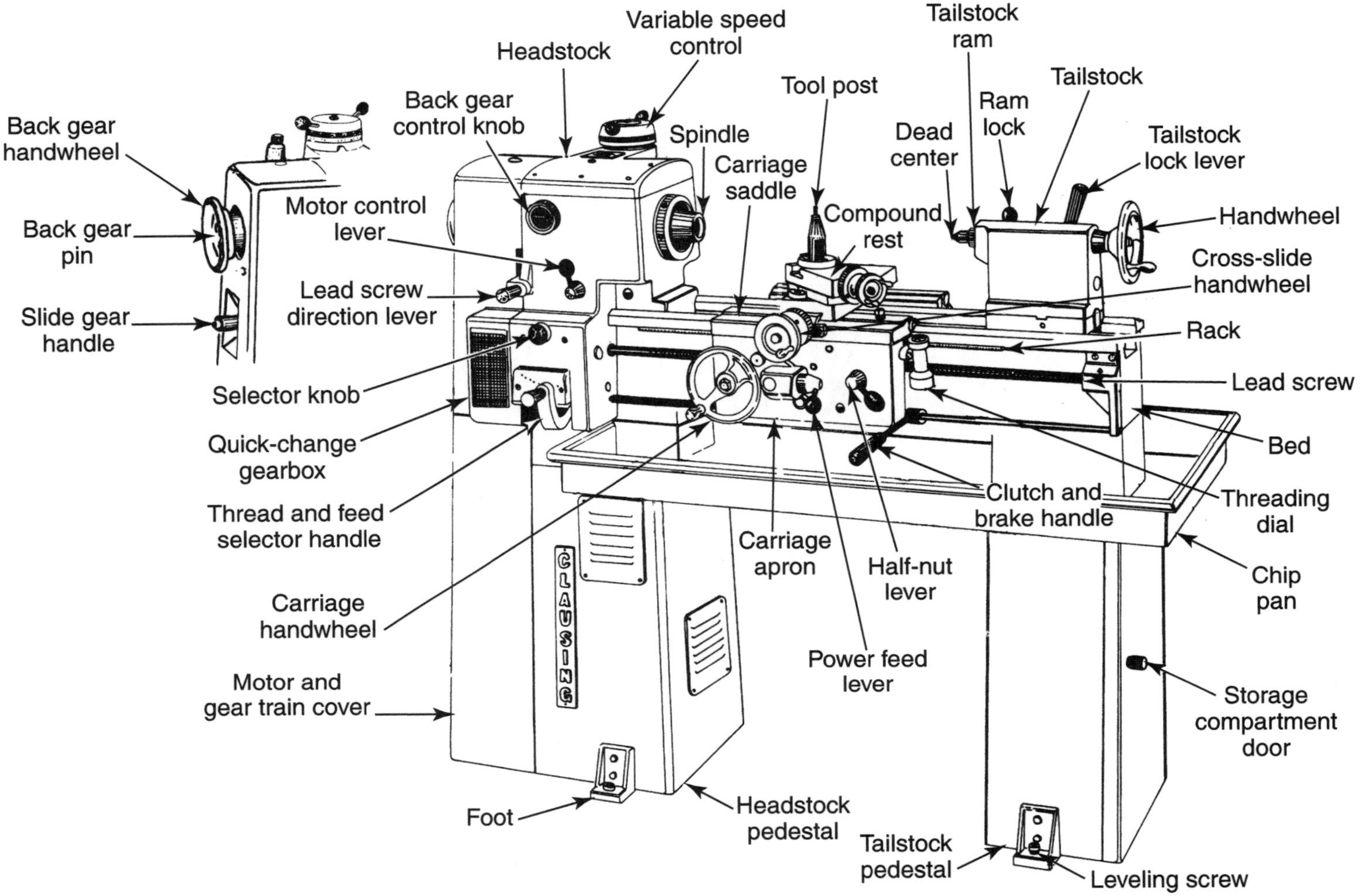

Back gear handwheel
Back gear pin
Slide gear handle
Headstock
Variable speed control
Back gear control knob
Motor control lever
Lead screw direction lever
Selector knob
Quick-change gearbox
Thread and feed selector handle
Carriage handwheel
Motor and gear train cover
Foot
Headstock pedestal
Spindle
Carriage saddle
Tool post
Compound rest
Dead center
Tailstock ram
Ram lock
Tailstock
Tailstock lock lever
Handwheel
Cross-slide handwheel
Rack
Lead screw
Bed
Threading dial
Clutch and brake handle
Chip pan
Carriage apron
Half-nut lever
Power feed lever
Storage compartment door
Tailstock pedestal
Leveling screw

26-3

Cutting Tool Shapes

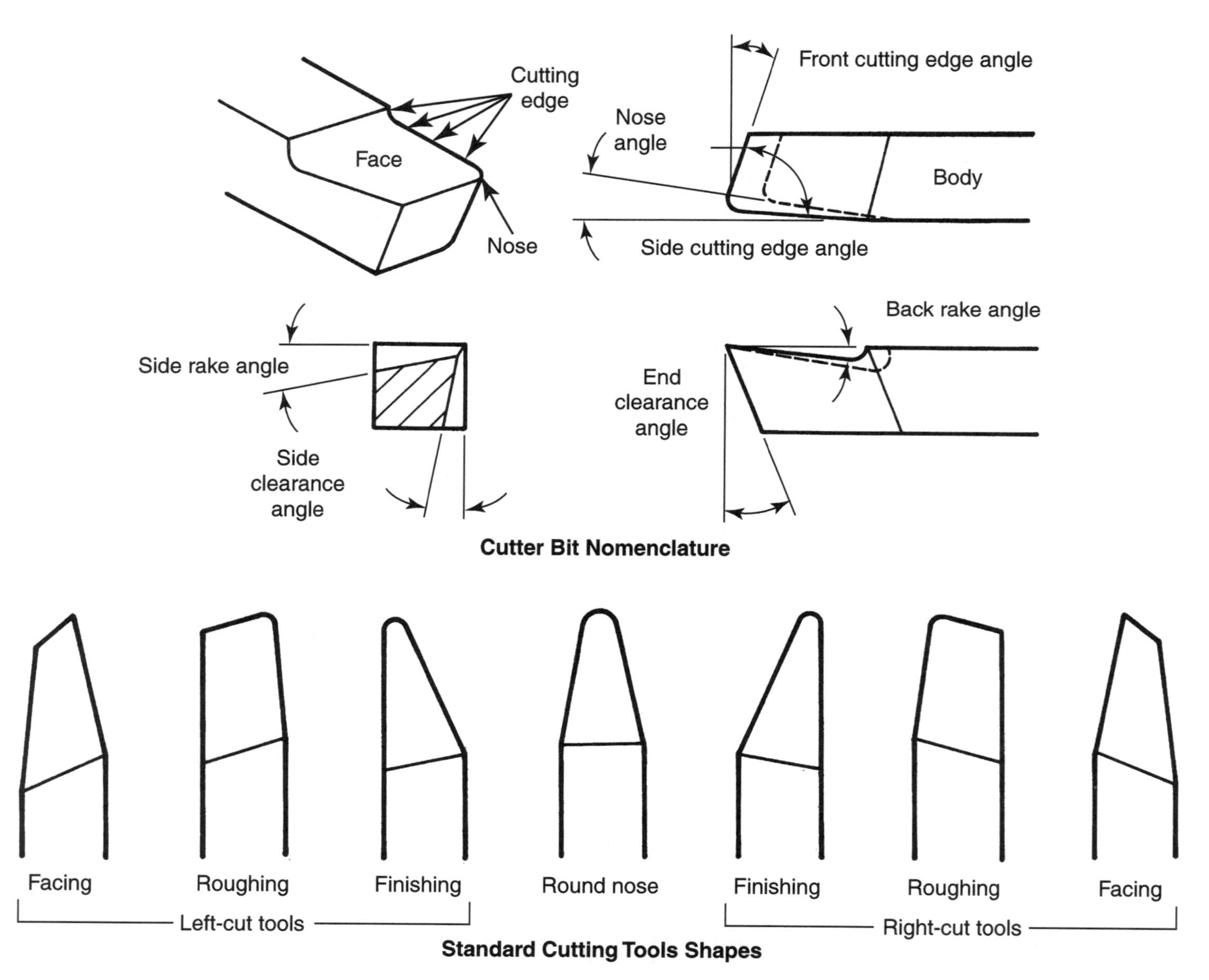

Cutter Bit Nomenclature

Standard Cutting Tools Shapes

26-4

Calculating Cutting Speeds

Cutting speeds (CS) are given in feet per minute (fpm), while the work speed is given in revolutions per minute (rpm). Thus, the peripheral speed of the work (CS) must be converted to rpm in order to determine the lathe speed required. The following formula can be used:

$$\text{rpm} = \frac{\text{CS} \times 4}{\text{D}}$$

rpm = revolutions per minute
CS = cutting speed of the particular metal being turned in feet per minute
D = diameter of the work in inches

Material to be cut	Roughing cut 0.01″ to 0.020″ 0.25 mm to 0.50 mm feed		Finishing cut 0.001″ to 0.010″ 0.025 mm to 0.25 mm feed	
	fpm	mpm	fpm	mpm
Cast iron 70	20	120	36	
Steel				
Low carbon	130	40	160	56
Medium carbon	90	27	100	30
High carbon	50	15	65	20
Tool steel (annealed)	50	15	65	20
Brass—yellow	160	56	220	67
Bronze	90	27	100	30
Aluminum*	600	183	1000	300

The speeds for rough turning are offered as a starting point. It should be all the machine and work will withstand. The finishing feed depends upon the finish quality desired.

* The speeds for turning aluminum will vary greatly according to the alloy being machined. The softer alloys can be turned at speeds upward of 1600 fpm (488 mpm) roughing to 3500 fpm (106 mpm) finishing. High silicon alloys require a lower cutting speed.

Cutting Speed Problems

Name: ______________________ **Date:**____________ **Period:**____________

Using the formula for cutting speeds, solve the following problems.

1. What spindle speed is required to finish turn 2″ diameter aluminum alloy? Round your answer to the nearest 50 rpm.

2. What spindle speed is required to finish turn 4″ diameter brass? Round your answer to the nearest 50 rpm.

3. What spindle speed is required to finish turn 1 1/2″ diameter tool steel (annealed)? Round your answer to the nearest 50 rpm.

4. What spindle speed is required to finish turn 3″ low-carbon steel? Round your answer to the nearest rpm.

Properly Drilled Center Hole

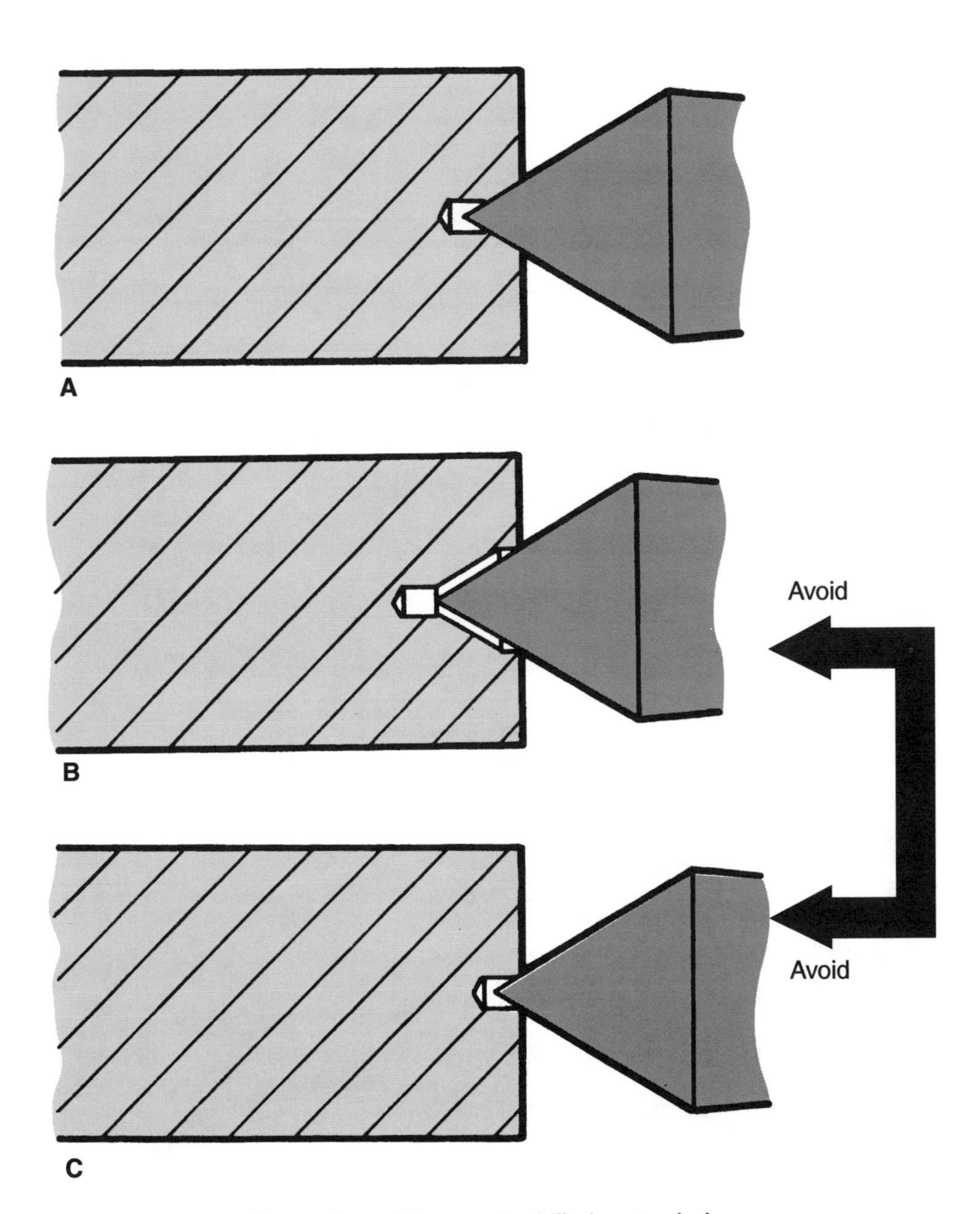

Correctly and incorrectly drilled center holes.
A—Properly drilled center hole. B—Hole drilled too deep.
C—Hole not drilled deep enough. Does not provide enough support; if used with a dead center, the center point will burn off.

Chapter 26 Quiz
Metal Lathe

Name: ______________________ **Date:**____________ **Period:**____________

Match each sentence with the appropriate word or phrase.

(a) Parting
(b) Tool holder
(c) Chatter
(d) Live center
(e) Universal chuck
(f) Facing
(g) Bed
(h) Dog
(i) Chucks
(j) Independent chuck
(k) Finish turning
(l) Half-nuts
(m) Roughing cut
(n) Face plate
(o) Cutter bit

________ 1. Reduces the work diameter to approximate size.

________ 2. Supports the cutting tool on many lathes.

________ 3. Jaws on this chuck operate at the same time.

________ 4. Easiest and fastest way of mounting work on a lathe.

________ 5. Brings work to exact size.

________ 6. Small piece of metal used to cut metal on a lathe.

________ 7. Jaws on this chuck operate individually.

________ 8. Operation on lathe that machines end of stock.

________ 9. Back bone of lathe.

________10. Rotating center.

________11. Vibration caused by cutting tool springing away from work. It produces small ridges on the machined surface.

________12. Device for clamping work so it can be machined between centers.

________13. Circular plate that mounts on headstock spindle.

________14. Mechanism that locks the lathe carriage to the lead screw for the purpose of cutting threads.

________16. Operation of cutting off work after it has been machined.

27 Cutting Tapers and Screw Threads on a Lathe

Learning Objectives

After studying this chapter, students should be able to:

- Describe how tapers are turned on a lathe.
- Calculate tailstock setover for turning a taper.
- Properly set up and operate a lathe for taper turning.
- Describe the various forms of screw threads.
- Cut screw threads on a lathe.

Chapter Resources

Text, pages 435–448
  Test Your Knowledge, pages 446–447
  Research and Development, page 447
Workbook, pages 127–130
Instructor's Manual
  Answer keys for:
    Test Your Knowledge Questions
    Workbook
    Chapter Quiz
  Reproducible Masters:
    27-1 Basic Taper Information
    27-2 Calculating Tailstock Setover (Taper Per Inch Given)
    27-3 Calculating Tailstock Setover (Taper Per Foot Given)
    27-4 Calculating Tailstock Setover (All Taper Dimensions Given)
    27-5 Screw Thread Applications
    27-6 Cutting Action of Threading Tools
    27-7 Three-Wire Method of Measuring Threads
    27-8 Thread Measuring Problems
    27-9 Chapter Quiz

Guide for Lesson Planning

Due to the amount of material covered, it would be advisable to divide the chapter into segments. Divide the chapter as best suits your program.

Part I—How to Turn a Taper

Prepare a lathe set up to demonstrate taper turning. Have students read and study Section 27.1. Review the assignment with them demonstrating the following:

- Methods available for turning tapers.
- Compound rest method. (Advantages and disadvantages)
- Offset tailstock method. (Advantages and disadvantages)
- Calculating tailstock setover. (Use Reproducible Masters 27-1 through 27-4.)
- Taper attachment method. (Advantages and disadvantages)
- How to measure tapers.

Part II—Cutting Screw Threads on a Lathe

Have students read and study Section 27.2. Review the assignment with them and discuss the following:

- Five uses of the screw thread.
- Screw thread forms and thread terminology.
- How to set up a lathe to cut threads. (Demonstrate)
- Cutting sharp "V" threads on a lathe. (Demonstrate)
- Methods used to measure screw threads.
- Using 3-wire method to measure screw threads.
- Copy and distribute Reproducible Masters 27-5 through 27-8. Have students practice calculating thread measuring problems. (Demonstrate)

- How to cut internal threads. (Demonstrate)

Part III—Lathe Safety

Have students read and study Section 27.3. Review the assignment and discuss the importance of following all safety rules.

Assignments

1. Assign the Test Your Knowledge questions.
2. Assign Chapter 27 in the *Modern Metalworking Workbook.*
3. Assign the chapter quiz. Copy and distribute Reproducible Master 27-9.
4. Permit students to volunteer for one or more of the Research and Development activities at the end of the chapter.

Test Your Knowledge

1. Any order: compound, offset tailstock, taper attachment, angular tool bit.
2. Setover
3. (a) 0.029″ (b) 1.500″ (c) 0.032″ (d) 1.850″ (e) 32.5 mm
4. taper attachment
5. Student answers will vary but may include any four of the following: making adjustments, assembling parts, transmitting motion, applying pressure, making measurements.
6. Lead
7. (b) pitch diameter
8. thread end groove
9. (a) M = 0.520″ (b) M = 0.270″ (c) M = 0.415″ (d) M = 0.509″
10. (b) toward the tailstock

Workbook

1. (a) Can be used to cut internal and external tapers. However, taper length is limited by compound rest movement.
   (b) Only external tapers can be turned. Work must be mounted between centers. Tailstock must be realigned after the job is complete.
   (c) Internal and external tapers can be cut. Work can be held in a chuck or between centers. Once the attachment is set, the same taper can be machined on any length of material.
   (d) Only very short tapers can be machined.
2. The distance a lathe tailstock has been offset from the normal centerline of the machine.
3. 0.100″ setover
4. 0.101″ setover
5. 0.280″ setover
6. 6.125 mm setover
7. 8.07 mm setover
8. Any order: making adjustments, assembling parts, transmitting motion, applying pressure, making measurements.
9. (e) Major diameter
10. (b) Pitch
11. (c) Root diameter
12. (d) Lead
13. (a) Pitch diameter
14. thread end
15. thread dial
16. Wire size 0.457″
17. Wire size 0.259″
18. Wire size 0.772″
19. Evaluate individually. Refer to Section 27.3.

Chapter Quiz

1. thread end
2. (d) one full
3. root
4. (a) Toward the tailstock.
5. major
6. Pitch
7. half-nuts
8. setover
9. tapered
10. pitch

Reproducible Masters

27-2
1. 0.0200″
2. 0.270″
3. 0.197″

27-3
1. 0.200″
2. 0.540″
3. 0.042″

27-4
1. 0.375″
2. 10 mm
3. 0.200″

27-5
1. 0.270″
2. 0.457″
3. 0.790″
4. 0.947″

Basic Taper Information

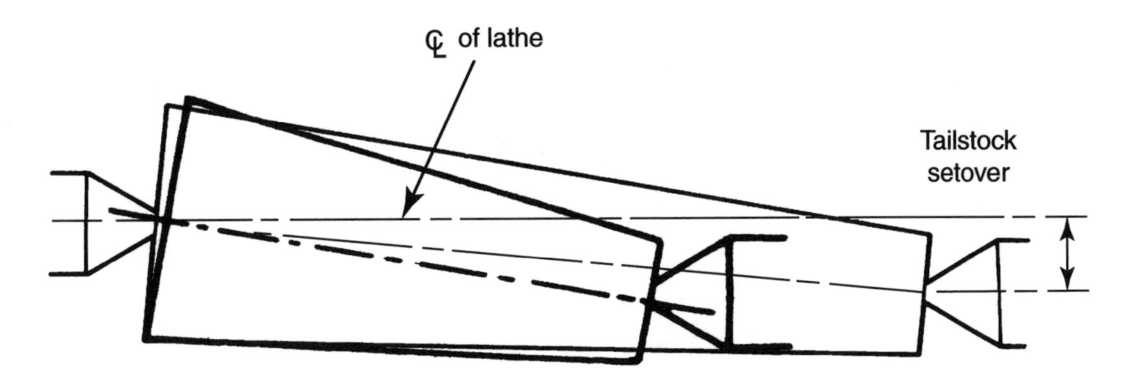

Length of work causes taper to vary even though tialstock offset remains the same.

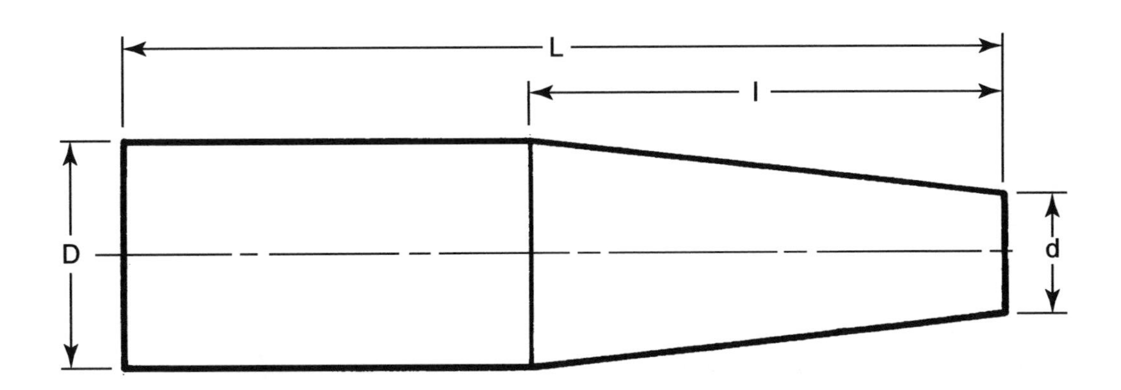

D = diameter at large end of taper; d = diameter at small end of taper; l = length of taper; L = total length of piece.

Calculating Tailstock Setover

(Taper Per Inch Given)

Name: ______________________ **Date:**____________ **Period:**____________

The tailstock offset must be calculated for each job because the work length plays an important part in the calculations.

TPI = Taper per inch
L = Total length of work

Formula: $\text{Offset} = \frac{L \times TPI}{2}$

1. What will be the setover for the following job?
   TPI = 0.050″
   L = 8.000″

2. What will be the setover for the following job?
   TPI = 0.060″
   L = 9.000″

3. What will be the setover for the following job?
   TPI = 0.045″
   L = 8.750″

Calculating Tailstock Setover

Taper Per Foot Given

Name: ______________________ **Date:**______________ **Period:**______________

The tailstock offset must be calculated for each job because the work length plays an important part in the calculations. When the taper per foot is given, it must first be converted to taper per inch (TPI). The following formula is used.

Formula: $\text{Offset} = \frac{\text{TPF} \times \text{L}}{24}$

1. What will be the setover for the following job?
   TPF = 0.600″
   L = 8.000″

2. What will be the setover for the following job?
   TPF = 0.720″
   L = 18.000″

3. What will be the setover for the following job?
   TPF = 0.100″
   L = 10.000″

Calculating Tailstock Setover

Dimensions of Tapered Section are Given but TPI, TPF, or TPmm are not

Name: ______________________ **Date:**____________ **Period:**____________

Often plans do not specify TPI, TPF, or TPmm but do give pertinent information. If inch dimensions are given in fractions, they must be converted to decimals.

Formula: $$\text{Offset} = \frac{L \times (D - d)}{2 \times \ell}$$

1. What will be the setover for the following job?
   D = 2 1/2″
   d = 1 3/4″
   ℓ = 6.00″
   L = 18.00″

2. What will be the setover for the following job?
   D = 60.0 mm
   d = 50.0 mm
   ℓ = 150.0 mm
   L = 300.0 mm

3. What will be the setover for the following job?
   D = 1.000″
   d = 0.750″
   ℓ = 5.000″
   L = 8.000″

Screw Thread Applications

Make Adjustments	Transmit Motion	Assemble Parts	Apply Pressure	Make Measurements

27-5

Cutting Action of Threading Tools

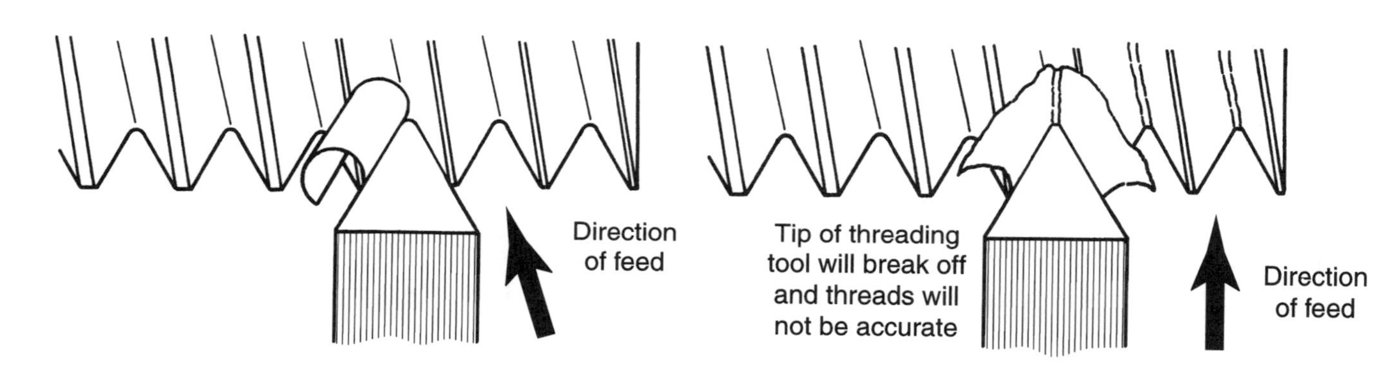

When the tool is fed in at a 29° angle, note that only one edge is cutting and that the cutting load is distributed evenly across the edge.

When fed straight in, note that both edges are cutting and the weakest part of the tool, the point, is doing the hardest work.

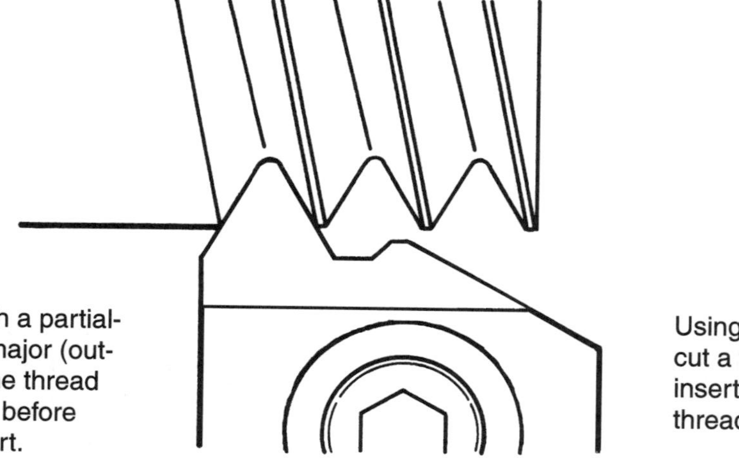

Cutting threads with a partial-profile insert. The major (outside) diameter of the thread must be cut to size before using this type insert.

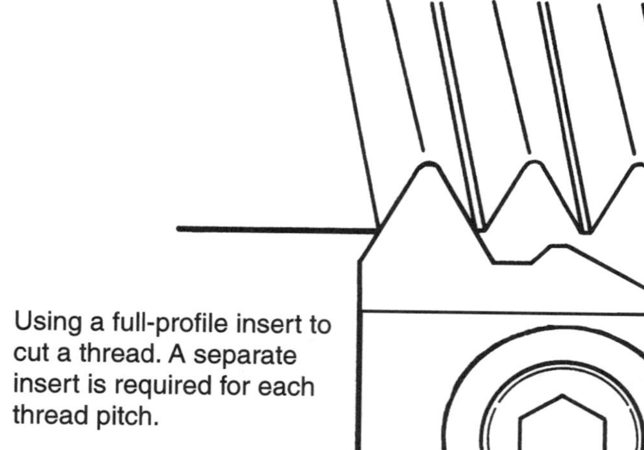

Using a full-profile insert to cut a thread. A separate insert is required for each thread pitch.

Three-Wire Method of Measuring Threads

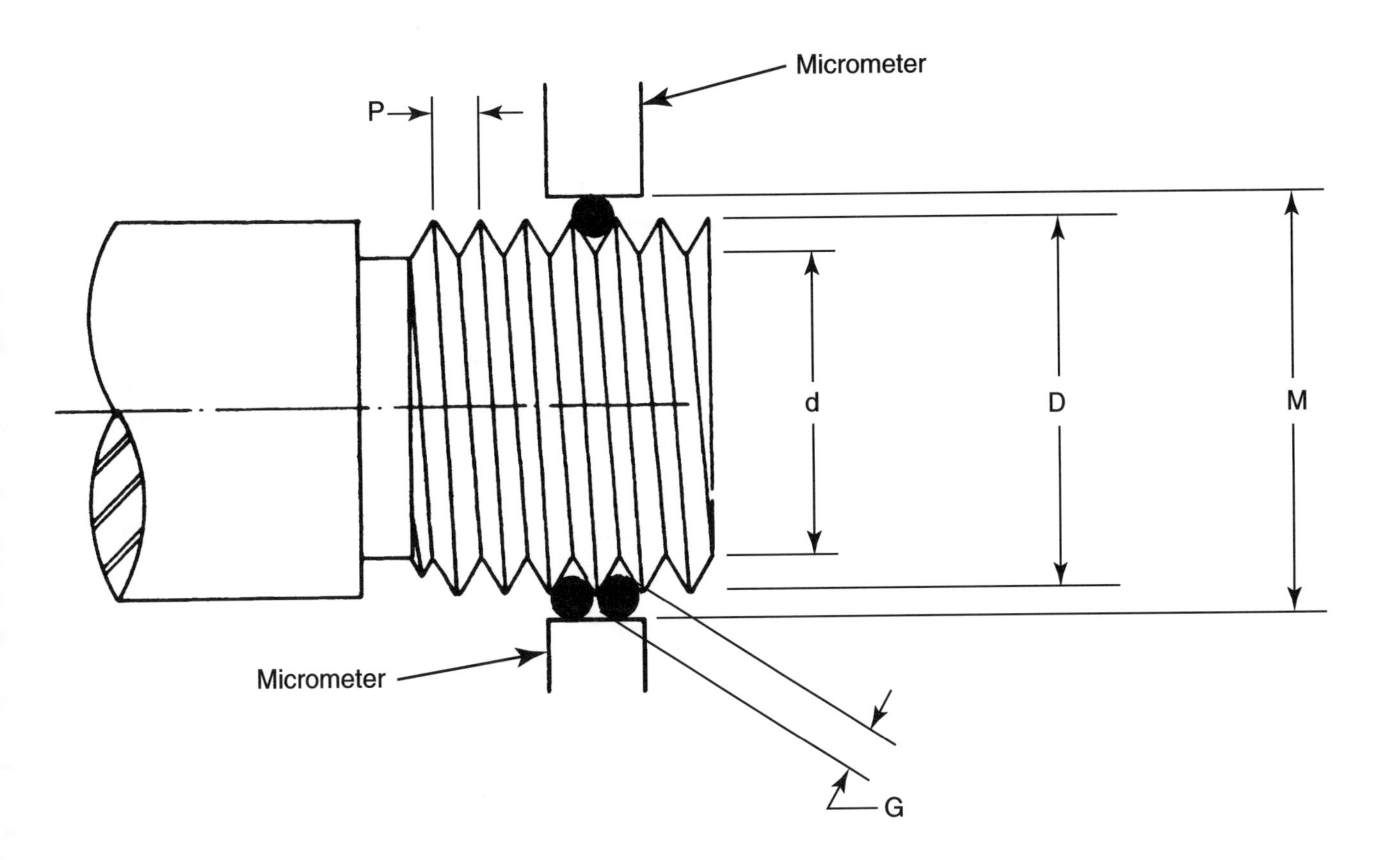

$$M = D + 3G - \frac{1.5155}{N}$$

Where: M = Measurement over the wires
D = Major diameter of thread
d = Minor diameter of thread
G = Diameter of wires

$$P = \text{Pitch} = \frac{1}{N}$$

N = Number of threads per inch

The smallest wire size that may be used for a given thread.

$$G = \frac{0.560}{N}$$

The largest wire size that may be used for a given thread.

$$G = \frac{0.900}{N}$$

The three-wire formula will work only if "G" is no larger or smaller than the sizes determined above. Any wire diameter between the two extremes may be used. All wires must be the same diameter.

Thread Measuring Problems

Name: ______________________ **Date:**____________ **Period:**____________

Using the three-wire method for measuring screw threads, calculate the correct measurement over the wires for the following thread sizes. Use the wire size given in the problem. Round your answers to three decimal places.

$$M = (D + 3G) - \frac{1.5155}{N}$$

M = Measurement over wires
G = Wire size
D = Thread diameter
N = Number of threads per inch

1. 1/4–20UNC (wire size 0.032″)

2. 7/16–20UNF (wire size 0.032″)

3. 3/4–16UNF (wire size 0.045″)

4. 7/8–14UNF (wire size 0.060″)

Chapter 27 Quiz

Cutting Tapers and Screw Threads on a Lathe

Name: ______________________ **Date:**__________ **Period:**__________

1. The _____ groove provides a place to stop the threading tool at the end of its cut. 1.__________
2. Lead is the distance a nut or threaded section will travel in ____ revolution(s) of the screw. 2.__________
   (a) one-quarter
   (b) one-half
   (c) two complete
   (d) one full
3. The smallest diameter of the thread is called the _____ diameter. 3.__________
4. When cutting left-hand threads, in which direction should the carriage travel? 4.__________
   (a) Toward the tailstock.
   (b) Away from the tailstock.
   (c) From right to left.
   (d) None of the above.
5. The largest diameter of the thread is the _____ diameter. 5.__________
6. _____ is the distance from one point on a thread to the corresponding point on the next thread. 6.__________
7. A thread dial is used to indicate when to engage the _____ to permit the cutting tool to follow in the original cut. 7.__________
8. The distance a lathe tailstock has been offset from the normal centerline of the machine is called _____. 8.__________
9. A piece of material is considered _____ when it increases or decreases in diameter at a uniform rate. 9.__________
10. The _____ diameter is the diameter of an imaginary cylinder that would pass through threads at such points to make the width of thread and the width of the spaces at these points equal. 10.__________

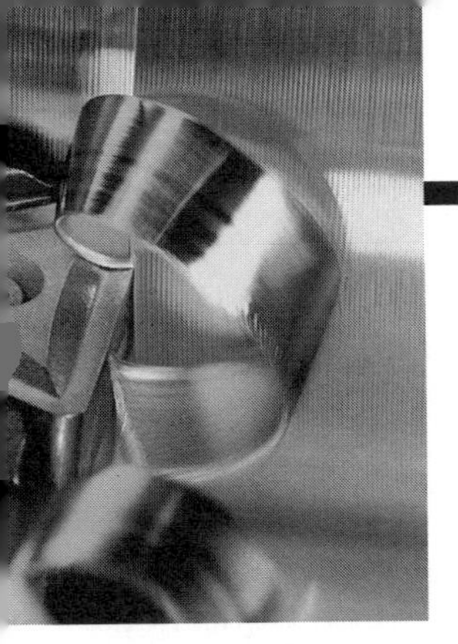

28 OTHER LATHE OPERATIONS

LEARNING OBJECTIVES

After studying this chapter, students should be able to:

- Set up and safely operate a lathe using various work-holding devices.
- Perform drilling, boring, knurling, grinding, and milling operations on a lathe.
- Demonstrate familiarity with industrial application of the lathe.

CHAPTER RESOURCES

Text, pages 449–462
 Test Your Knowledge, page 461
 Research and Development, page 461
Workbook, pages 131–132
Instructor's Manual
 Answer keys for:
  Test Your Knowledge Questions
  Workbook
  Chapter Quiz
 Reproducible Masters:
  28-1 Drilling on the Lathe
  28-2 Boring on the Lathe
  28-3 Knurling on the Lathe
  28-4 Chapter Quiz

GUIDE FOR LESSON PLANNING

As with all machining operations demonstrated in class or performed on the job, the equipment must be in safe operating condition with all guards and safety devices in place. Make sure students wear eye protection and have a clear view of the demonstration.

PART I—HOW TO DRILL AND REAM ON A LATHE

Set up a lathe to demonstrate drilling and reaming. Have the following equipment available:

- Jacobs chuck
- Center drill
- Straight shank drill
- Taper shank drill
- Cutting fluid

Have students read and study Section 28.1. Copy and distribute Reproducible Master 28-1. Review the assignment and discuss the following:

- Why cutting fluids are needed when drilling most metals.
- Why a center drill is used to start the hole being drilled.
- The proper way to mount drills so they will run true.
- Why drill size and material being drilled determines lathe speed.
- The reason for a pilot hole when drilling holes larger than 1/2″ (12.5 mm).
- When and how reamers are used.
- Safety precautions to be observed when drilling and reaming.

PART II—HOW TO BORE ON A LATHE

Set up the lathe for boring operations with a drilled section of stock mounted in place. Have a selection of boring bars and holders available for class examination. Have students read and study Section 28.2. Copy and distribute Reproducible Master 28-2. Review the assignment and discuss the following:

- Why boring is done.
- How boring techniques differ from conventional turning.
- How to select the proper boring bar for the job.

- Positioning the boring bar for cutting.
- How chatter can be prevented when using long slender boring bars.
- Precautions to be taken when boring on the lathe.

Demonstrate the following:

- How to mount the boring bar for adequate clearance.
- Using power feed for boring operations.
- How to use a telescoping gage and micrometer caliper to determine hole size.

Part III—Knurling

Set up the lathe for knurling operations. Have an assortment of jobs with knurled sections for student examination. Examples of poorly made knurled sections should also be available.

Have students read and study Section 28.3. Copy and distribute Reproducible Master 28-3. Review the assignment and discuss the following:

- Reason some work is knurled.
- Different types of knurls.
- How to set up the lathe for knurling.
- Knurling problems that may be encountered and how they can be corrected.
- Why cutting fluids are used when knurling.
- Safety precautions to be observed when knurling.

Demonstrate the following:

- How to set up the lathe and work for knurling.
- Knurl a section of metal.

Briefly review the demonstration. Provide students with an opportunity to ask questions.

Part IV—Filing and Polishing

Set up a lathe for filing and polishing with all of the necessary equipment available. For a change of pace, have students demonstrate filing and polishing on the lathe. Have students read and study Section 28.4.1. Review the assignment and discuss the following:

- Reasons filing and polishing on the lathe should be kept to a minimum.
- The correct file to use when filing is necessary.
- Why the left-hand method of filing is recommended.
- The proper way to file on a lathe.
- Why the carriage and ways should be protected when polishing on the lathe.
- The various abrasives that can be used to polish on a lathe.
- Precautions that must be observed when filing and polishing on a lathe.

Allow students to do the actual filing and polishing demonstrations. Briefly review the demonstrations. Provide students with the opportunity to ask questions. Permit students who did the demonstrations to answer the questions.

Part V—Other Lathe Operations.

Have students read and study the remainder of the chapter. Review the assignment and discuss the following:

- Why and how lathe mandrels are used.
- How the tool post grinder is used.
- Why the carriage and ways must be protected when grinding on a lathe.
- Milling on the lathe.
- Safety precautions that must be observed when grinding and milling on a lathe.
- Industrial applications of the lathe.

If equipment is available, demonstrate the correct way to grind and mill on a lathe. Briefly review the assignment. Allow time for students to ask questions.

Assignments

1. Assign the Test Your Knowledge questions.
2. Assign Chapter 28 in the *Modern Metalworking Workbook.*
3. Assign the chapter quiz. Copy and distribute Reproducible Master 28-1.
4. Permit students to volunteer for one or more of the Research and Development activities at the end of the chapter.

Test Your Knowledge

1. pilot
2. The left-hand method. It involves holding the file handle in the left hand. The right hand is then clear of the revolving chuck or faceplate.
3. (b) when the outside diameter must be machined concentric to a reamed or bored hole
4. tool post grinder
5. Turret
6. (d) Knurling
7. (b) Filing

8. (a) Reaming
9. (e) Polishing
10. (c) Boring

Workbook

1. Student answers will vary but may include any four of the following: threading, boring, drilling, reaming, knurling, grinding, polishing, milling, filing.
2. (b) two-thirds
3. outside
4. The knurls will not track and will quickly dull. The work may also be destroyed.
5. With the left-hand method, your right hand is clear of the revolving chuck or faceplate. The right-hand method places your left arm over the revolving chuck or faceplate.
6. Any abrasive chips left from the cloth can cause rapid wear of the machine's moving parts.
7. mandrel
8. arbor press
9. A turret lathe is equipped with a six-sided tool holder called a turret, to which a number of different cutting tools are fitted.
10. Evaluate individually. Refer to Section 28.9.

Chapter Quiz

1. Student answers will vary but may include any four of the following: threading, boring, drilling, reaming, knurling, grinding, polishing, milling, filing.
2. Boring
3. It involves holding the file handle in the left hand. The right hand is then clear of the revolving chuck or faceplate.
4. knurling
5. Polishing
6. mandrel
7. tool post grinder
8. The operation used to make a hole accurate in diameter and finish.
9. Filing
10. The knurls will not track and will quickly dull. The work may also be destroyed.

Notes

Drilling on the Lathe

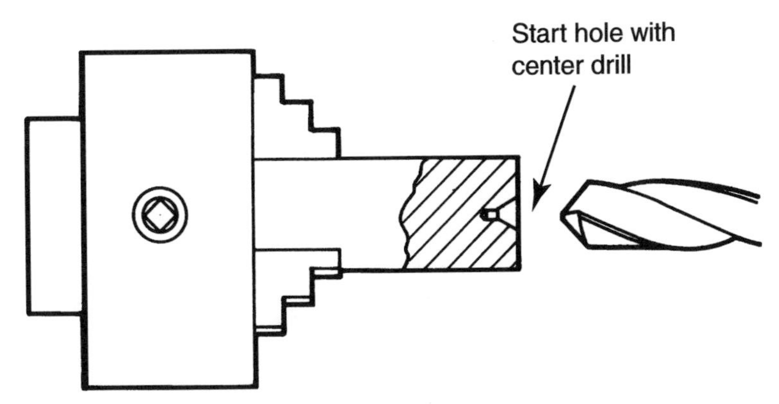

The drill will cut exactly on the center if the hole is started with a center drill.

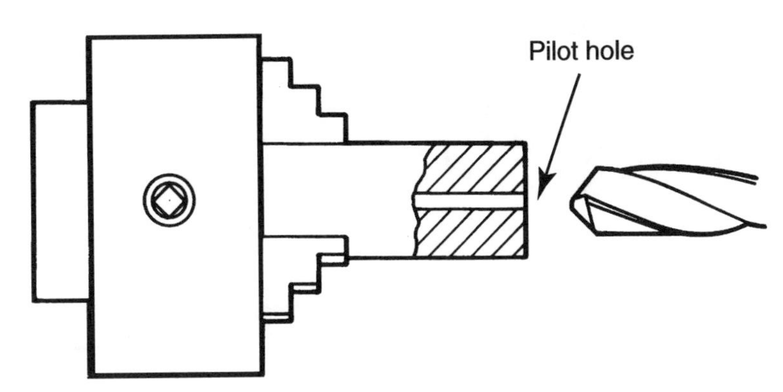

Holes larger than 1/2″ (12.5 mm) in diameter require drilling of a pilot hole.

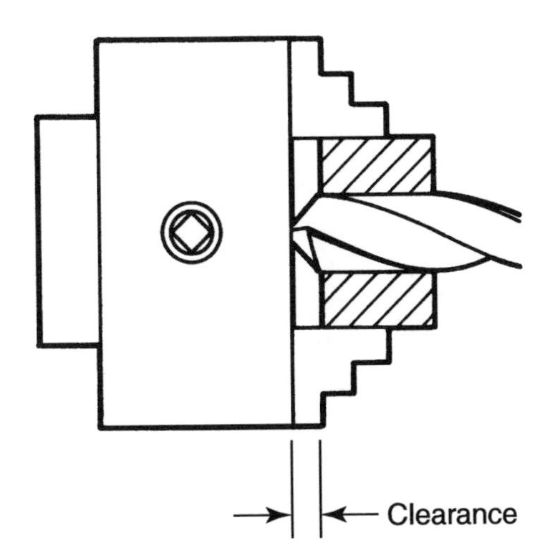

There must be enough clearance between the back of the work and the chuck face to permit the drill to break through the work without damaging the chuck.

Boring on the Lathe

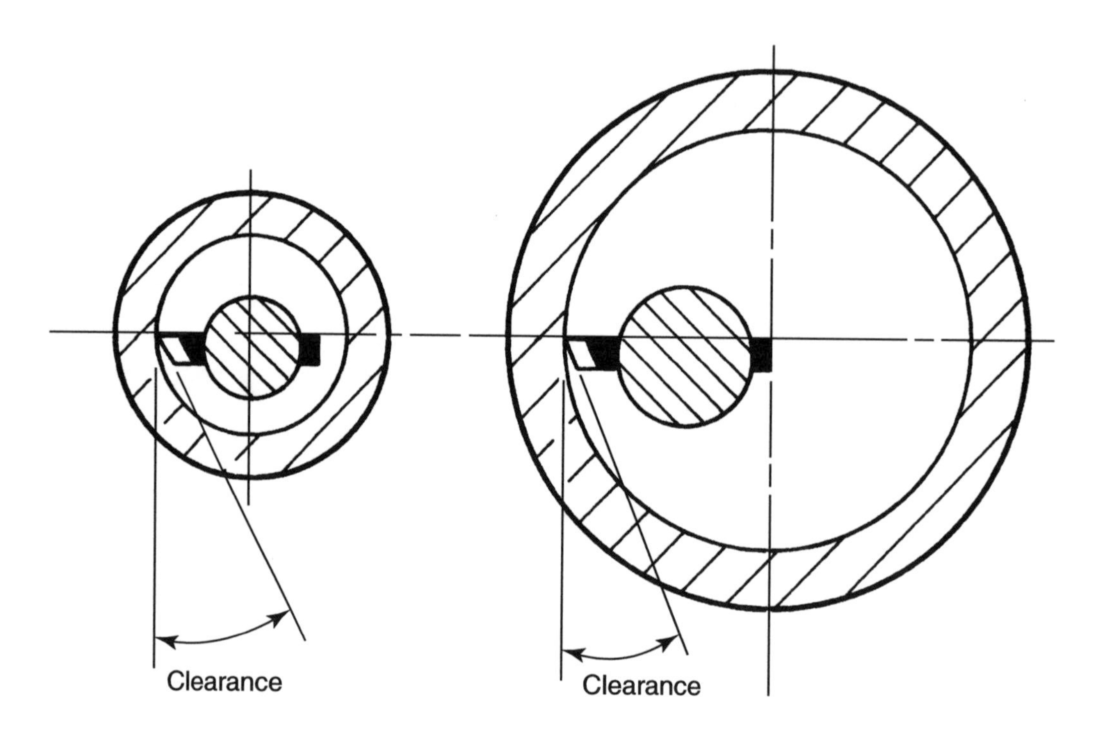

Tool used to bore small diameter holes requires greater front clearance to prevent rubbing.

Knurling on the Lathe

Procedure

1. Mark of section to be knurled.
2. Adjust the lathe to a slow back-geared speed and fairly rapid feed.
3. Place the knurling tool in the post. Bring it up to work. Both wheels must bear evenly on the work with their faces parallel with the centerline of the piece. See figure below.

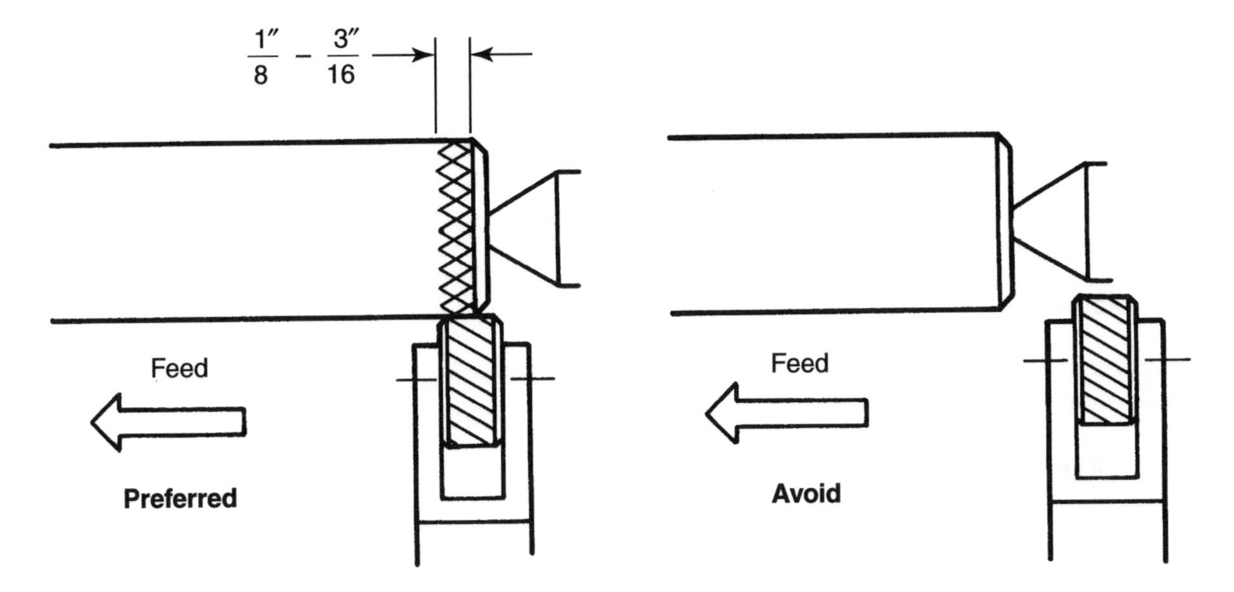

4. Start the lathe and slowly force the knurls into the work surface until a pattern begins to form. Tool travel should be *toward* the headstock whenever possible. Engage the automatic feed and let the tool travel across the work. Flood the work with cutting fluid.
5. When the knurling tool reaches the proper position, reverse spindle rotation and allow the tool to move back across the work to the starting point. Apply additional pressure to force the knurls deeper into the work.

Chapter 28 Quiz

Other Lathe Operations

Name: ______________________ **Date:** ______________ **Period:** ______________

1. List four machining operations (besides turning) that can be done on the lathe.

   ____

   ____

2. _____ is the operation used to produce a hole that is concentric with the outside diameter of the work. 2. ______________

3. Which should the left-hand method of filing be used when filing on a lathe?

   ____

   ____

   ____

4. Horizontal or diamond-shaped serrations are machined on the circumference of the work through the machining process known as _____. 4. ______________

5. _____ is the operation used to produce a fine finish on a piece of work. 5. ______________

6. Work is pressed on a(n) _____ with a mechanical arbor press. 6. ______________

7. Through the use of a(n) _____, both internal and external grinding can be performed on a lathe. 7. ______________

8. What is *reaming?*

   ____

   ____

9. _____ is the operation used to remove burrs and round off sharp edges. 9. ______________

10. What will happen if the knurling tool setup is *not* made properly?

    ____

    ____

    ____

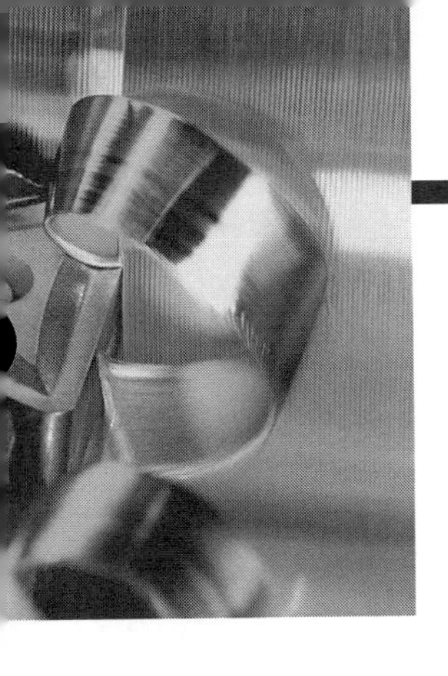

Broaching Operations

Learning Objectives

After studying this chapter, students should be able to:

- Describe how broaching operates.
- Explain the advantages of broaching.
- Set up and cut a keyway using a keyway broach and arbor press.
- Observe safety precautions when cutting a keyway.

Chapter Resources

Text, pages 463–468
    Test Your Knowledge, page 468
    Research and Development, page 468
Workbook, pages 133–134
Instructor's Manual
    Answer keys for:
        Test Your Knowledge Questions
        Workbook
        Chapter Quiz
    Reproducible Masters:
        29-1 How a Broaching Tool Cuts
        29-2 Chapter Quiz

Guide for Lesson Planning

Have students read and study the chapter. They should pay particular attention to the illustrations. Review the assignment using Reproducible Master 29-1 as an overhead transparency or handout. Discuss the following:

- The broaching process.
- Types of broaching machines.
- How a broaching tool cuts.
- Advantages of broaching.
- Demonstrate how to broach a keyway.

Assignments

1. Assign the Test Your Knowledge questions.
2. Assign Chapter 29 in the *Modern Metalworking Workbook.*
3. Assign the chapter quiz. Copy and distribute Reproducible Master 29-2.
4. Permit students to volunteer for one or more of the Research and Development activities at the end of the chapter.

Test Your Knowledge

1. Any order: internal, external, pot. Evaluate descriptions individually.
2. Evaluate individually.
3. Evaluate individually.
4. Any three of the following: high productivity; capability of maintaining close tolerances; production of good surface finishes; economy; long tool life, since only a small amount of metal is removed by each tooth; capability of using semiskilled workers, since once set up, the equipment is automated.
5. Adding burnishing elements to the finishing end of the tool.

Workbook

1. (d) All of the above.
2. So the cutting tool (broach) can be inserted.
3. pull
4. slab
5. pot
6. Roughing, semifinishing, finishing.
7. only a small amount of metal is removed by each tooth
8. (d) All of the above.
9. (d) All of the above.
10. burnishing
11. Evaluate student work individually.

Chapter Quiz

1. Internal
2. external
3. pot
4. Roughing, semifinishing, finishing
5. (d) All of the above.
6. So the cutting tool (broach) can be inserted.
7. (d) All of the above.

How a Broaching Tool Cuts

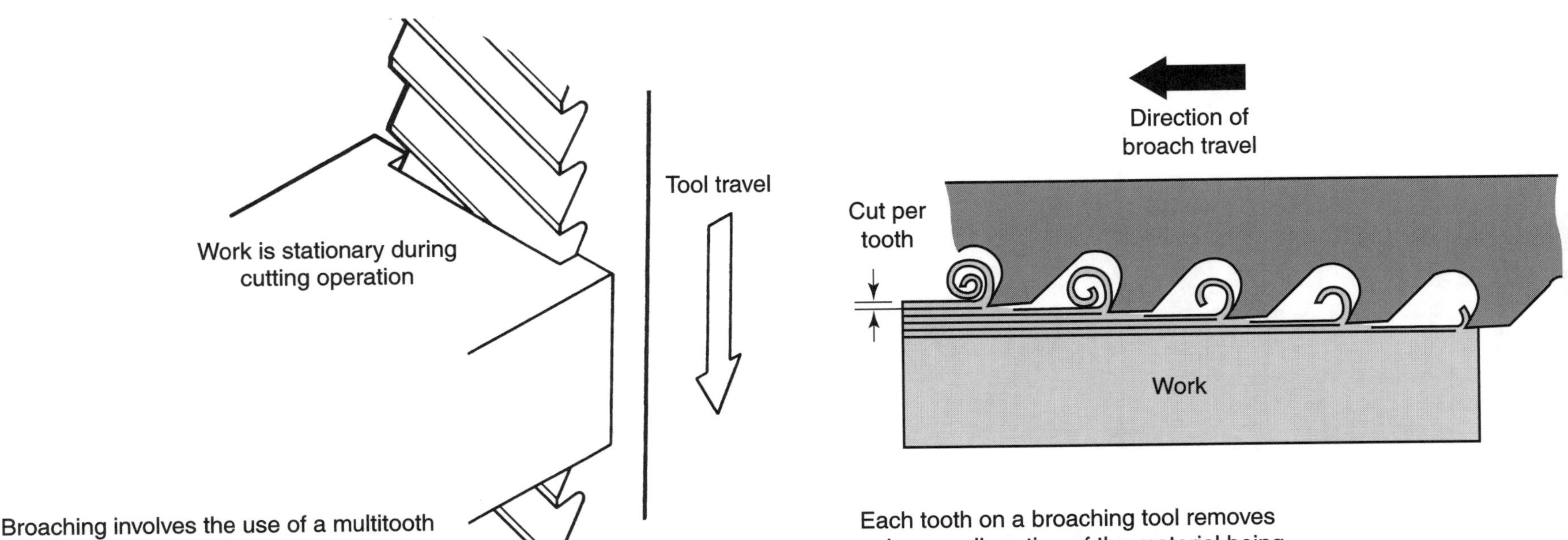

Broaching involves the use of a multitooth cutting tool (the broach) that moves against the stationary work. The operation may be on a vertical or horizontal plane, and may involve making internal or external cuts.

Each tooth on a broaching tool removes only a small portion of the material being machined.

29-1

Chapter 29 Quiz

Broaching Operations

Name: ______________________ **Date:**__________ **Period:**__________

1. _____ broaching makes use of a pull broach. 1.__________

2. The slab broach is used for _____ broaching. 2.__________

3. In _____ broaching, the tool is stationary and the work is pulled through or over it. 3.__________

4. What are the three kinds of teeth on a broaching tool? ______________________

____

____

5. Broaching is a manufacturing process for machining _____. 5.__________
   (a) internal and external surfaces
   (b) flat, rounded, and contoured surfaces
   (c) helical splines
   (d) All of the above.
   (e) None of the above.

6. Why does internal broaching require a starting hole? ______________________

____

____

7. The broaching operation _____. 7.__________
   (a) is usually completed in a single pass of the cutting tool
   (b) can remove metal faster in one pass than any other machining technique
   (c) can maintain consistently close tolerances
   (d) All of the above.
   (e) None of the above.

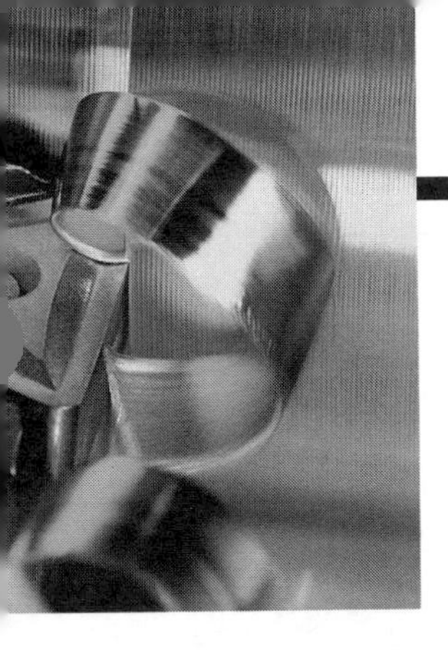

30 Milling Machines

Learning Objectives

After studying this chapter, students should be able to:

- Explain how milling machines operate.
- Identify the various types of milling machines.
- Select, mount, and care for milling cutters.
- Calculate cutting speeds and feeds.
- Prepare machine and mount work for a variety of milling operations.
- Practice proper safety precautions when operating milling machines.

Chapter Resources

Text, pages 469–506
  Test Your Knowledge, pages 504–505
  Research and Development, page 505
Workbook, pages 135–140
Instructor's Manual
  Answer keys for:
    Test Your Knowledge Questions
    Workbook
    Chapter Quiz
  Reproducible Masters:
    30-1 The Horizontal Milling Machine
    30-2 Conventional Milling and Climb Milling
    30-3 Efficiency of Small Diameter Cutters
    30-4 Cutting Speeds and Feeds
    30-5 Rules for Determining Cutting Speeds and Feeds
    30-6 Cutting Speed and Feed Problems
    30-7 The Vertical Milling Machine
    30-8 Using the Edge Finder
    30-9 Chapter Quiz

Guide for Lesson Planning

Divide the chapter into several segments. The number of segments taught will be determined by the time and equipment at your disposal.

To aid in teaching this chapter, have a horizontal and a vertical milling machine set up with a cut completed.

Part I—Types of Milling Machines

Have students read and study Section 30.1. They should pay particular attention to the illustrations. Review the assignment with them and discuss, question, and demonstrate the following:

- Types of milling machines.
- Methods of machine controls.

Review milling safety practices in Section 30.2.

Part II—Milling Operations

Students should read Sections 30.3 through 30.6, paying particular attention to the illustrations. Review the assignment with them and discuss, question, and demonstrate the following:

- Types of milling operations.
- How cutters are classified.
- Types of milling cutters and their uses.
- Methods of cutting.
- Care of milling cutters.
- Safety precautions when handling milling cutters.
- Cutter holding and driving devices.
- Care of cutter holding and driving devices.

Part III—Cutting Speeds and Feeds

Have students read and study Section 30.7. Using overhead transparencies and handouts (Reproducible Masters 30-4, 30-5, and 30-6) instruct them on how to calculate cutting speeds and feeds.

Part IV—Cutting Fluids and Work-Holding Attachments

Have students read and study Sections 30.8 and 30.9. They should pay particular attention to the illustrations. Review the assignment with them and discuss:

- Reasons why cutting fluid is used in milling operations.
- Types of vises available.
- The dividing head, its operation, and why it is used.
- How to calculate and set up a dividing head. Provide a few simple dividing head problems. Also explain how the dividing head is used in gear cutting.

Part V—Milling Operations on a Horizontal Milling Machine

Have students read and study Section 30.10. They should pay particular attention to the illustrations. Review the assignment with them. Discuss and demonstrate:

- Milling flat surfaces.
- Techniques used to align a vise.
- Safety precautions that must be observed when handling milling cutters.
- Selecting the proper size cutter(s).
- Side milling and the type cutter used.
- Straddle milling and how to set up for the operation.
- Gang milling.
- Techniques for accurately positioning a side cutter to mill a slot.
- How to center a cutter on round stock.
- Other operations that can be performed on a horizontal milling machine—drilling, reaming, boring, etc.
- Care and cleaning of a horizontal milling machine.

Part VI—The Vertical Milling Machine

Have students read and study Section 30.11. They should pay particular attention to the illustrations. Review the assignment discussing, questioning, and demonstrating the following:

- The vertical milling machine and its operation.
- Cutters for vertical milling machines.
- How to set up and mill angular surfaces.
- Positioning a cutter to mill a slot or keyway.
- Techniques for milling an internal opening in stock.
- How to drill and bore on a vertical milling machine.
- Using a "wiggler" or "edge finder" to accurately position a hole for drilling or boring.
- Care and cleaning of a vertical milling machine.

Part VII—Milling Safety

Have students review Section 30.2. Review the assignment and impress upon them the necessity for observing safety procedures when operating a milling machine, or any machine tool.

Part VIII—Industrial Applications for Milling Machines

Review the industrial applications of milling machines discussed in Section 30.12. Emphasize the skills necessary to operate these machines.

Machines and equipment used to demonstrate milling operations should be in safe operating condition with all safety devices in place. Cutters must be sharp and the work mounted solidly.

Position students so they all can see the demonstrations but clear of potential flying chips. Approved eye protection must be worn.

Assignments

1. Assign the Test Your Knowledge questions at the end of the chapter.
2. Assign Chapter 30 in the *Modern Metalworking Workbook.*
3. Assign the chapter quiz. Copy and distribute Reproducible Master 30-9.
4. Permit students to volunteer for one or more of the Research and Development activities at the end of the chapter.

Test Your Knowledge

1. Any order: vertical, cross (traverse), longitudinal.
2. fixed-bed, knee and column
3. Any order: manual, semi-automatic, fully automatic, computerized (CNC).
4. Wrap them in cloth and handle with extreme care.

5. machine accessories, large work
6. solidly
7. Evaluate individually. Refer to Figure 30-11 A and B.
8. Any order: solid, inserted-tooth.
9. Any order: arbor cutter, shank cutter, facing cutter.
10. two
11. plain
12. staggered
13. slitting saw
14. shortest
15. draw-in bar
16. The distance (measured in feet or meters) a point (tooth) on the circumference of the cutter moves in one minute. It is expressed in feet per minute (fpm) or meters per minute (mpm).
17. 480 rpm
18. 800 rpm
19. Any order: carry away heat, prevent chips from sticking to the cutter, flush away chips.
20. swivel
21. (b) divide the circumference of a workpiece into any number of equal parts
22. Any order: slab, inserted tooth, shell face.
23. dial indicator
24. (c) two or more cutters cut at the same time
25. Evaluate individually. Refer to Section 30.10.7.
26. Jacobs chuck
27. boring
28. end, face
29. Any order: tilt the spindle head at the desired angle; position the work at the desired angle in a vise; use a universal vise.
30. two-flute

Workbook

1. Evaluate individually. Refer to text Figure 30-1.
2. Any order: vertical, traverse, longitudinal.
3. (a) Universal
   (b) Vertical spindle
   (c) Plain
4. Evaluate individually. Refer to Section 30.2.
5. (c) Solid cutter
6. (e) Arbor cutter
7. (a) Face milling
8. (f) Shank cutter
9. (d) Inserted tooth cutter
10. (b) Peripheral milling
11. (g) Facing cutter
12. Evaluate sketches individually. Refer to text Figure 30-11.
13. (a) Arbor cutter
    (b) Facing cutter
    (c) Straight fluted end mill
    (d) Staggered-tooth side milling cutter
14. Angle
15. Thin milling cutters that resemble the circular saw blades used in woodworking. They are used for cutting narrow slots and for cutoff operations.
16. Evaluate individually. Refer to text Section 30.6.4.
17. 100–150 rpm
18. 150–400 rpm
19. 2100–3800 rpm
20. They carry away the heat generated during the machining operation, act as a lubricant to prevent chips from sticking or fusing to the cutter teeth, and flush away chips. The lubricating qualities also influence the quality of the finish of the machined surface.
21. (f) Index table
22. (a) Vise
23. (d) Toolmaker's vise
24. (g) Dividing head
25. (c) Swivel vise
26. (b) Flanged vise
27. (e) Rotary table
28. Gang milling
29. Evaluate individually. Refer to text Section 30.10.7.
30. Evaluate individually. Refer to text Section 30.11.6.

Reproducible Master 30-6

1. Recommended cutting speed for aluminum = 775 fpm
   Recommended feed per tooth = 0.015″
   rpm = 750
   Feed = 180″
2. Recommended cutting speed for low-carbon steel = 212 fpm
   Recommended feed per tooth = 0.012″
   rpm = 250
   Feed = 24″

Chapter Quiz

1. (n) Manual controls
2. (f) Semi-automatic controls
3. (q) Face milling
4. (r) Solid cutter
5. (m) Inserted-tooth cutter
6. (s) Arbor cutter
7. (t) Shank cutter
8. (e) Facing cutter
9. (b) Side cutter
10. (d) Metal slitting saw
11. (g) Two-flute end mill
12. (c) Conventional milling
13. (j) Climb milling
14. (p) Feed
15. (l) Dividing head
16. (i) Straddle mill
17. (k) Gang milling
18. (o) Slitting
19. (a) Slotting
20. (h) Edge finder

The Horizontal Milling Machine

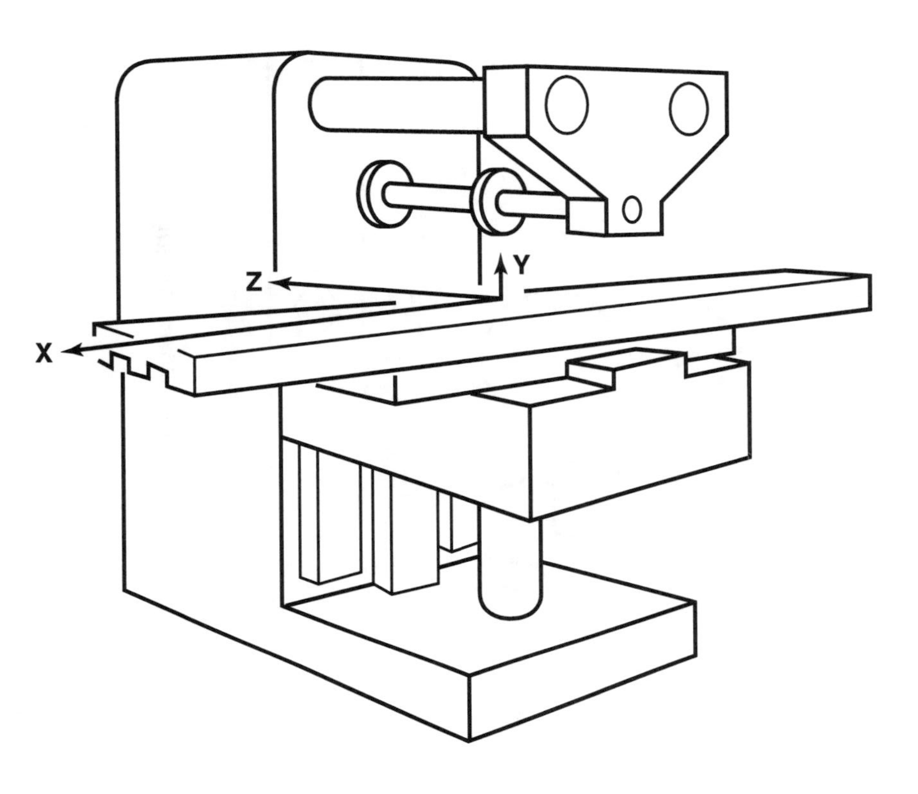

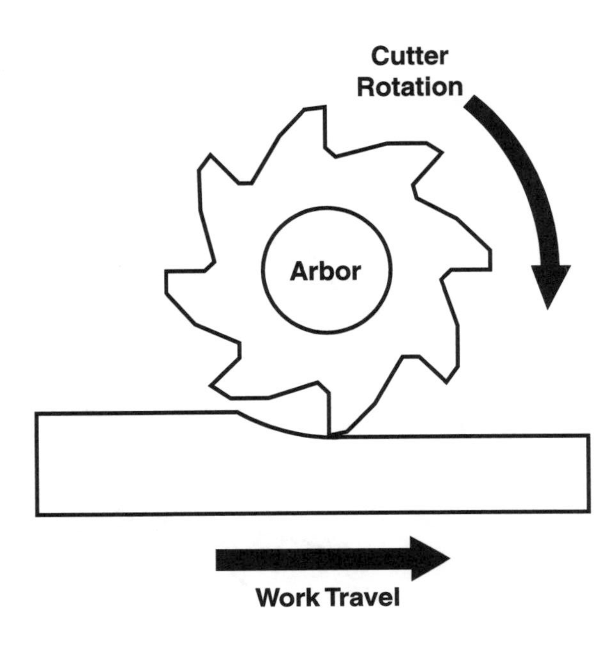

30-1

Conventional Milling and Climb Milling

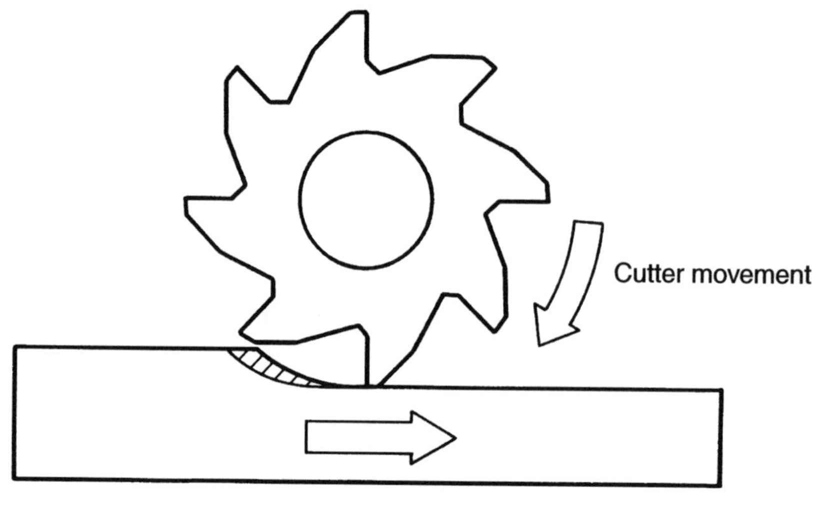

Conventional (up) Milling

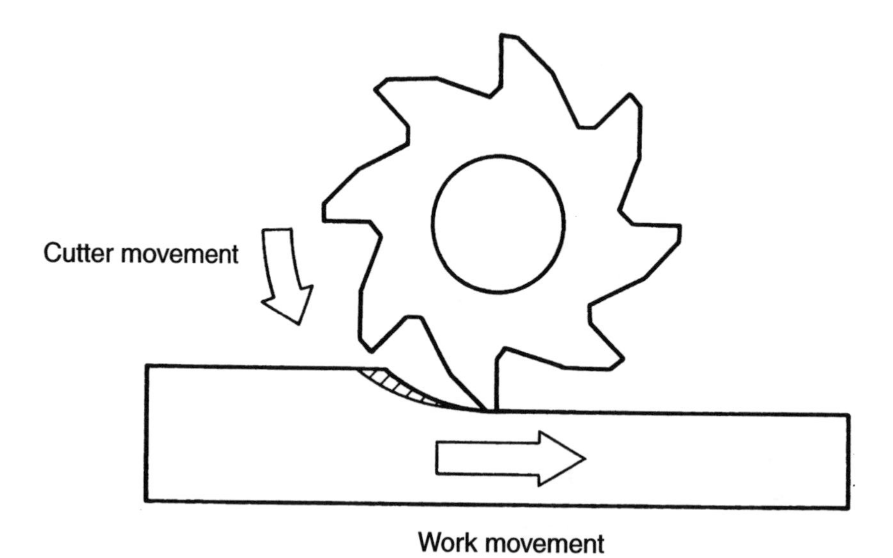

Climb (down) Milling

30-3

Efficiency of Small Diameter Cutters

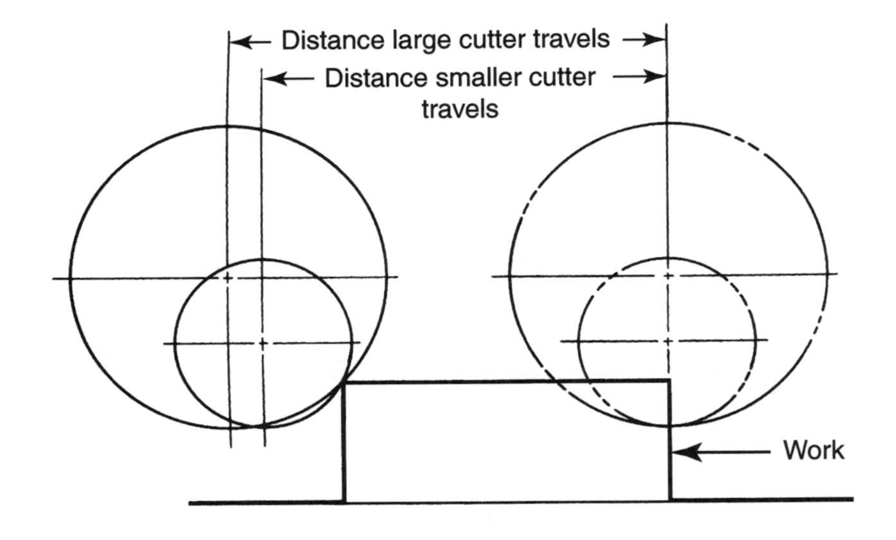

Cutting Speeds and Feeds

Material	High-speed steel cutter		Carbide cutter	
	Feet per minute	Meters per minute*	Feet per minute	Meters per minute*
Aluminum	550–1000	170–300	2200–4000	670–1200
Brass	250–650	75–200	1000–2600	300–800
Low carbon steel	100–325	30–100	400–1300	120–400
Free cutting steel	150–250	45–75	600–1000	180–300
Alloy steel	70–175	20–50	280–700	85–210
Cast iron	45–60	15–20	180–240	55–75

Reduce speeds for hard materials, abrasive materials, deep cuts, and high alloy materials. Increase speeds for soft materials, better finishes, light cuts, frail work, and setups. Start at midpoint on the range and increase or decrease speed until best results are obtained.
*Figures rounded off.

Type of cutter	Material				
	Aluminum	Brass	Cast iron	Free cutting steel	Alloy steel
End mill	0.009 (0.22) 0.022 (0.55)	0.007 (0.18) 0.015 (0.38)	0.004 (0.10) 0.009 (0.22)	0.005 (0.13) 0.010 (0.25)	0.003 (0.08) 0.007 (0.18)
Face mill	0.016 (0.40) 0.040 (1.02)	0.012 (0.30) 0.030 (0.75)	0.007 (0.18) 0.018 (0.45)	0.008 (0.20) 0.020 (0.50)	0.005 (0.13) 0.012 (0.30)
Shell end mill	0.012 (0.30) 0.030 (0.75)	0.010 (0.25) 0.022 (0.55)	0.005 (0.13) 0.013 (0.33)	0.007 (0.18) 0.015 (0.38)	0.004 (0.10) 0.009 (0.22)
Slab mill	0.008 (0.20) 0.017 (0.43)	0.006 (0.15) 0.012 (0.30)	0.003 (0.08) 0.007 (0.18)	0.004 (0.10) 0.008 (0.20)	0.001 (0.03) 0.004 (0.10)
Side cutter	0.010 (0.25) 0.020 (0.50)	0.008 (0.20) 0.016 (0.40)	0.004 (0.10) 0.010 (0.25)	0.005 (0.13) 0.011 (0.28)	0.003 (0.08) 0.007 (0.18)
Saw	0.006 (0.15) 0.010 (0.25)	0.004 (0.10) 0.007 (0.18)	0.001 (0.03) 0.003 (0.08)	0.003 (0.08) 0.005 (0.13)	0.001 (0.03) 0.003 (0.08)

US Customary value expressed in inches per tooth. Metric value (shown in parentheses) expressed in millimeters per tooth.
Increase or decrease feed until the desired surface finish is obtained.
Feeds may be increased 100 percent or more depending upon the rigidity of the machine and the power available, if carbide tipped cutters are used.

Rules for Determining Cutting Speeds and Feeds

Rules for determining speed and feed			
To find	**Having**	**Rule**	**Formula**
Speed of cutter in feet per minute (fpm)	Diameter of cutter and revolutions per minute	Diameter of cutter (in inches) multiplied by 3.1416 (π) multiplied by revolutions per minute, divided by 12	$fpm = \frac{\pi D \times rpm}{12}$
Speed of cutter in meters per minute	Diameter of cutter and revolutions per minute	Diameter of cutter multiplied by by 3.1416 (π) multiplied by revolutions per minute, divided by 1000	$mpm = \frac{D(mm) \times \pi \times rpm}{1000}$
Revolutions per minute (rpm)	Feet per minute and diameter of cutter	Feet per minute, multiplied by 12, divided by circumference of cutter (πD)	$rpm = \frac{fpm \times 12}{\pi D}$
Revolutions per minute (rpm)	Meters per minute and diameter of cutter in millimeters (mm)	Meters per minute multiplied by 1000, divided by the circumference of cutter (D)	$rpm = \frac{mpm \times 1000}{\pi D}$
Feed per revolution (FR)	Feed per minute and revolutions per minute	Feed per minute, divided by revolutions per minute	$FR = \frac{F}{rpm}$
Feed per tooth per revolution (ftr)	Feed per minute and number of teeth in cutter	Feed per minute (in inches or millimeters) divided by number of teeth in cutter × revolutions per minute	$ftr = \frac{F}{T \times rpm}$
Feed per minute (F)	Feed per tooth per revolution, number of teeth in cutter, and rpm	Feed per tooth per revolution multiplied by number of teeth in cutter, multiplied by revolutions per minute	$F = ftr \times T \times rpm$
Feed per minute (F)	Feed per revolution and revolutions per minute	Feed per revolution multiplied by revolutions per minute	$F = FR \times rpm$
Number of teeth per minute (TM)	Number of teeth in cutter and revolutions per minute	Number of teeth in cutter multiplied by revolutions per minute	$TM = T \times rpm$

rpm = Revolutions per minute
T = Teeth in cutter
D = Diameter of cutter
π = 3.1416 (pi)
frm = Speed of cutter in feet per minute

TM = Teeth per minute
F = Feed per minute
FR = Feed per revolution
ftr = Feed per tooth per revolution
mpm = Speed of cutter in meters per minute

30-5

Milling Machines—Calculating Cutting Speeds and Feeds

Name: ______________________________ **Date:**______________ **Period:**______________

Refer to the rules for determining cutting speeds and feeds to calculate the cutting speed and feed for specific materials. Round cutting speed to the nearest 50 rpm.

1. Determine the proper speed and feed for a 4″ diameter side cutter (HSS) with 16 teeth milling aluminum.
   Information available:
      Recommended cutting speed for aluminum (midpoint on range) = __________
      Recommended feet per tooth (midpoint on range) = __________
      Cutter diameter = 4″
      Number of teeth on cutter = 16

2. Determine the proper speed and feed for a 3″ diameter slab cutter (HSS) with 8 teeth milling low-carbon steel.
   Information available:
      Recommended cutting speed for low-carbon steel (midpoint on range) = __________
      Recommended feet per tooth (midpoint on range) = __________
      Cutter diameter = 3″
      Number of teeth on cutter = 8

The Vertical Milling Machine

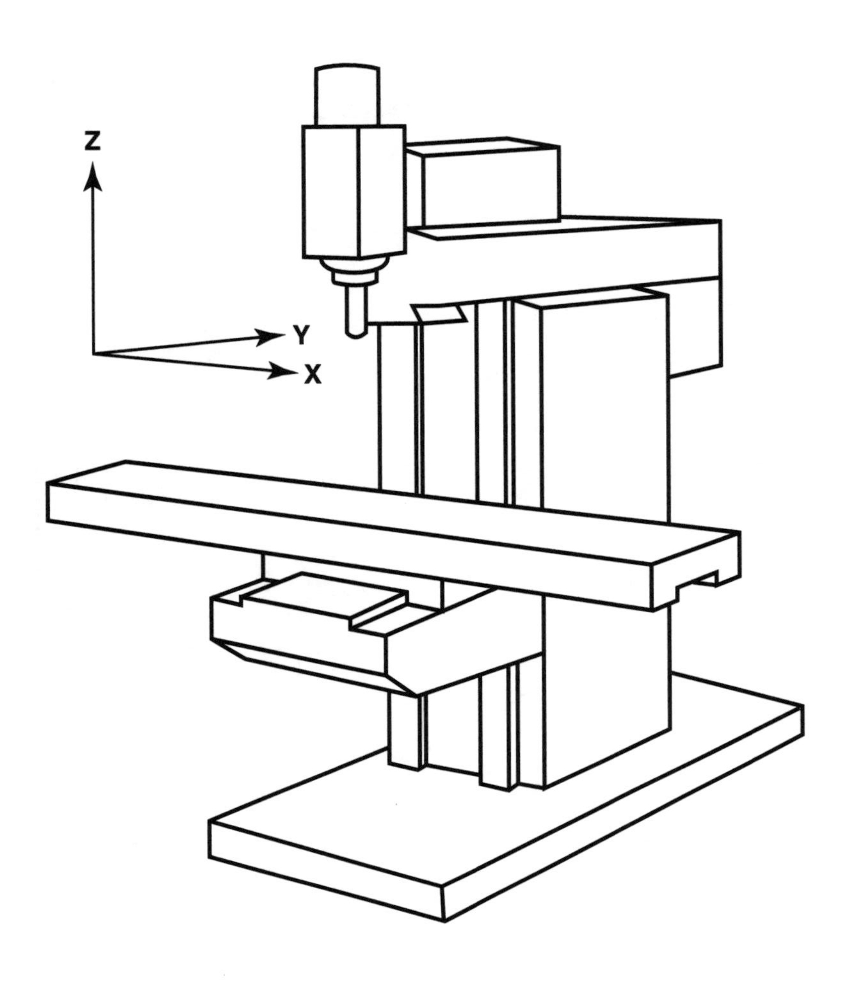

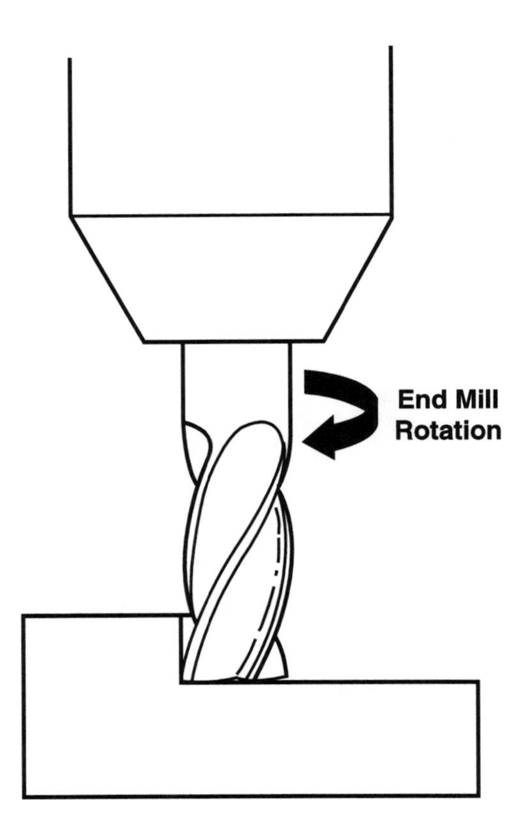

30-7

Using the Edge Finder

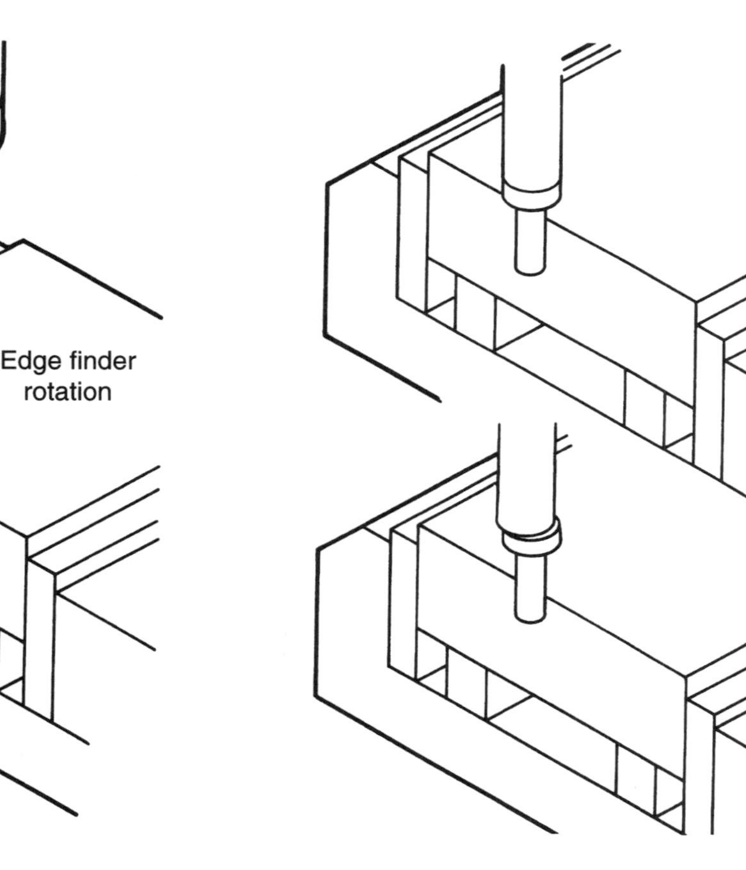

With spindle rotating at moderate speed, and with edge finder tip as shown, slowly feed tip of tool against work.

Edge finder tip will gradually become centered with its shank.

When the tip becomes exactly centered, it will aabruptly jump sideways about 1/32″ (0.8mm). When this occurs, stop table movement immediately. Center of the spindle will be exactly one-half tip diameter away from edge of work. Set the micrometer dial to "0" and, with edge finder clear of work, move table longitudinally the required distance plus one-half the tip diameter. Follow the same procedure to get traverse measurement.

Chapter 30 Quiz
Milling Machines

Name: ______________________ **Date:**______________ **Period:**______________

Match the sentence with the appropriate term or phrase.

_________ 1. All machine movements are made by hand-operated controls.

_________ 2. Machine movements are controlled by hand and/or power feeds.

_________ 3. Surface being machined is parallel with the cutter face.

_________ 4. Shank and body of the cutter are made in one piece.

_________ 5. Teeth of cutter are made of special cutting material and are brazed or clamped in place.

_________ 6. Cutter with a suitable hole for mounting it on a machine arbor.

_________ 7. Cutters with either straight or taper shank that is an integral part of the cutter.

_________ 8. Cutters that can be mounted directly to the spindle nose or stub arbor.

_________ 9. Cutters with cutting edges on their circumference and on one or both sides.

_________10. Thin milling cutter similar to circular saw blade in woodworking.

_________11. Can be fed into work like a drill.

_________12. Work is fed into the rotation of the cutter.

_________13. Work moves in the same direction as cutter rotation.

_________14. Rate at which the work moves into the cutter.

_________15. Device used to divide the circumference of a circular piece into equally spaced divisions.

_________16. Cutters are used in pairs to machine two opposite sides of a job at the same time.

_________17. Use of several cutters permitting several surfaces to be machined in one pass.

_________18. Sawing thin metal stock into various widths and lengths.

_________19. Cuts the groove in the head of a screw.

_________20. Used to align and position work for milling.

(a) Slotting
(b) Side cutter
(c) Conventional milling
(d) Metal slitting saw
(e) Facing cutter
(f) Semi-automatic controls
(g) Two-flute end mill
(h) Edge finder
(i) Straddle mill
(j) Climb milling
(k) Gang milling
(l) Dividing head
(m) Inserted-tooth cutter
(n) Manual controls
(o) Slitting
(p) Feed
(q) Face milling
(r) Solid cutter
(s) Arbor cutter
(t) Shank cutter

METAL SPINNING

LEARNING OBJECTIVES

After studying this chapter, students should be able to:
- Define metal spinning.
- Set up and prepare a spinning lathe for use.
- Demonstrate the spinning process.
- Describe the shear spinning process.
- Apply spinning safety rules.

CHAPTER RESOURCES

Text, pages 507–520
Test Your Knowledge, page 520
Research and Development, page 520
Workbook, pages 141–144
Instructor's Manual
Answer keys for:
Test Your Knowledge Questions
Workbook
Chapter Quiz
Reproducible Masters:
31-1 The Spinning Process
31-2 Determining Disc Size for Spinning
31-3 How Shear Spinning Differs from Conventional Spinning
31-4 Chapter Quiz

GUIDE FOR LESSON PLANNING

Samples of spun products should be on hand for students to examine. Before making the chapter assignment, question students on how they think the items were made.

CLASS DISCUSSION AND DEMONSTRATION

Have students read and study all or part of the chapter. They should pay particular attention to the illustrations. Review the assignment with them. Discuss and demonstrate (if spinning equipment is available) the following:
- The spinning process.
- Spinning equipment.
- How to spin a simple form.
- How to determine disc size. Assign a simple problem or two for student practice.
- Techniques used for spinning complex forms.
- Spinning a bead on the edge of a spun shape.
- Spinning deep forms.
- Industrial applications of spinning.
- Shear spinning and how it differs from conventional spinning.
- Safety precautions that must be observed when spinning.

When demonstrating spinning operations, position students so all can see and hear, but should not be in direct line with the rotating disc. The spinning lathe should be in safe operating condition with all safety features and devices in place. The machine must also be mounted solidly to the floor. Students must be wearing approved eye protection.

ASSIGNMENTS

1. Assign the Test Your Knowledge questions.
2. Assign Chapter 31 in the *Modern Metalworking Workbook.*
3. Assign the chapter quiz. Copy and distribute Reproducible Master 31-4.
4. Permit students to volunteer for one or more of the Research and Development activities at the end of the chapter.

TEST YOUR KNOWLEDGE

1. three-dimensional
2. tool rest

3. hardened
4. softened
5. tongue
6. solid
7. A segmented chuck is used when spinning complex shapes.
8. follow block
9. It prevents galling of the metal.
10. yellow bar soap, beeswax
11. physically, economically
12. It can be used to strengthen the piece and to decorate it.
13. Evaluate individually. Refer to Section 31.7.
14. It is a cold extrusion process in which the parts are shaped by rollers exerting tremendous pressures on a rotating starting blank or preform. This displaces the metal parallel to the centerline of the workpiece. The metal is taken from the thickness of the blank.
15. Evaluate individually.

Workbook

1. (c) working metal into three dimensional shape
2. It is made in sections (segments) so it can be disassembled for removal from the spun form.
3. (e) None of the above.
4. (c) Solid chuck
5. (f) Joined chuck
6. (a) Spinning tools
7. (e) Segmented chuck
8. (b) Tool rest
9. (d) Follow block
10. Evaluate individually. Refer to Figure 31-14.
11. By placing the back stick against the left side of the disc and forcing the metal to pass between the back tool and the forming tool. It may also be removed from the chuck and flattened with a mallet.
12. In shear spinning, the metal is taken from the thickness of the blank, whereas in spinning the metal is taken from the diameter of the blank. Instead of displacing the metal as in the shear spinning technique, spinning folds the metal over and places it on or close to the spinning form.
13. Evaluate individually. Refer to Section 31.7.
14. Evaluate individually. Refer to Section 31.8.
15. Evaluate students' simple forms individually.
16. Evaluate student designs and segmented chucks individually.
17. Evaluate student ability to spin objects using a segmented chuck individually.

Chapter Quiz

1. three-dimensional
2. inside
3. solid
4. segmented
5. (b) prevent the forming tool from damaging the metal's surface
6. spinning form
7. (a) rollers that exert tremendous pressure on a rotating metal blank or preform

The Spinning Process

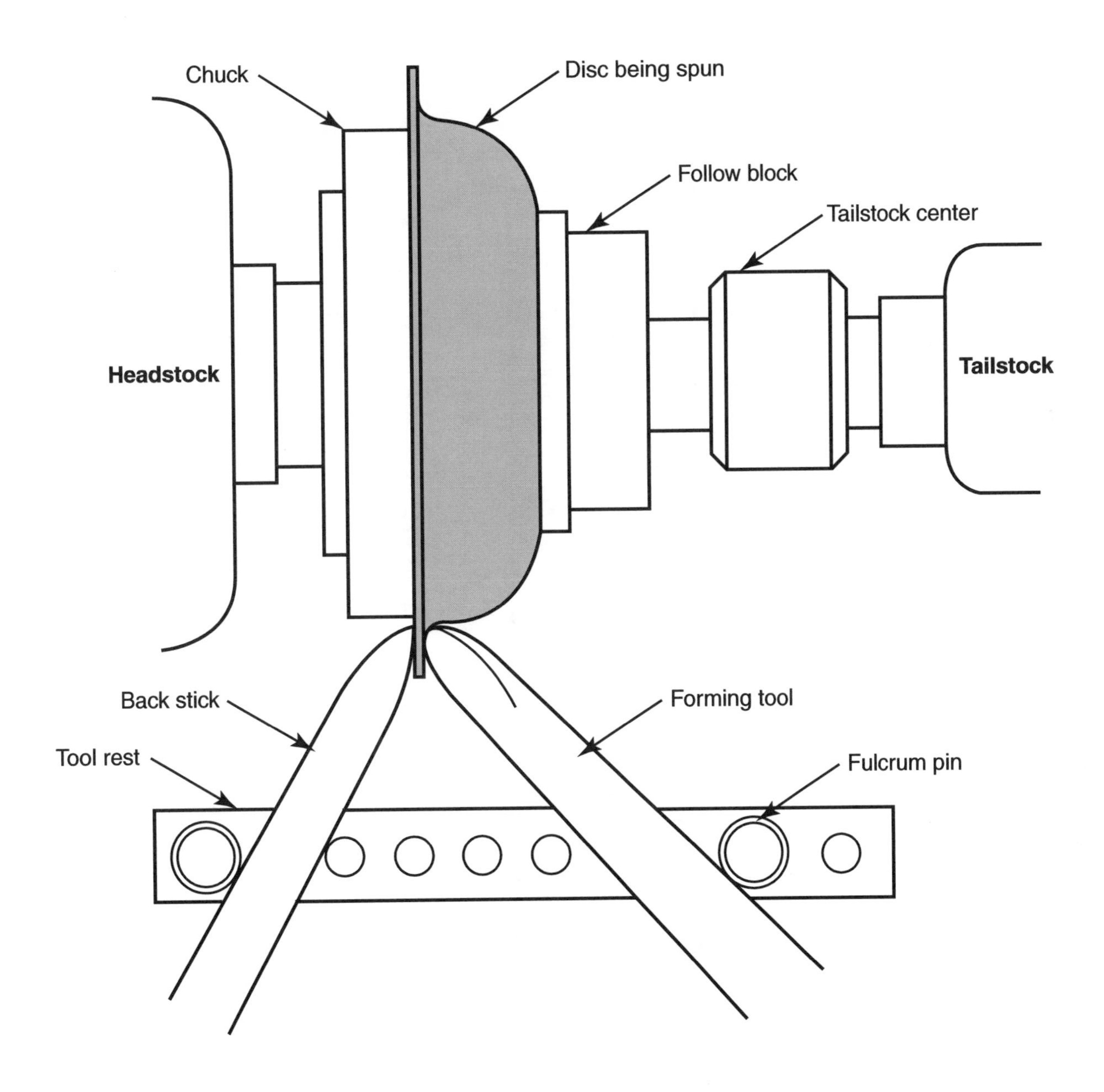

Determining Disc Size for Spinning

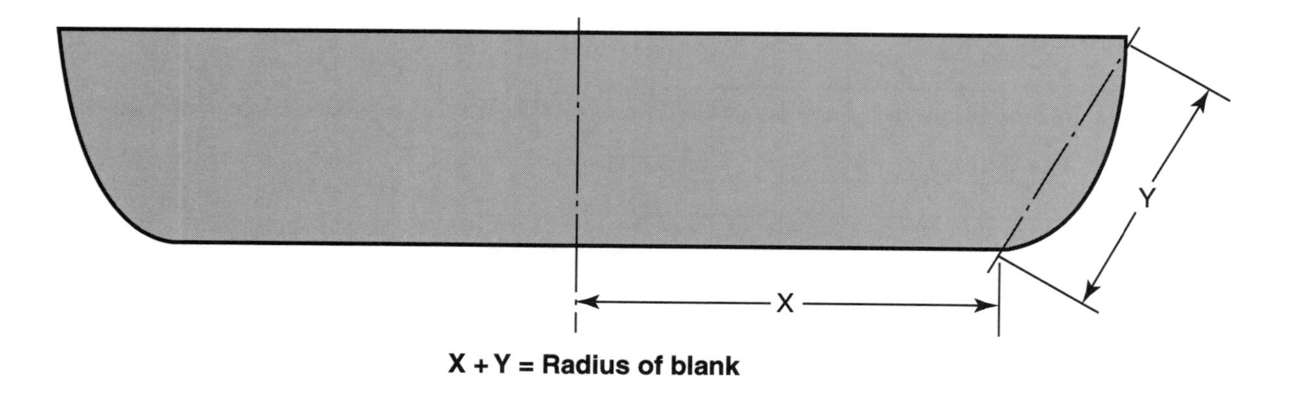

31-2

31-3

How Shear Spinning Differs from Conventional Spinning

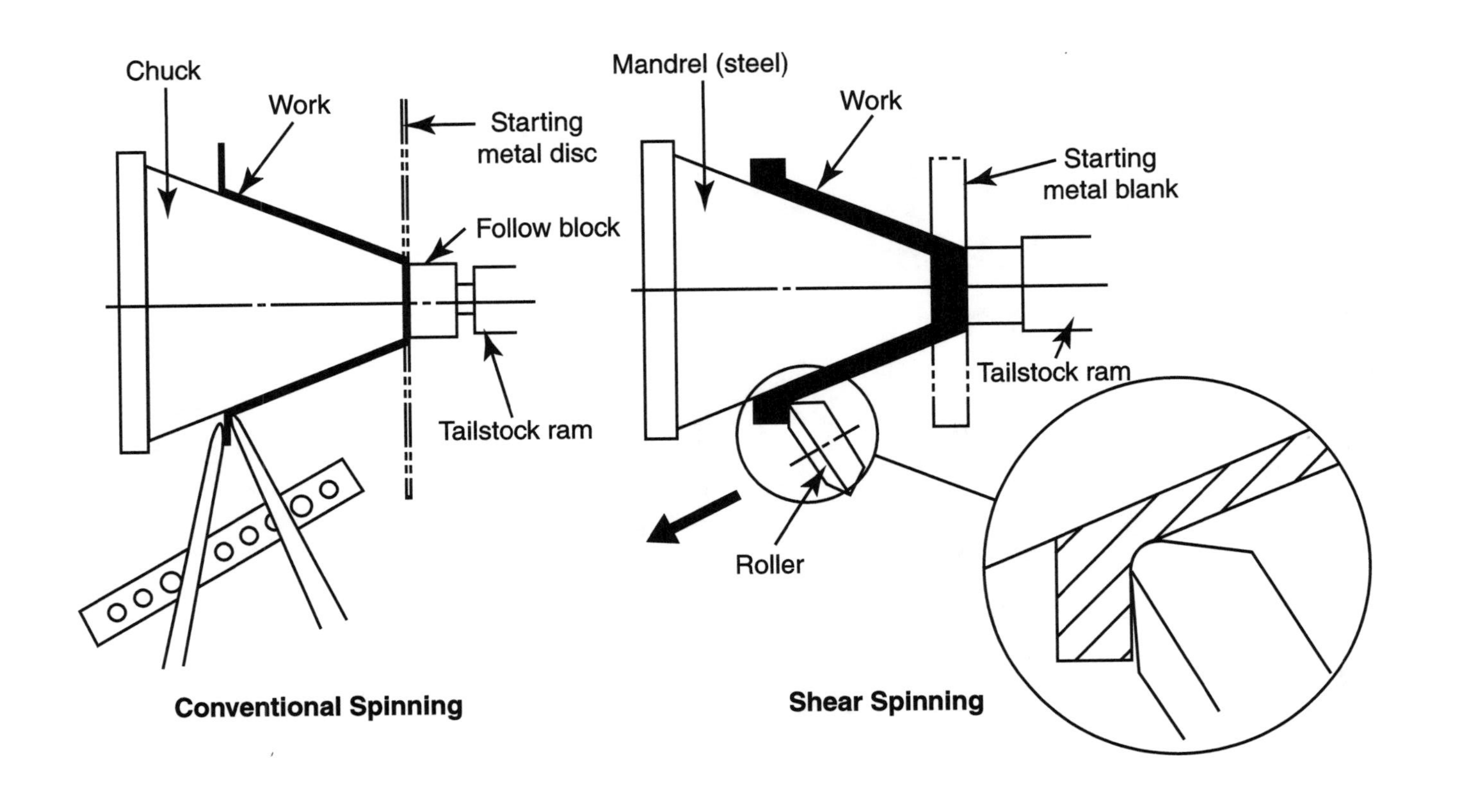

Chapter 31 Quiz

Metal Spinning

Name: ______________________ **Date:** ____________ **Period:** ____________

1. Metal spinning is a method of working sheet metal into three-dimensional shape. It involves, in its simplest form, working sheet metal into _____ form. 1.__________
2. The forming chuck used in spinning must be made to the dimensions specified for the _____ of the desired object. 2.__________
3. Simple forms can be spun on a(n) _____ chuck. 3.__________
4. A(n) _____ chuck is required if the object to be spun is more complex in shape. 4.__________
5. A lubricant must be used when spinning to _____. 5.__________
   (a) prevent damage to the tailstock center
   (b) prevent the forming tool from damaging the metal's surface
   (c) keep the forming tool cool
   (d) All of the above.
   (e) None of the above.
6. Spinning folds the metal over and places it on or close to the _____. 6.__________
7. In shear spinning the metal is shaped by _____. 7.__________
   (a) rollers that exert tremendous pressure on a rotating metal blank or preform
   (b) folding the metal down on a spinning form
   (c) using a technique similar to conventional spinning
   (d) All of the above.
   (e) None of the above.

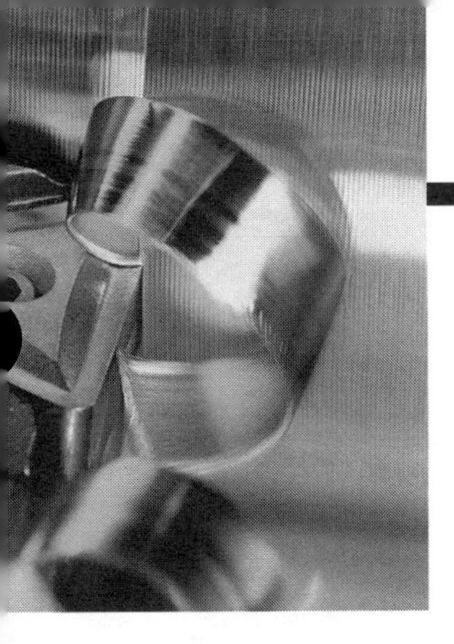

32 Cold Forming Metal Sheet

Learning Objectives

After studying this chapter, students should be able to:
- Describe several cold forming operations.
- Explain how shearing, blanking, and piercing operations differ.
- Describe how many of the cold forming techniques are accomplished.

Chapter Resources

Text, pages 521–532
    Test Your Knowledge, pages 530–531
    Research and Development, page 531
Workbook, pages 145–146
Instructor's Manual
    Answer keys for:
        Test Your Knowledge Questions
        Workbook
        Chapter Quiz
    Reproducible Masters:
        32-1 Shearing Operation
        32-2 Blanking Operation
        32-3 Progressive Die Punching Operation
        32-4 Drawing Operation
        32-5 Guerin Process
        32-6 Marform Process
        32-7 Hydroform Process
        32-8 Stretch Forming Operation
        32-9 Tube Bending
        32-10 Bending Operation
        32-11 Roll Forming
        32-12 Chapter Quiz

Guide for Lesson Planning

Samples of products shaped and formed by the operations described in this chapter will aid in explaining cold forming metal sheet techniques. They could include one-piece food containers, cooking pots, a knife and fork, a section of rain gutter and down spout, some auto parts, and similar items.

Class Discussion and Demonstration

Have students read and study all or part of the chapter paying special attention to the illustrations. Review the assignment with them using the Reproducible Masters as overhead transparencies or handouts as resource material.

Discuss the following:
- Stamping
- Cutting operations
- Shearing
- Blanking
- Progressive dies
- Drawing process
- Forming operations
- Bending operations
- Roll forming

Assignments

1. Assign the Test Your Knowledge questions.
2. Assign Chapter 32 in the *Modern Metalworking Workbook.*
3. Assign the chapter quiz. Copy and distribute Reproducible Master 32-12.
4. Permit students to volunteer for one or more of the Research and Development activities at the end of the chapter.

Test Your Knowledge

1. Cutting
2. Shearing
3. plasma
4. (c) Blanking
5. A progressive die has all the cutting and forming dies required to make a complete part lined up in proper order in one die.

Each stage in making the part cuts a portion of the blank, but leaves small connections so the metal section can be transferred to the next stage.

6. three-dimensional shapes
7. Evaluate individually.
8. Evaluate individually.
9. Bending
10. Roll forming is the process in which a flat strip is passed through a series of rolls that progressively form it into final shape.

Workbook

1. (f) Stamping
2. (g) Cutting
3. (k) Shearing
4. (b) Squaring shears
5. (m) Blanking
6. (l) Drawing
7. (c) Piercing
8. (d) Guerin process
9. (i) Marform process
10. (e) Hydroforming
11. (h) Bulging
12. (a) Stretch forming
13. (j) Roll forming
14. Evaluate student designs and work individually.
15. Evaluate student designs and work individually.

Chapter Quiz

1. (j) Stamping
2. (g) Blanking
3. (d) Cutting
4. (h) Squaring shears
5. (i) Guerin process
6. (c) Hydroforming
7. (e) Marform process
8. (k) Bulging
9. (m) Drawing
10. (a) Stretch forming
11. (l) Roll forming
12. (b) Punching

Shearing Operation

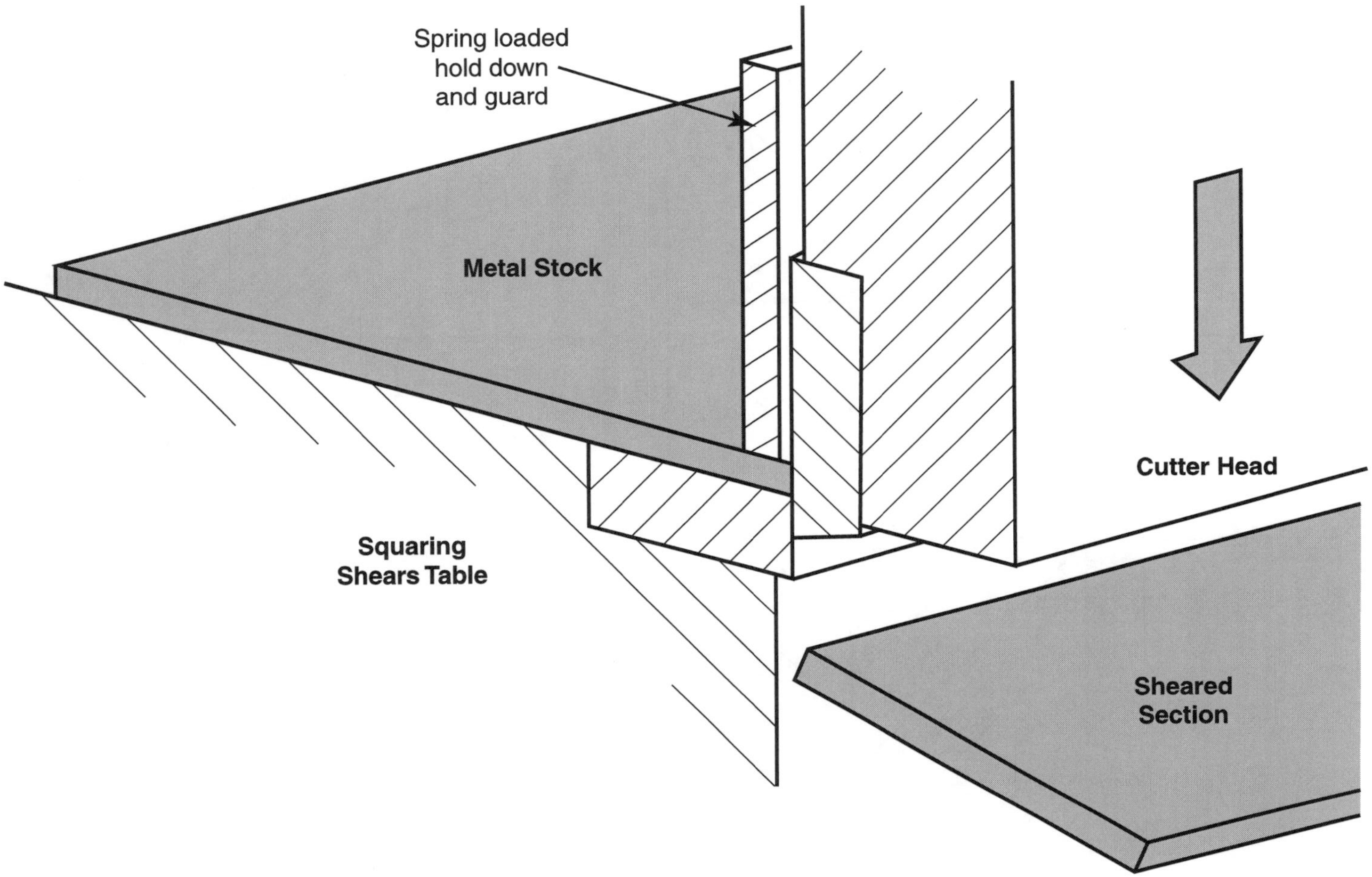

Spring loaded hold down and guard

Metal Stock

Squaring Shears Table

Cutter Head

Sheared Section

32-1

Blanking Operation

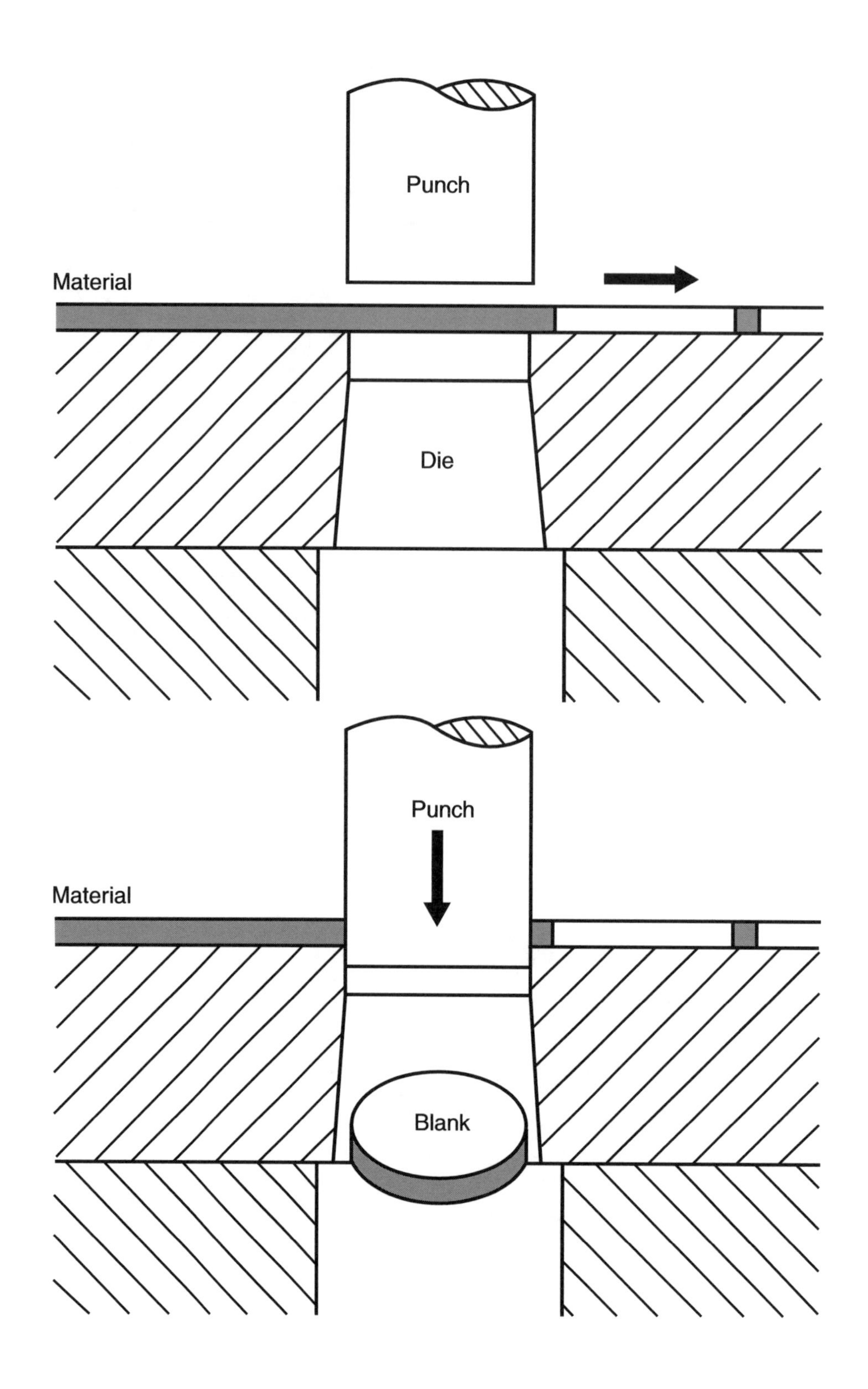

Progressive Die Punching Operation

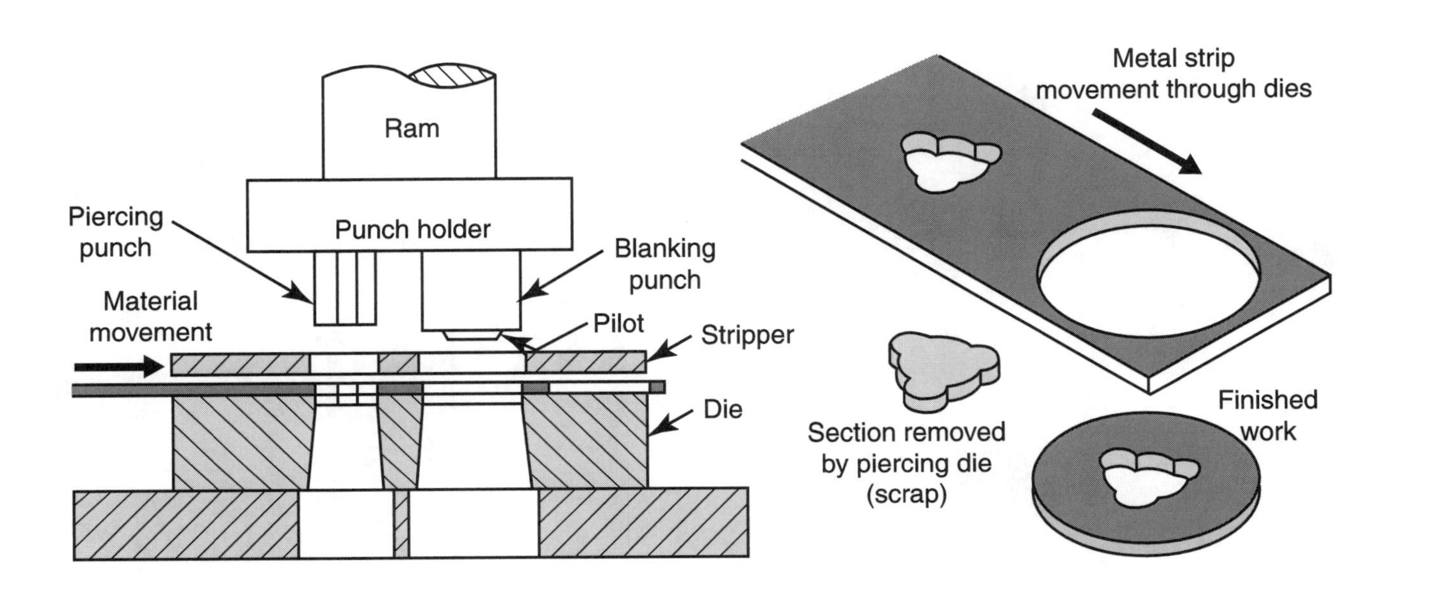

32-3

Drawing Operation

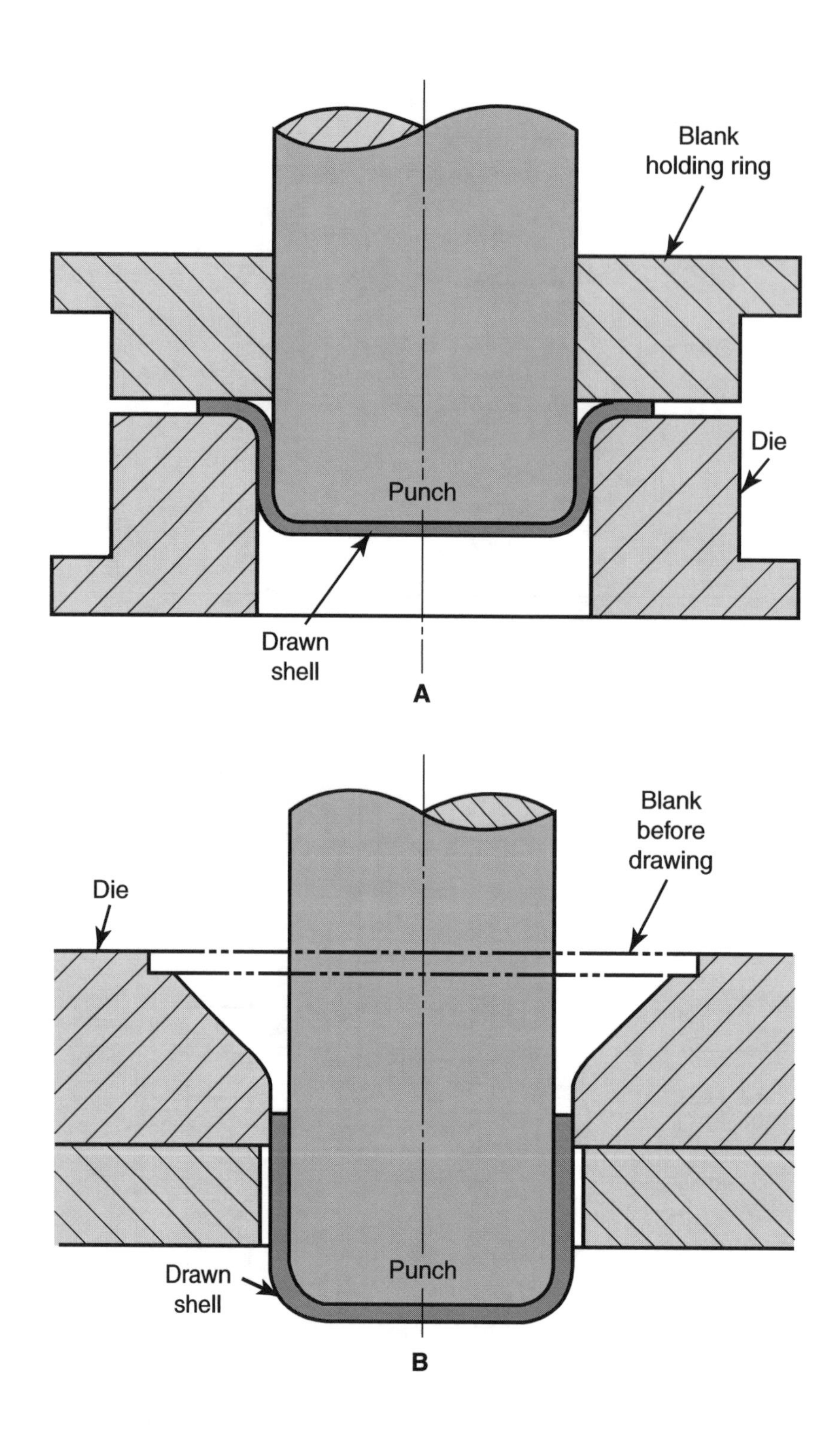

Guerin Process

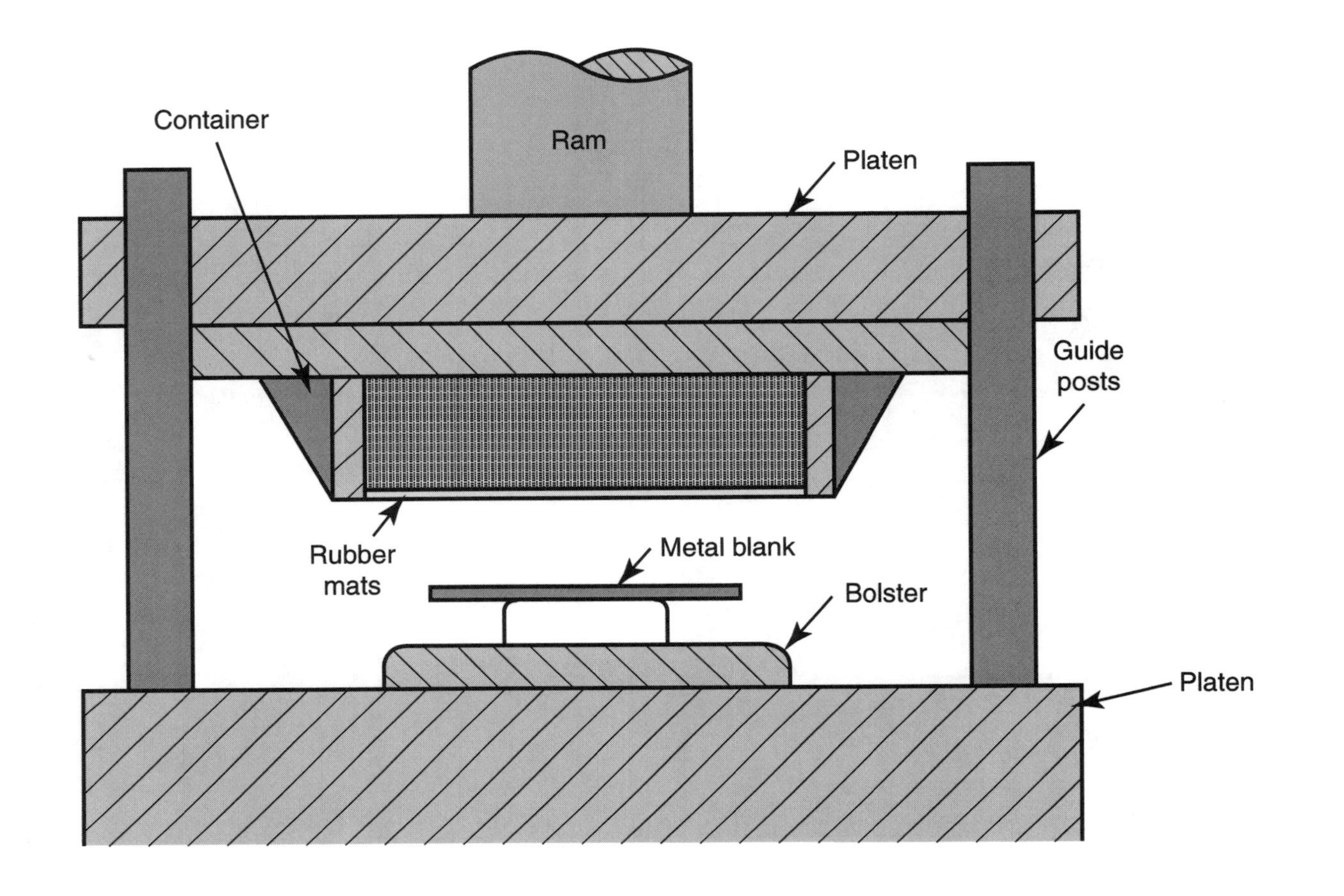

32-5

Marform Process

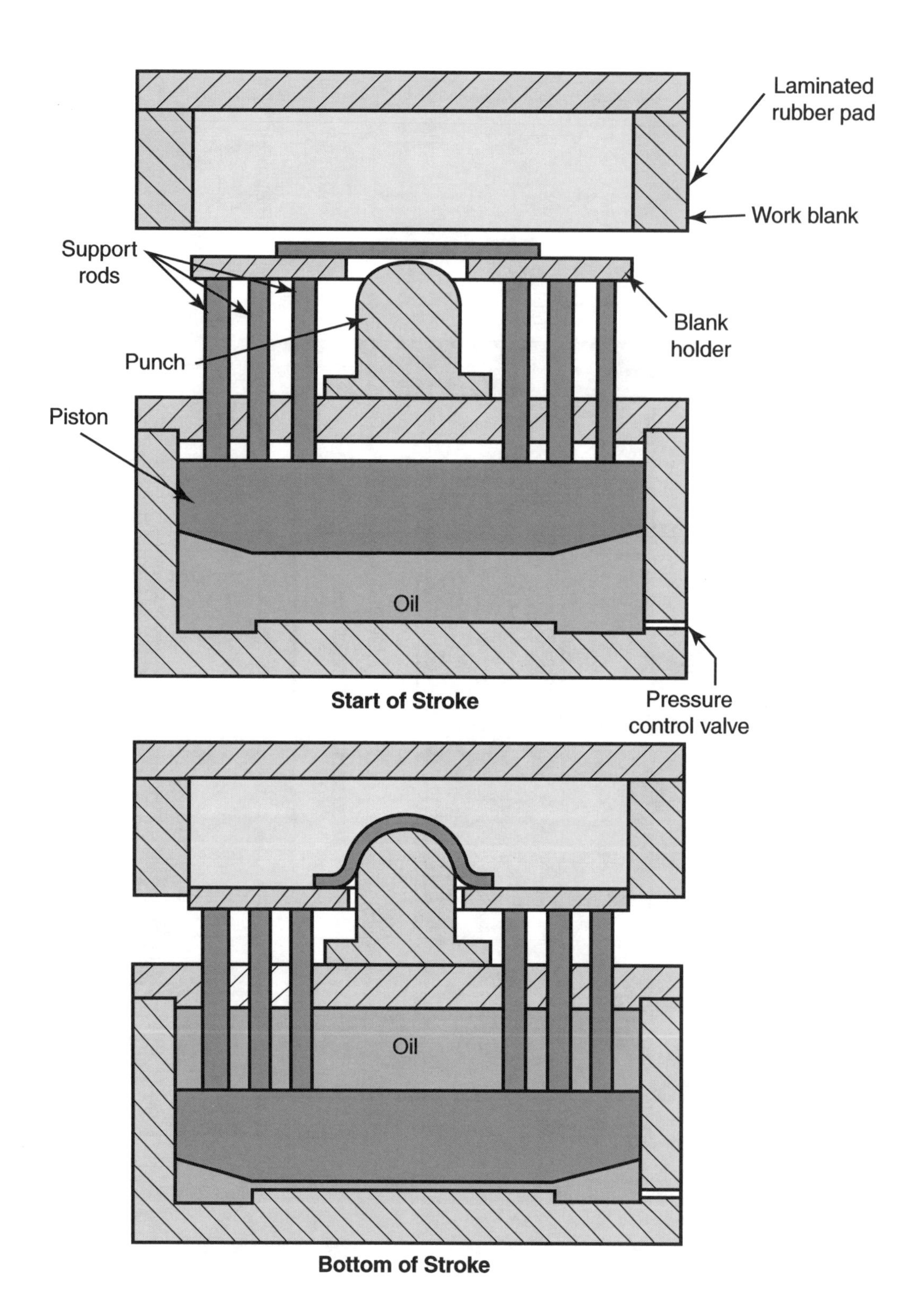

Start of Stroke

Bottom of Stroke

Hydroform Process

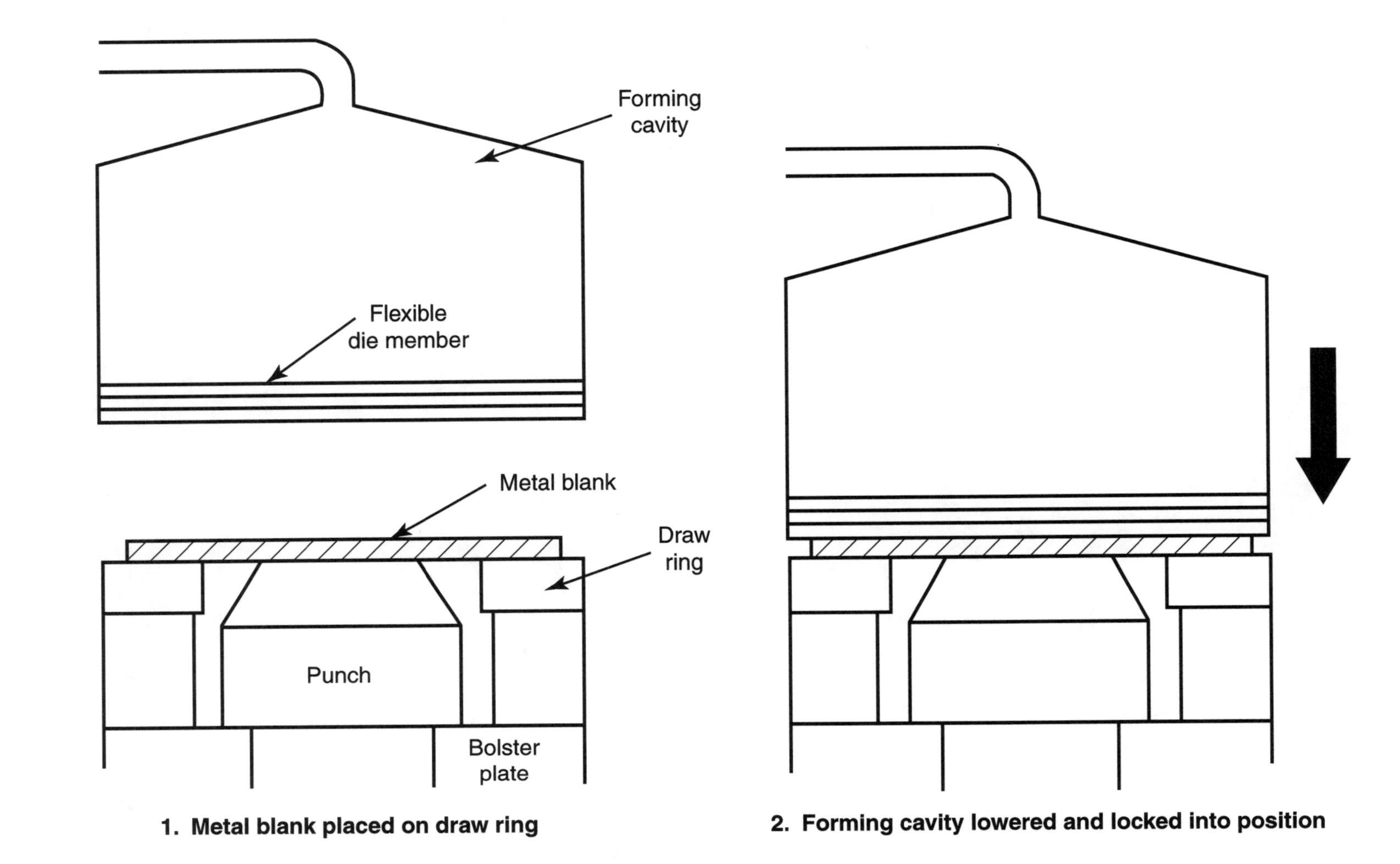

1. Metal blank placed on draw ring

2. Forming cavity lowered and locked into position

32-7
(Continued)

Hydroform Process
(Continued)

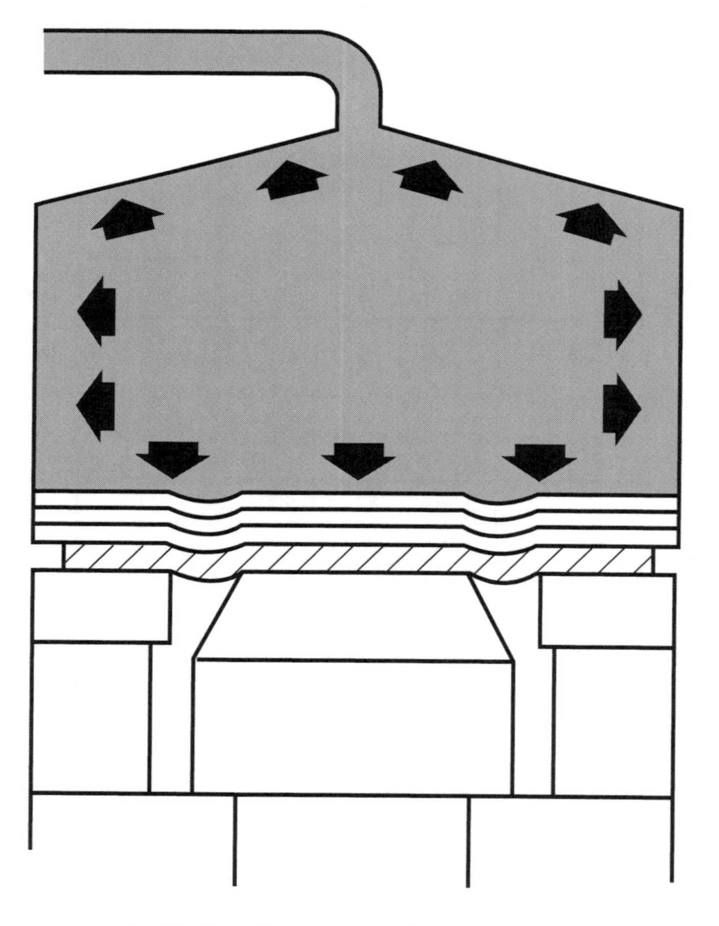

3. Hydraulic pressure built up to a predetermined setting

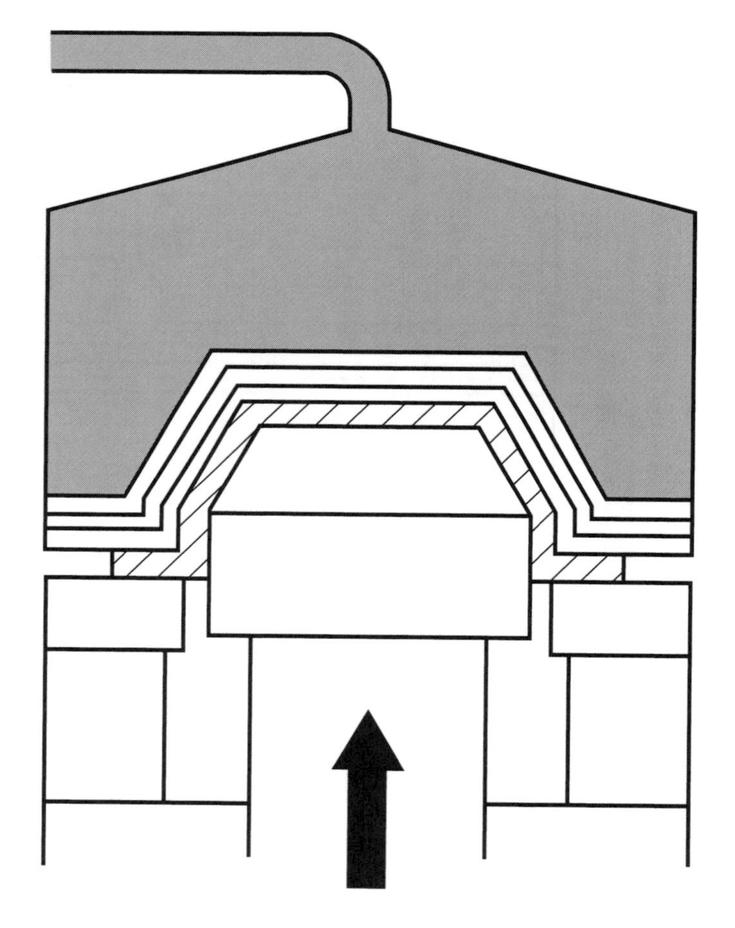

4. Punch is moved up into flexible die member

Hydroform Process
(Continued)

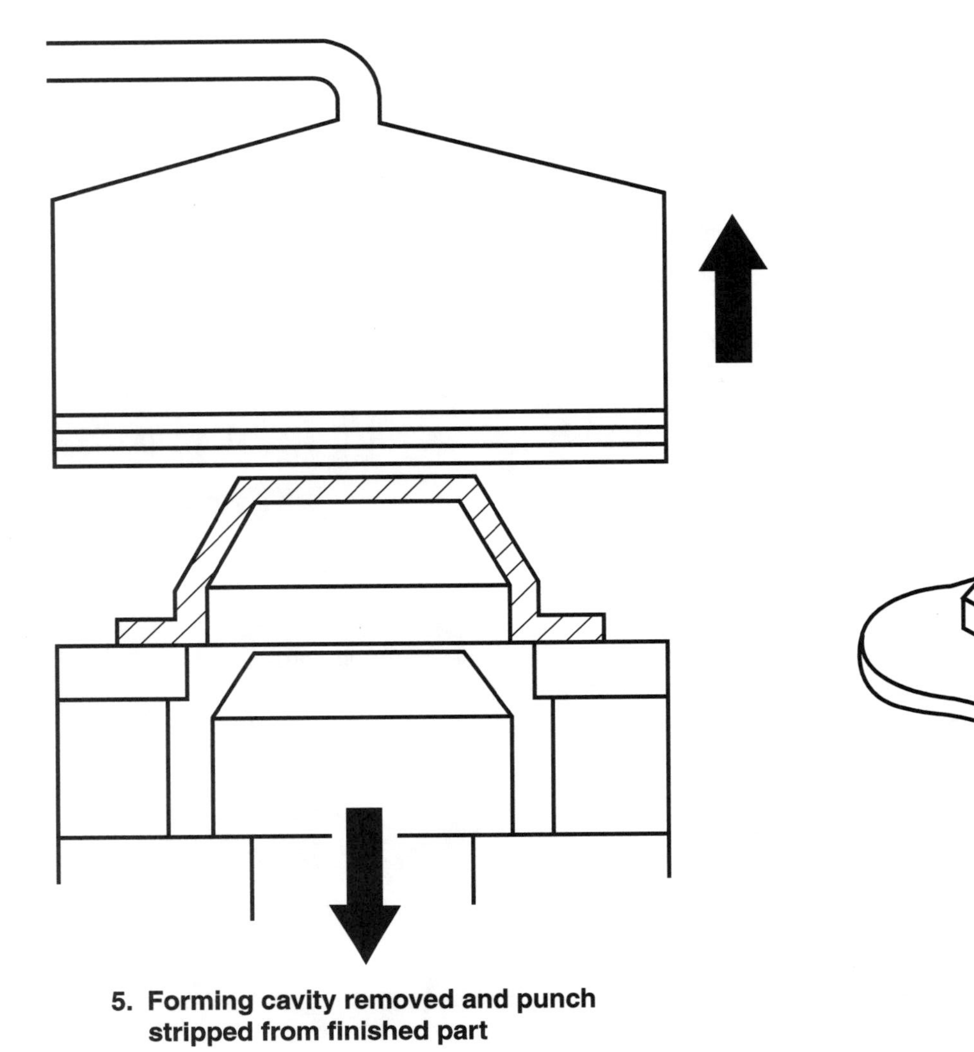

5. Forming cavity removed and punch stripped from finished part

6. Finished part ready for trimming

32-7

32-8

Stretch Forming Operation

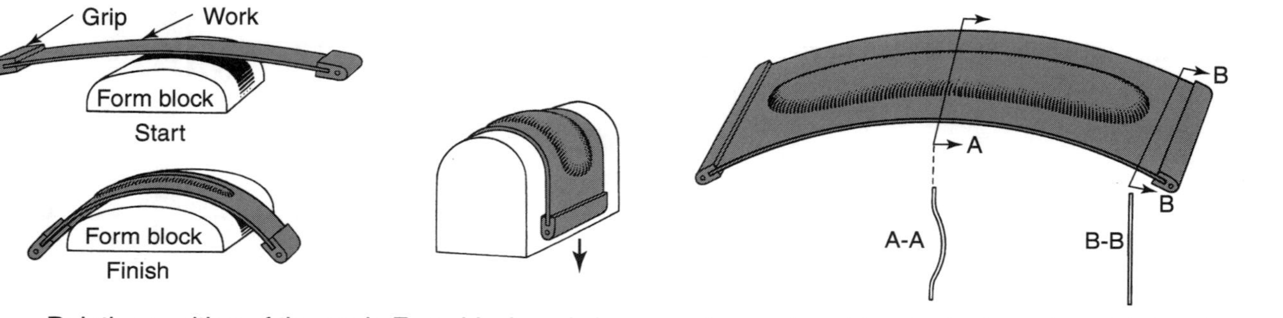

Relative position of the work. Form block and clamps at the start and finish of a typical stretch forming operation produce a raised rib. Typical shape produced by stretch forming over a form block. A stretch-formed shape is similar to the bottom of a canoe. The section will be riveted or yields to other sections to make the canoe.

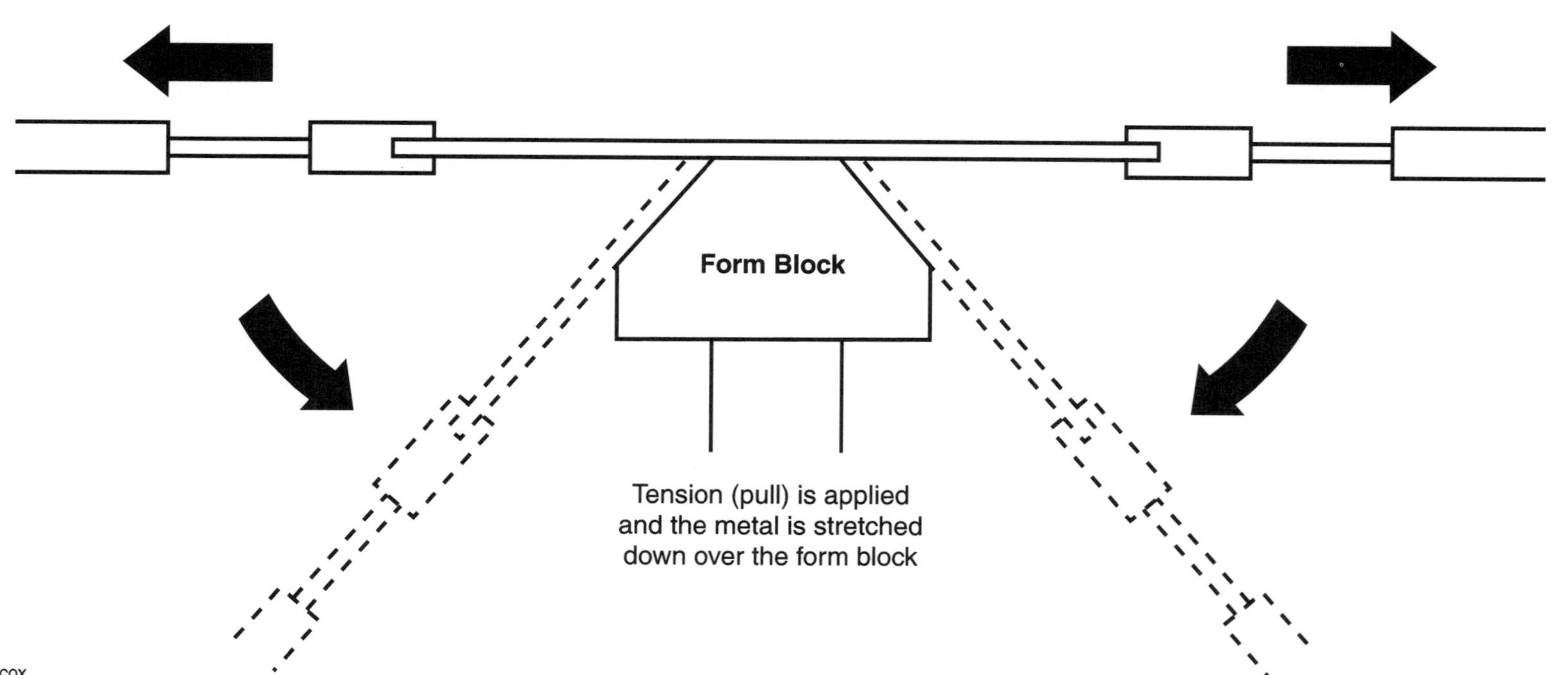

Tube Bending

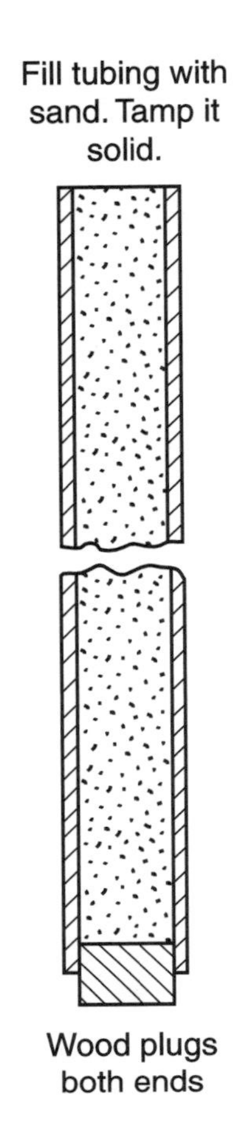

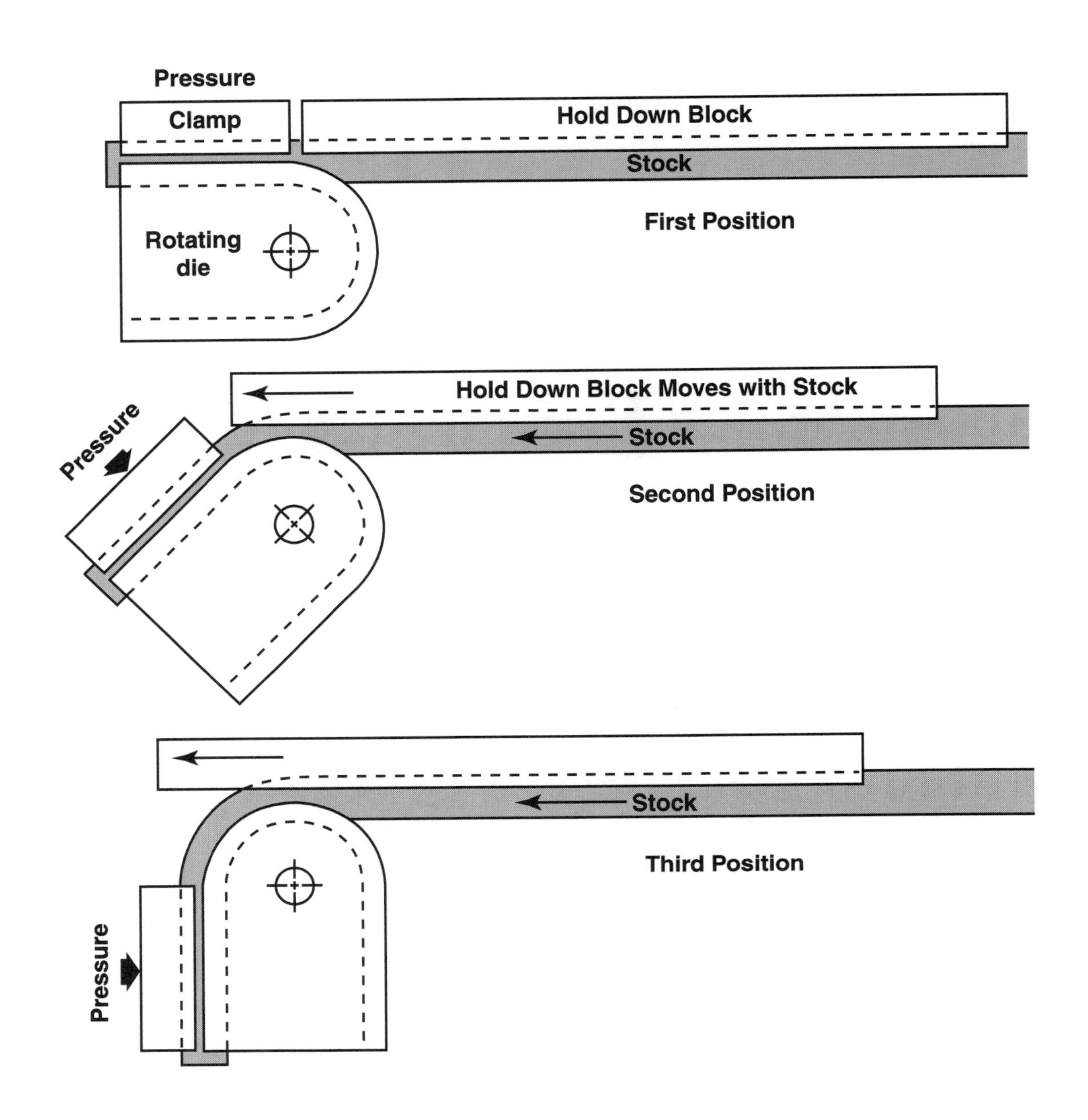

32-9

Bending Operation

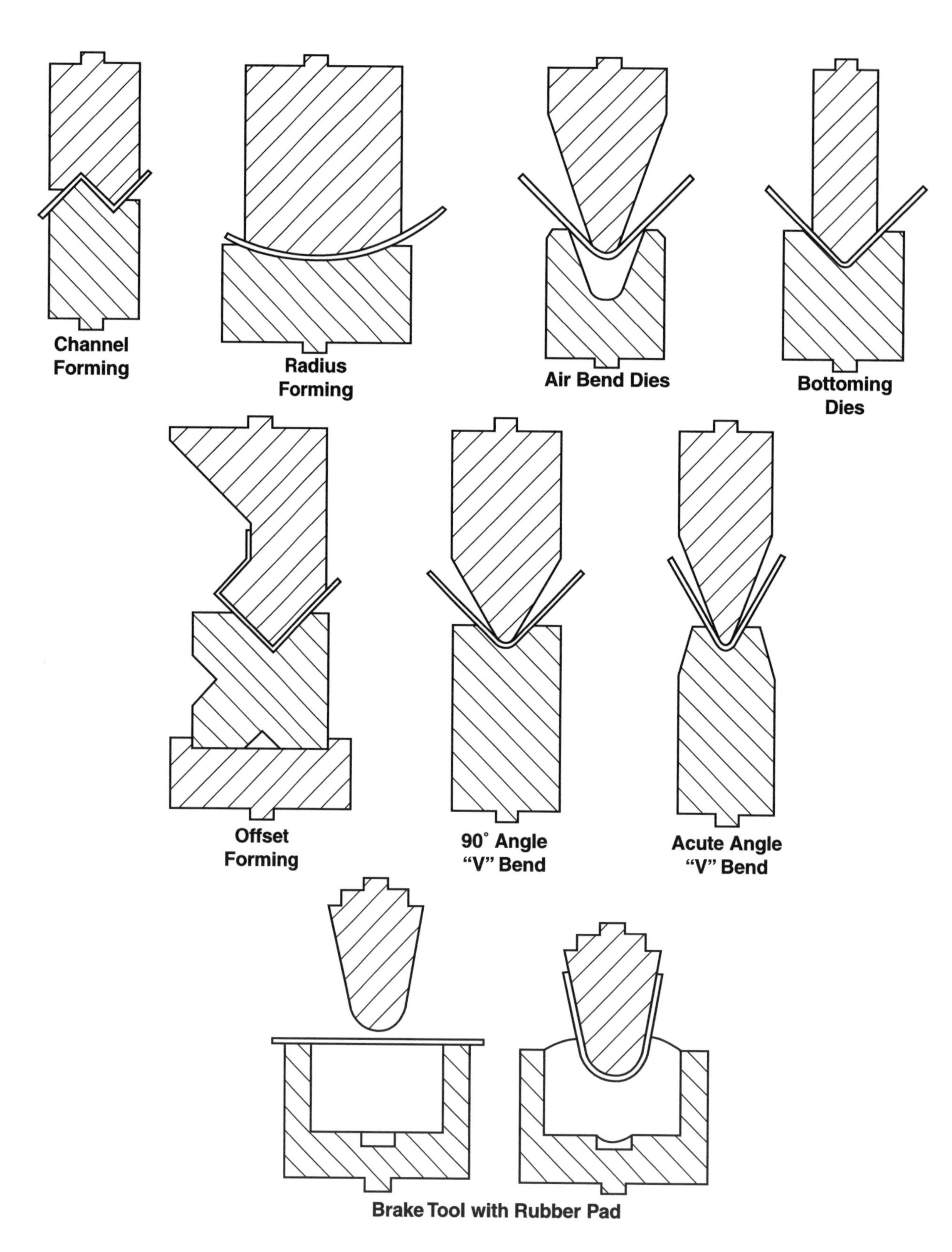

Roll Forming

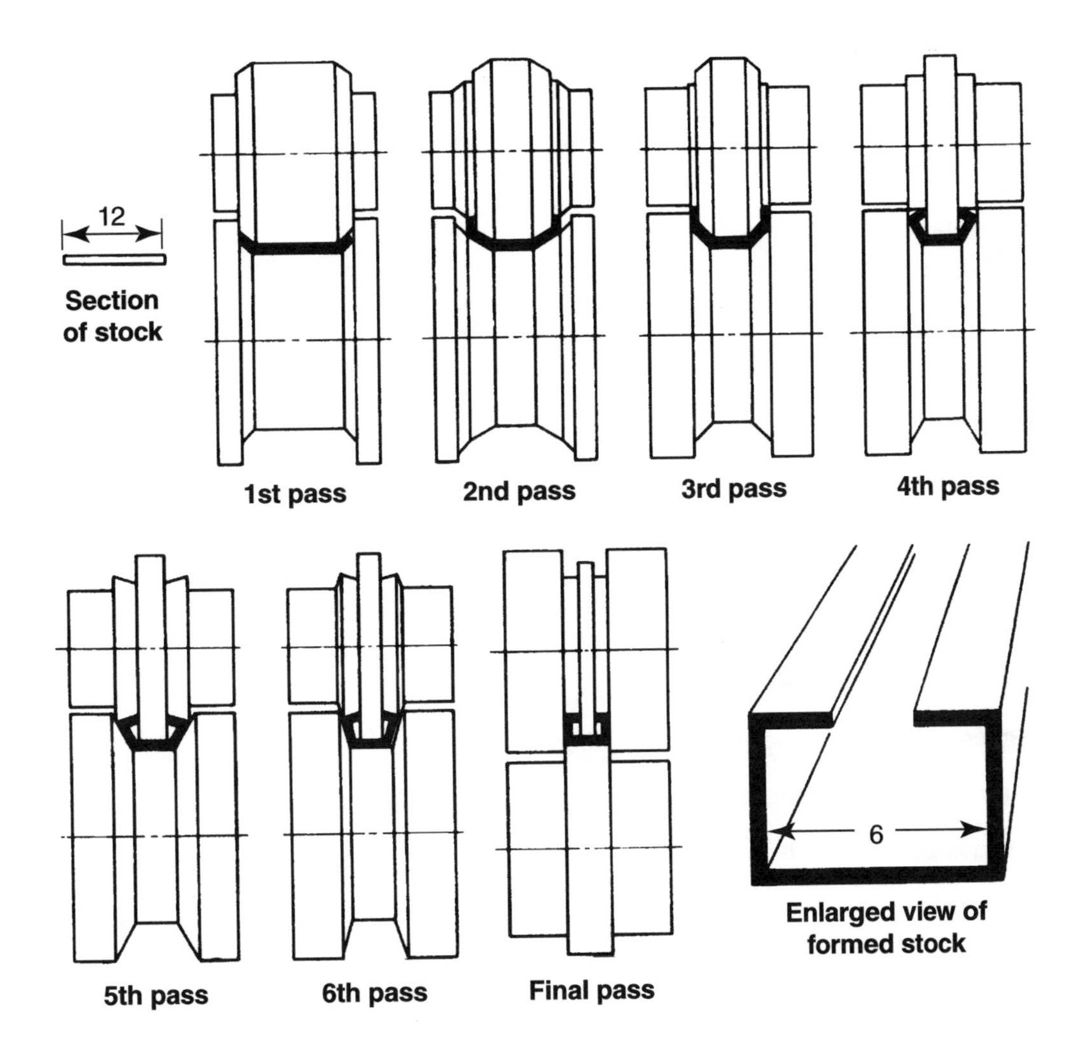

Chapter 32 Quiz

Cold Forming Metal Sheet

Name: ______________________ **Date:**____________ **Period:**____________

Match the following words or phrases with the appropriate sentence.

(a) Stretch forming
(b) Punching
(c) Hydroforming
(d) Cutting
(e) Marform process
(f) Shearing
(g) Blanking
(h) Squaring shears
(i) Guerin process
(j) Stamping
(k) Bulging
(l) Roll forming
(m) Drawing

_________ 1. Term used for many press forming operations.

_________ 2. Process where the metal remaining is the waste.

_________ 3. Used to cut metal to size for machining.

_________ 4. Equipment for cutting straight edges.

_________ 5. Process that forms shallow parts in rubber dies.

_________ 6. Process in which a rubber diaphragm is backed up by hydraulic pressure.

_________ 7. Technique used for deep drawing operations.

_________ 8. Rubber is used to transmit the pressure needed to expand the metal blank or tube against the die.

_________ 9. Metal-shaping process performed on a mechanical or hydraulic draw press using a matched punch and die set.

_________ 10. Metal blank is gripped on opposite edges with clamps, then lightly pulled, forcing the metal to wrap around a form of the desired shape.

_________ 11. Shapes flat metal sheet by passing it through a series of rollers.

_________ 12. Process in which the metal removed becomes the waste or scrap.

33 Extrusion Processes

Learning Objectives

After studying this chapter, students should be able to:

- Explain the extrusion process.
- Identify the three types of extrusion used to shape metal.
- Describe impact extrusion.

Chapter Resources

Text, pages 533–538
  Test Your Knowledge, pages 537–538
  Research and Development, page 538
Workbook, pages 147–148
Instructor's Manual
  Answer keys for:
    Test Your Knowledge Questions
    Workbook
    Chapter Quiz
  Reproducible Masters:
    33-1 Sectional View of Extrusion Press
    33-2 Impact Extrusion Process
    33-3 Chapter Quiz

Guide for Lesson Planning

Samples of products made by the extrusion process should be available for student examination.

Class Discussion and Demonstration

Challenge students to devise a method to demonstrate the extrusion process. Modeling clay and Play Dough® can be used as substitutes for metal.

Have students read and study the chapter paying careful attention to the illustrations. Review and discuss the chapter.

Assignments

1. Assign the Test Your Knowledge questions.
2. Assign Chapter 33 in the *Modern Metalworking Workbook.*
3. Assign the chapter quiz. Copy and distribute Reproducible Master 33-3.
4. Permit students to volunteer for one or more of the Research and Development activities at the end of the chapter.

Test Your Knowledge

1. (c) direct extrusion
2. (a) indirect extrusion
3. (b) impact extrusion
4. squeezed
5. Cold extrusion
6. press capacity
7. Impact (cold)

Workbook

1. (c) Indirect
2. (a) Impact
3. (b) direct
4. The press capacity.
5. (b) the ingot is usually left at room temperature
6. Evaluate individually. Refer to Figure 33-06 in the text.
7. Evaluate individually.

Chapter Quiz

1. (a) the ingot is usually left at room temperature
2. (c) Impact
3. (b) direct
4. (a) Indirect
5. (a) extrusion press capacity

6. complex
7. (d) All of the above.
8. Evaluate student sketches individually.

Sectional View of Extrusion Press

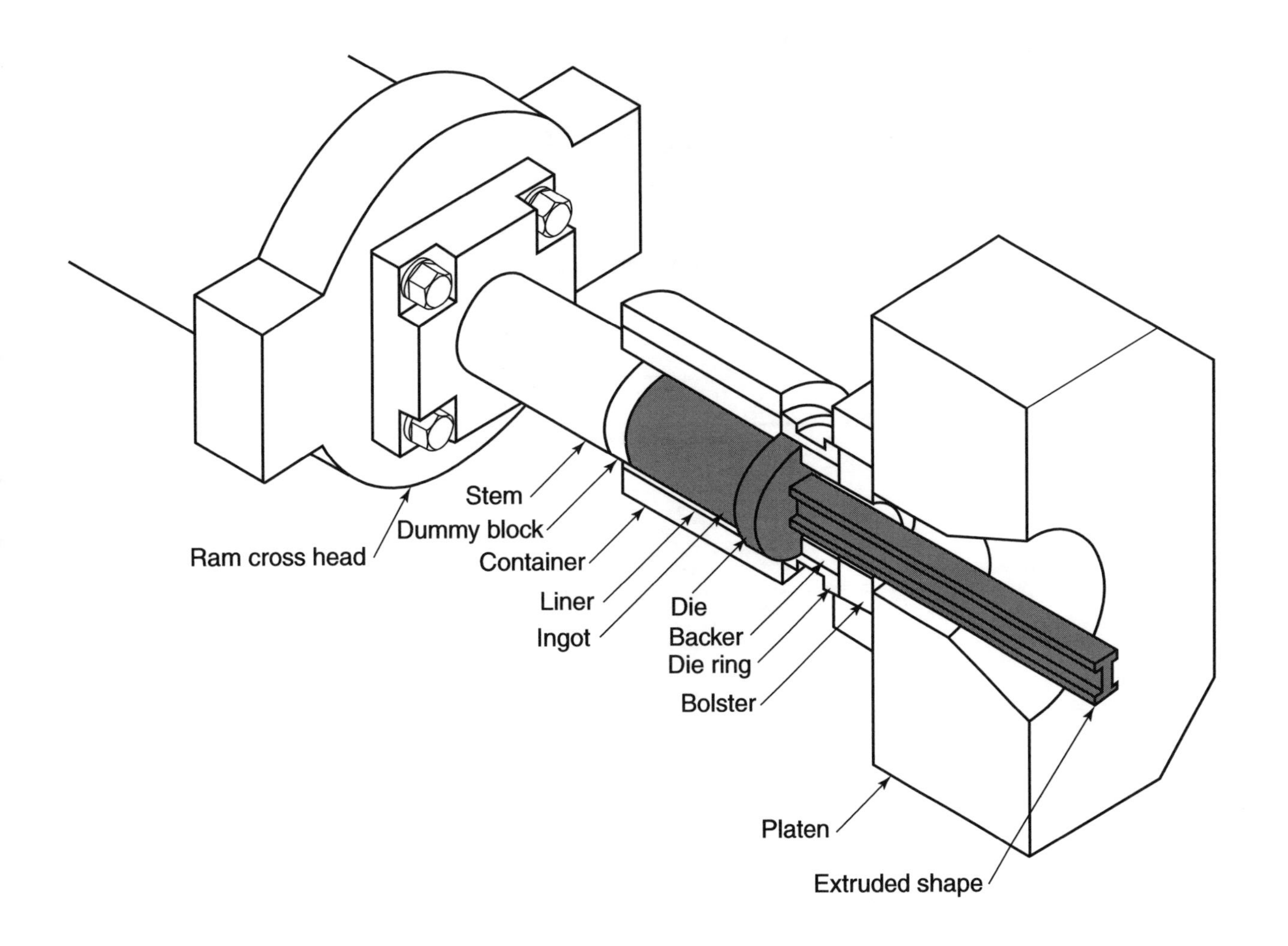

Impact Extrusion Process

33-2

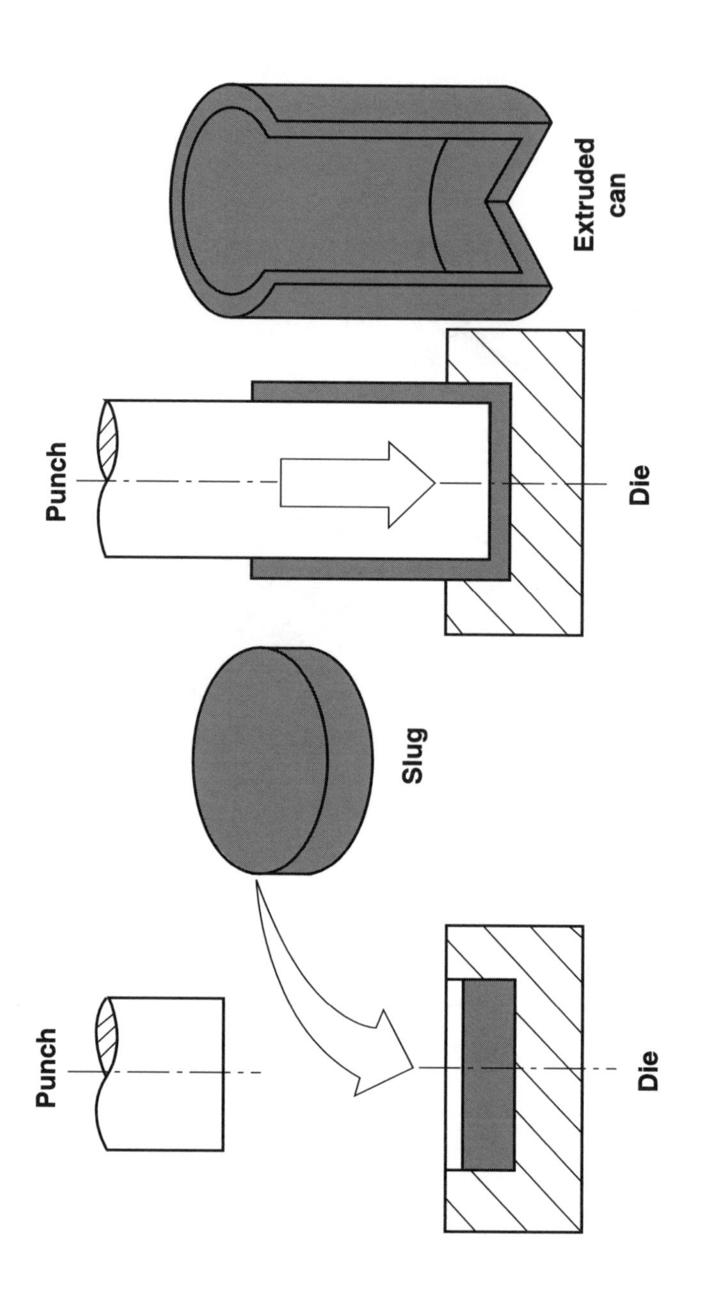

Chapter 33 Quiz

Extrusion Processes

Name: ______________________ **Date:**______________ **Period:**______________

1. In cold extrusion operations, _____. 1.______________
   (a) the ingot is usually left at room temperature
   (b) the punch is cooled to near freezing temperatures
   (c) the billet is cooled to near freezing temperatures
   (d) All of the above.
   (e) None of the above.

2. _____ is an extrusion process in which parts are formed by striking a slug of metal with a punch moving at high velocity. 2.______________
   (a) Indirect
   (b) Direct
   (c) Impact
   (d) Hot

3. In _____ extrusion, the ram and product move in the same direction against the die. 3.______________
   (a) indirect
   (b) direct
   (c) impact
   (d) hot

4. _____ extrusion is the process in which the piece of metal remains stationary while a hollow die stand forces the die back into the cylinder. 4.______________
   (a) indirect
   (b) direct
   (c) impact
   (d) hot

5. Extrusion size is limited by _____. 5.______________
   (a) extrusion press capacity
   (b) length of the extrusion
   (c) design of the extrusion
   (d) All of the above.
   (e) None of the above.

6. The extrusion process is an economical method of producing _____ shapes. 6.______________

7. After extrusion, the extruded section must be _____. 7.______________
   (a) straightened
   (b) cut to length
   (c) some alloys must be heat treated
   (d) All of the above.
   (e) None of the above.

Chapter 33 Quiz *(Continued)*

Name: ______________________________ **Date:**______________ **Period:**______________

8. Make a sketch showing how a die 24″ in diameter can be used to extrude a section 30″ wide.

Powder Metallurgy

Learning Objectives

After studying this chapter, students should be able to:

- Explain the powder metallurgy process.
- Give examples of products made from powder metallurgy.
- Describe the steps in producing a product by the powder metallurgy process.

Chapter Resources

Text, pages 539–544
Test Your Knowledge, page 543
Research and Development, page 543
Workbook, pages 149–150
Instructor's Manual
Answer keys for:
Test Your Knowledge Questions
Workbook
Chapter Quiz
Reproducible Masters:
34-1 Powder Metallurgy Sequence
34-2 Chapter Quiz

Guide for Lesson Planning

A selection of products made by the powder metallurgy process should be available for students to examine. Samples can include self-lubricating bearings, auto fuel filter, auto parts, Alnico magnet, etc.

Have students read and study the chapter. Review the assignment. Discuss the steps in making a powder metallurgy product. Some teachers have demonstrated the process using aluminum and bronze powders in student-made dies.

Assignments

1. Assign the Test Your Knowledge questions at the end of the chapter.
2. Assign Chapter 34 in the *Modern Metalworking Workbook.*
3. Assign the chapter quiz. Copy and distribute Reproducible Master 34-2.
4. Permit students to volunteer for one or more of the Research and Development activities at the end of the chapter.

Test Your Knowledge

1. powders
2. To ensure uniform behavior during processing and the output of a consistent product.
3. A briquette or green compact. It is fragile and brittle and will crumble if not handled carefully.
4. Sintering transforms the briquette into a strong, useful unit.
5. (d) All of the above.

Workbook

1. A technique used to shape parts from metal powders by heating them to a predetermined temperature and compacting them into a predetermined shape.
2. Evaluate individually. Refer to Section 34.2.
3. The careful manufacture of the metal powders.
4. sintered
5. Coining and sizing
6. Evaluate individually.

Chapter Quiz

1. metal powders
2. They are self-lubricating.
3. sintered
4. green compact
5. Evaluate individually. Refer to Section 34.2.
6. Shrinkage and distortion caused by the heating operation often require the pieces to go through a sizing or coining operation to restore or improve the finished dimensions of the piece.

Powder Metallurgy Sequence

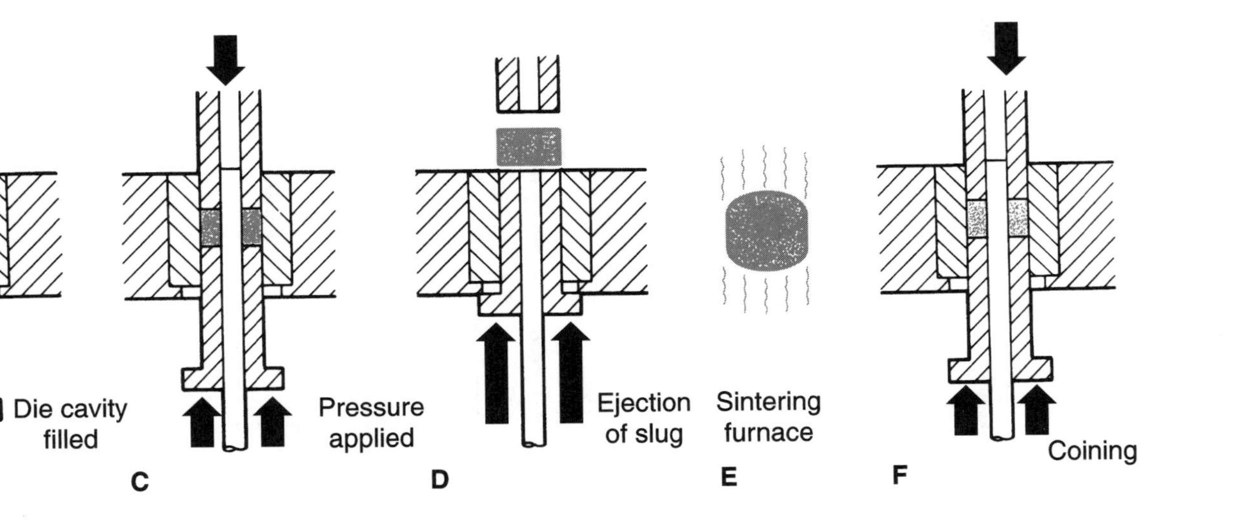

34-1

Chapter 34 Quiz

Powder Metallurgy

Name: ______________________ **Date:**____________ **Period:**____________

1. The technique of fabricating parts from _____ is known as powder metallurgy. 1.____________

2. What is unique about bearings made by the powder metallurgy process?

____

____

3. The part formed from metal powder, as ejected from the die, is quite brittle and fragile. To transform this briquette into a strong and usable unit, it must be _____. 3.____________

4. A briquette is also known as a(n) _____. 4.____________

5. List four uses of products made by the powder metallurgy technique.

____

____

____

____

6. Why are powder metallurgy parts subject to sizing, coining, or forging operations?

____

____

____

____

____

____

35 Nontraditional Machining Techniques

Learning Objectives

After studying this chapter, students should be able to:

- Describe several nontraditional machining techniques.
- Explain how nontraditional techniques differ from conventional machining processes.
- Relate the need for nontraditional machining techniques.
- List the advantages and disadvantages of several nontraditional techniques.

Chapter Resources

Text, pages 545–562
Test Your Knowledge, pages 561–562
Research and Development, page 562
Workbook, pages 155–156
Instructor's Manual
Answer keys for:
Test Your Knowledge Questions
Workbook
Chapter Quiz
Reproducible Masters:
35-1 Waterjet Cutting
35-2 Electrical Discharge Wire Cutting (WireCutEDM)
35-3 Electro Chemical Machining (ECM)
35-4 Electron Beam Machining (EBM)
35-5 Laser Beam Machining (LBM)
35-6 Ultrasonic Machining
35-7 Explosive Forming
35-8 Magnetic Forming
35-9 Impact Machining
35-10 Chapter Quiz

Guide for Lesson Planning

Some of the machining techniques described in this chapter are only in limited industrial use. Securing examples of items produced by using them may prove difficult for some metalworking shops/labs. However, local industries or parents may be a source of information and samples of these and more conventional machining techniques. Explore local sources personally or through your students.

In areas with limited industrial resources, additional information on nontraditional machining techniques can be found in trade publications, scientific publications, periodicals, and on the Internet. Encourage students to research the various techniques and report on them.

The chapter may be presented in several segments with students adding their research material during the discussion.

Part I—Electrical Discharge Machining (EDM)

Have students read and study Section 35.1. Review the segment and discuss:

- How Electrical Discharge Machining works.
- Benefits of EDM.
- EDM applications.

Part II—Electrical Discharge Wire Cutting (WireCutEDM)

Have students read and study the assignment. Review and discuss the following:

- How WireCutEDM operates.
- Why it was developed.

Advanced metalworking students may choose to construct demonstration models of electrical discharge machines. The machines must be constructed under close supervision and the completed job inspected by a licensed electrician and electronic technician before being operated.

Part III—Electrochemical Machining (ECM)

Have students read and study Section 35.3. Review and discuss the following:

- Why is ECM referred to as electroplating in reverse.
- The principle behind its operation
- Some of ECM's advantages.

Part IV—Chemical Machining

Have students read and study Section 35.4. Review the material with them and discuss the following:

- Two classifications of chemical machining and how they differ.
- Principle steps in chem-milling.
- Principle steps in chem-blanking.

Part V—Hydrodynamic Machining

Have students read and study Section 35.5. Review the material and discuss:

- The waterjet cutting process.
- Why the process was developed. (Some material caused conventional cutting tools to dull rapidly. The heat generated during conventional machining often damaged the material. The fine, sharp particles that were produced often penetrated the skin or were breathed in, causing medical problems.)
- Waterjet cutting can be held to close tolerances, reduce waste, and produce fewer particulate (fragments) than other processes.
- The technique had problems until a nozzle material was developed that did not erode quickly from the high pressure fluid passing through it.

Part VI—Electron Beam and Laser-Beam Machining

Have students read and study Sections 35.6 and 35.7. Review and discuss the following:

- How electron beam cutter/welder operates.
- Work the process is capable of performing.
- How laser machining operates.
- Safety precautions that must be observed when EBM and laser machining is done.

Part VII—Ultrasonic and Impact Machining

Have students read and study Section 35.8. Review and discuss the following:

- How ultrasonic techniques are used to machine and weld.

Part VIII—High-Energy-Rate Forming (HERF)

Have students read and study Section 35.9. Review and discuss HERF operations.

Caution: Do *not* attempt to demonstrate explosive forming in the school shop/lab.

Part IX—Magnetic Forming

Have students read and study Section 35.9.2. Review and discuss magnetic forming operations.

Assignments

1. Assign the Test Your Knowledge questions.
2. Assign Chapter 35 in the *Modern Metalworking Workbook.*
3. Assign the chapter quiz. Copy and distribute Reproducible Master 35-10.
4. Permit students to volunteer for one or more of the Research and Development activities at the end of the chapter.

Test Your Knowledge

1. Evaluate individually. Refer to Section 35.1.
2. It uses a small diameter wire electrode instead of a mirror-image electrode.
3. The process produces a part that is the reverse shape of the electrode.
4. carefully controlled
5. (a) not to be etched
6. Evaluate individually. Refer to text Section 35.4.3.
7. Evaluate individually. Refer to Section 35.5.
8. Evaluate individually. Refer to Section 35.7.
9. Evaluate individually. Refer to text Section 35.8.2.
10. High-energy-rate forming
11. Evaluate individually.
12. If the explosive is too close, the pressure pulse could perforate the metal. If the explosive is too far away, the preform would not be forced against the die.
13. (f) Infrasonic
14. (d) Laser beam machining

15. (e) Ultrasonic
16. (c) EBM
17. (a) EDM
18. (b) ECM
19. (g) Impact machining
20. (h) HERF

Workbook

1. (d) All of the above.
2. electrically-conductive
3. Electrical discharge wire cutting (WireCutEDM)
4. Electrochemical machining (ECM)
5. Chemical machining
6. In chemical milling, selected areas of the metal are etched away to produce an accurately contoured surface. In chemical blanking, metal is totally removed from selected areas by chemical action.
7. (c) too much metal may be removed
8. Chemical blanking
9. Evaluate individually. Refer to text Section 35.4.3.
10. A relatively new cutting process developed to cut composites using a high-powered waterjet.
11. Electronic beam (EBM)
12. Laser-beam
13. Ultrasonic machining
14. Impact (slurry)
15. When the metal tends to try to regain its original shape.
16. High-energy-rate forming (HERF)
17. Explosive forming
18. The time needed to get the part into production and produce the first piece.
19. magnetic
20. Evaluate student demonstrations individually.

Chapter Quiz

1. (f) Waterjet cutting
2. (g) Chemical blanking
3. (j) Chemical machining
4. (h) Electrical discharge machining
5. (a) Electrochemical machining
6. (i) Electron beam machining
7. (b) Laser beam machining
8. (k) Impact machining
9. (d) High-energy-rate-forming
10. (e) Electromagnetic forming

Waterjet Cutting

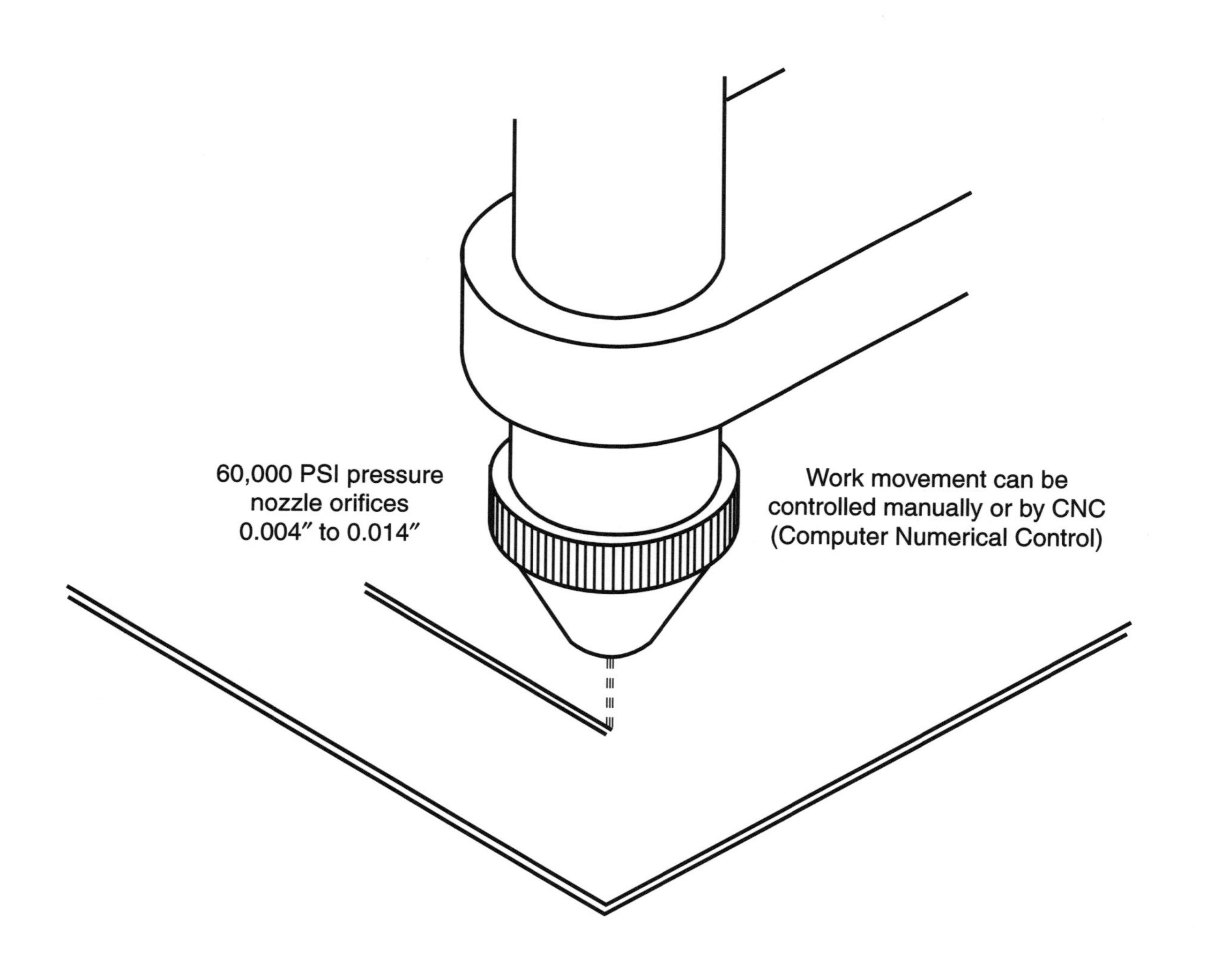

35-1

Electrical Discharge Wire Cutting (WireCutEDM)

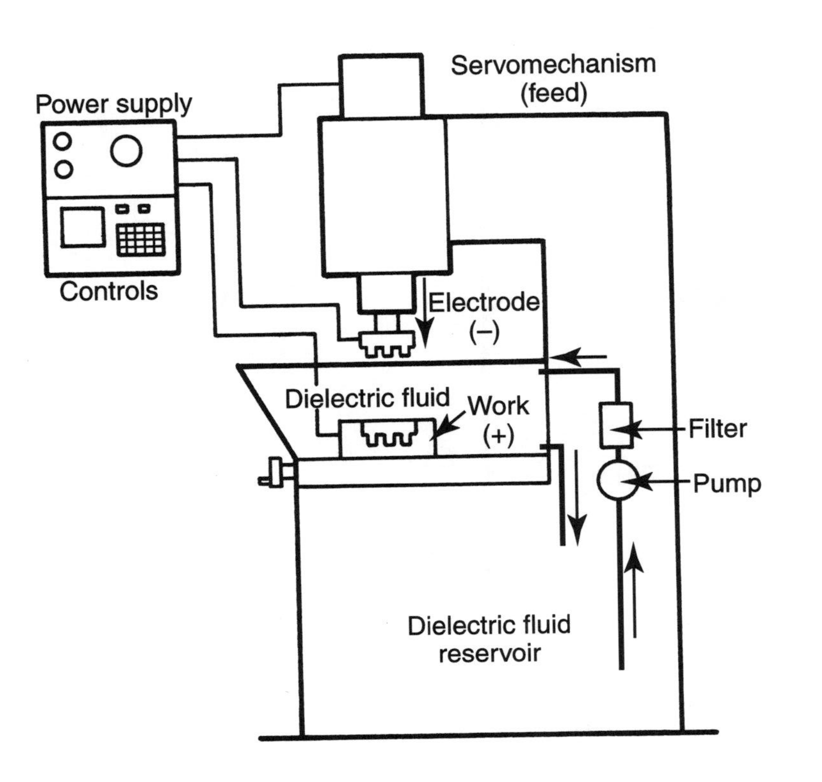

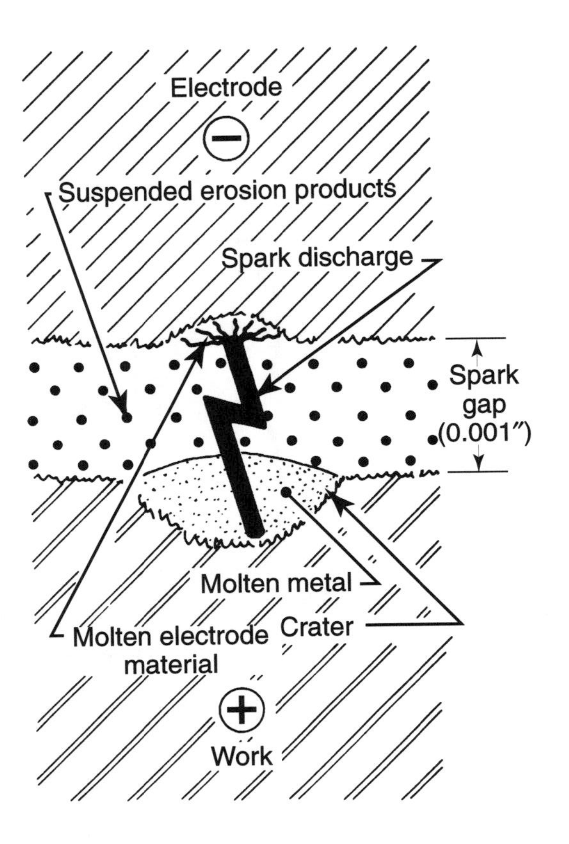

35-2

Electro Chemical Machining (ECM)

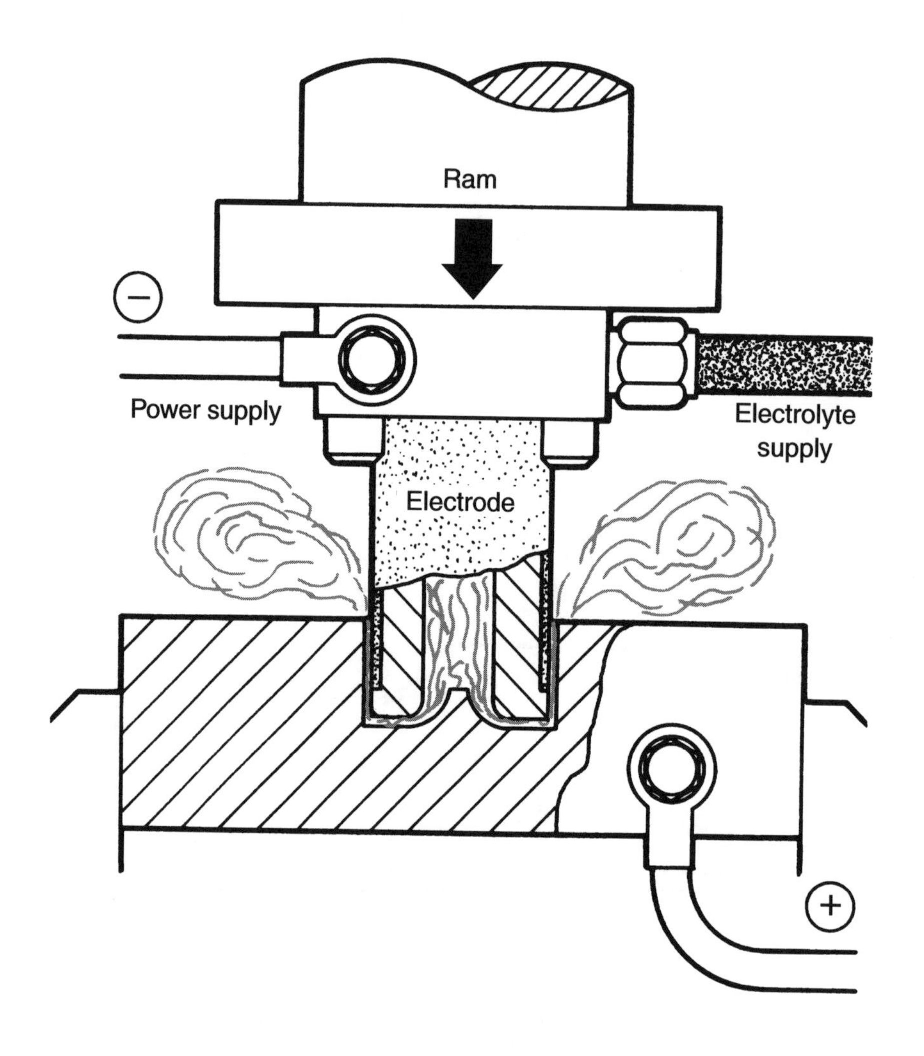

Electron Beam Machining (EBM)

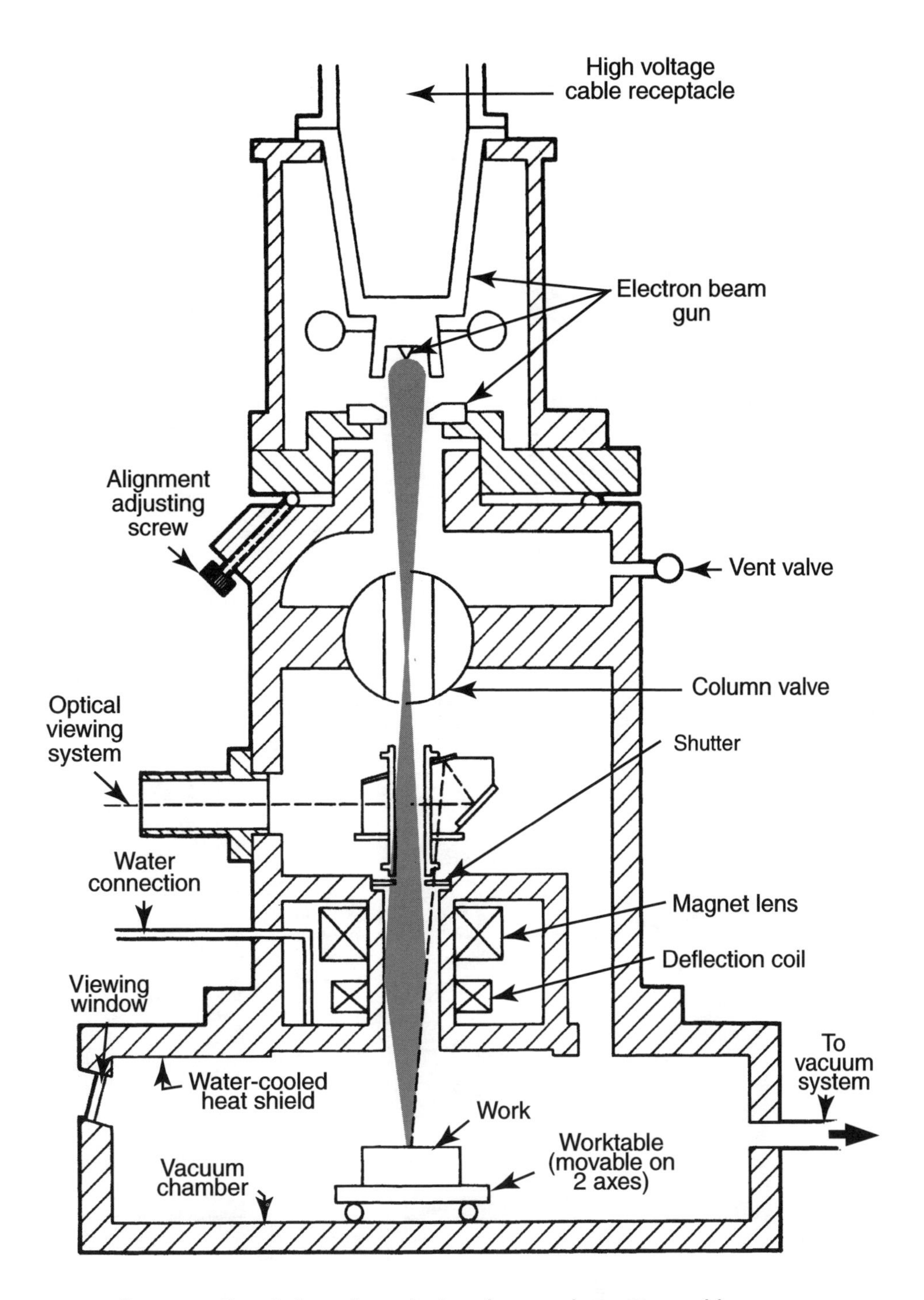

Cross-sectional view of an electron beam microcutter-welder.

Laser Beam Machining (LBM)

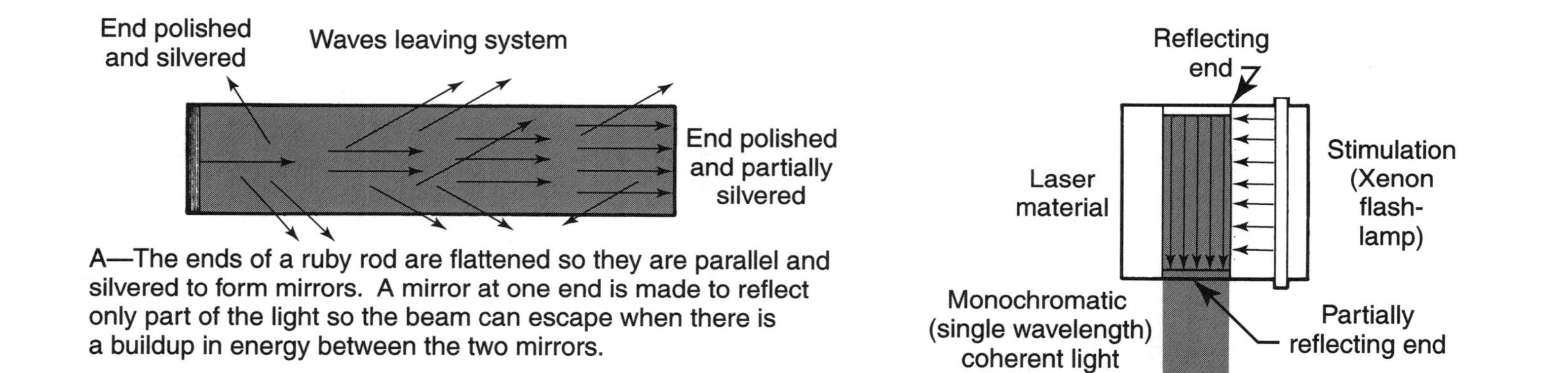

A—The ends of a ruby rod are flattened so they are parallel and silvered to form mirrors. A mirror at one end is made to reflect only part of the light so the beam can escape when there is a buildup in energy between the two mirrors.

B—Soon after chromium atoms in the ruby crystal are pumped by a flashlamp to a higher energy level, they drop to another level, and stimulated emission takes place. Waves moving at angles to the crystal's axis leave the system, but those traveling along the axis grow by stimulated emission of photons.

A flashlamp capable of producing an intense light is employed to "pump" a laser into a high level of excitement.

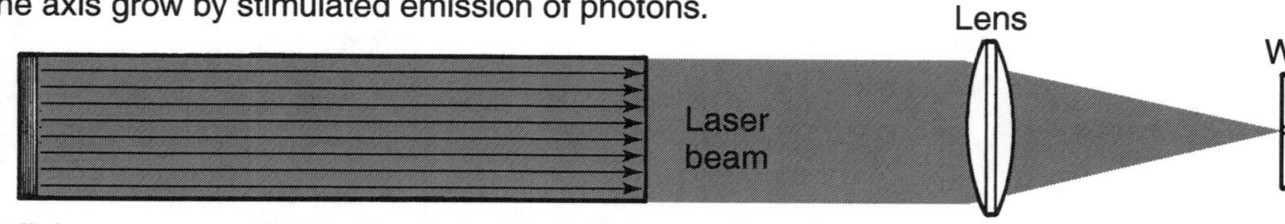

C—Parallel waves are reflected back and forth between the mirrors and the wave system grows in intensity. A pale red glow indicates a certain amount of light being lost at the mirror, but beyond a critical point, the waves intensify enough to overcome this loss. An intense red beam flashes out of the partially silvered end of the crystal.

Ultrasonic Machining

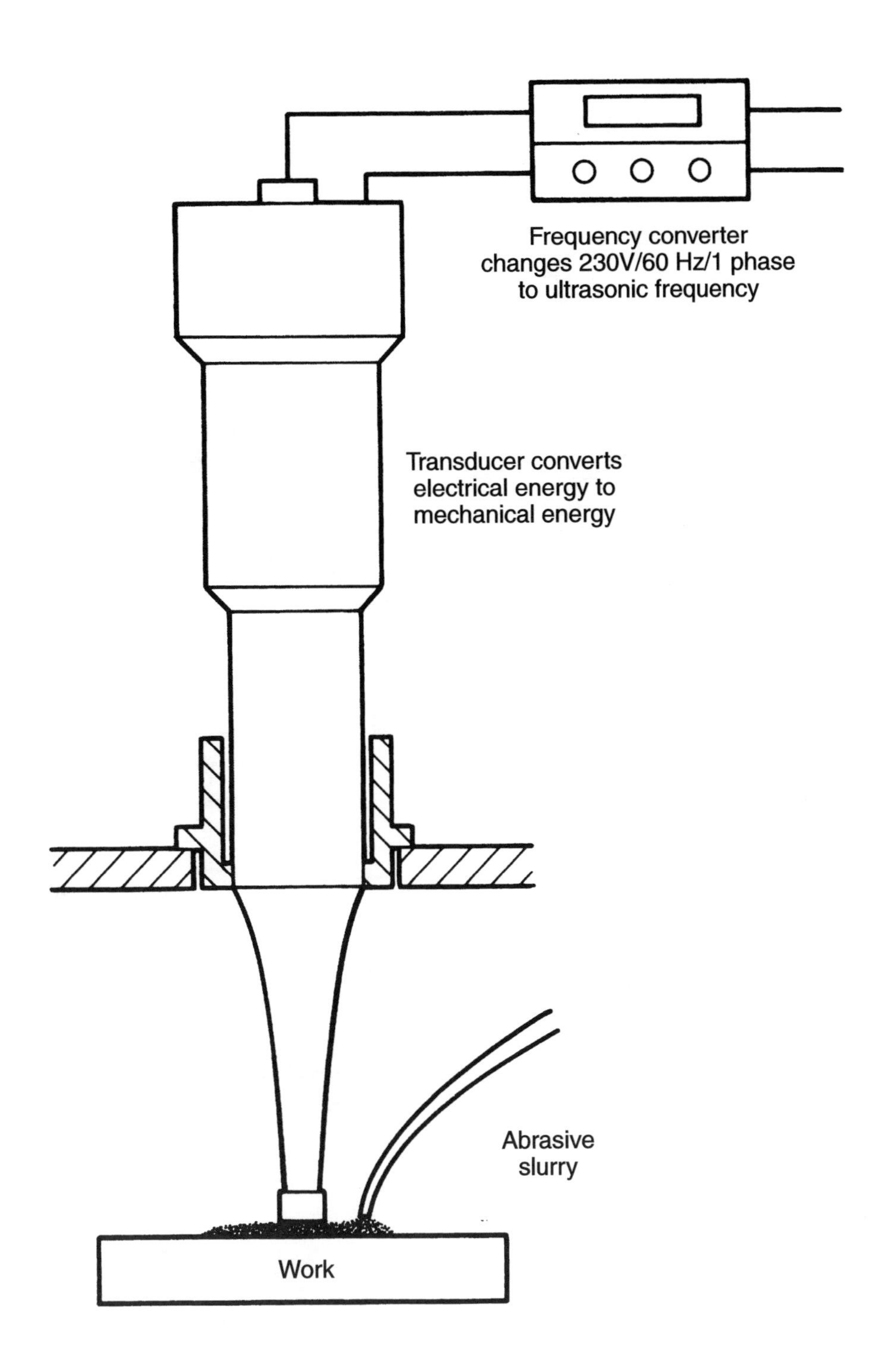

Explosive Forming

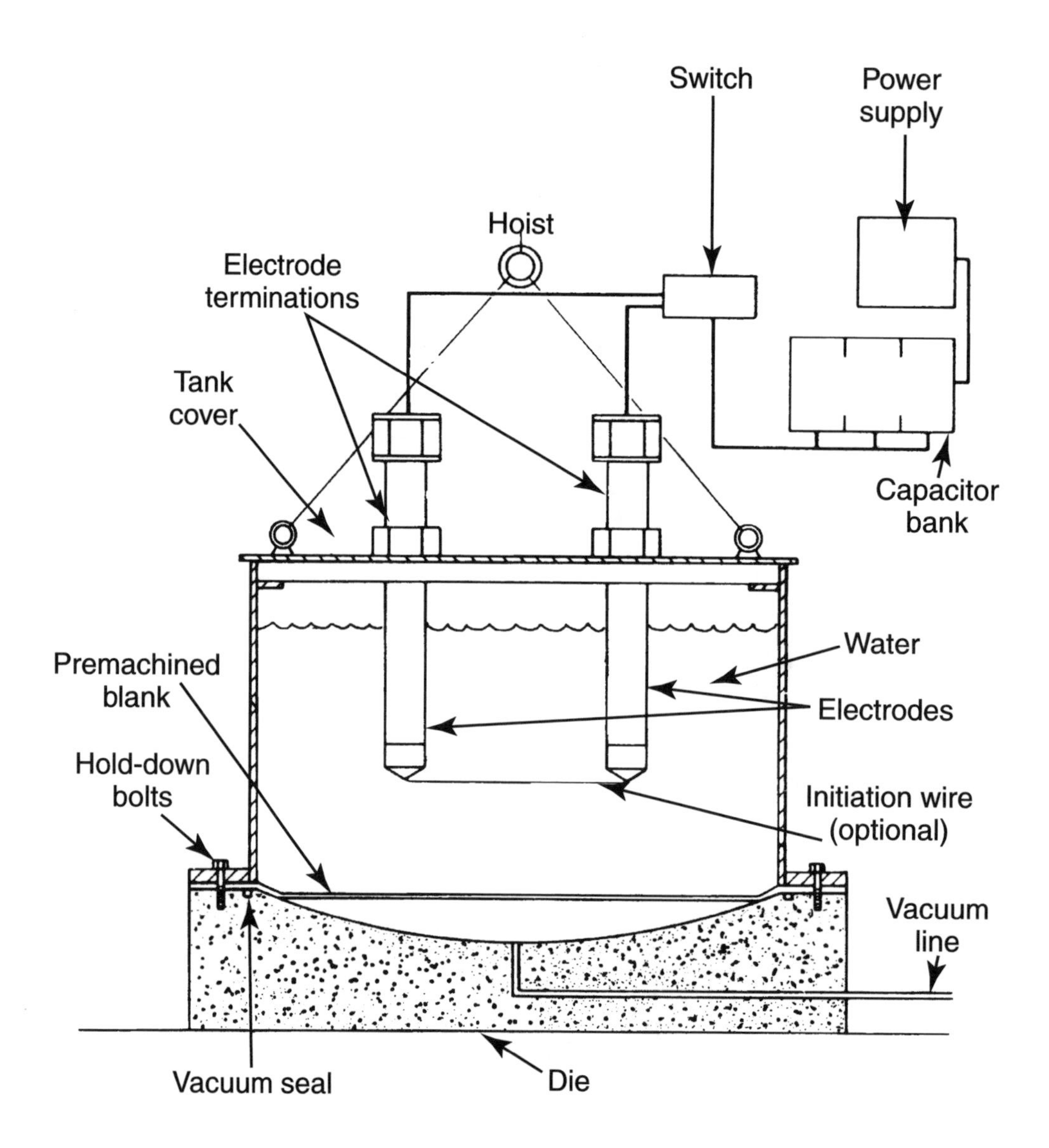

Magnetic Forming

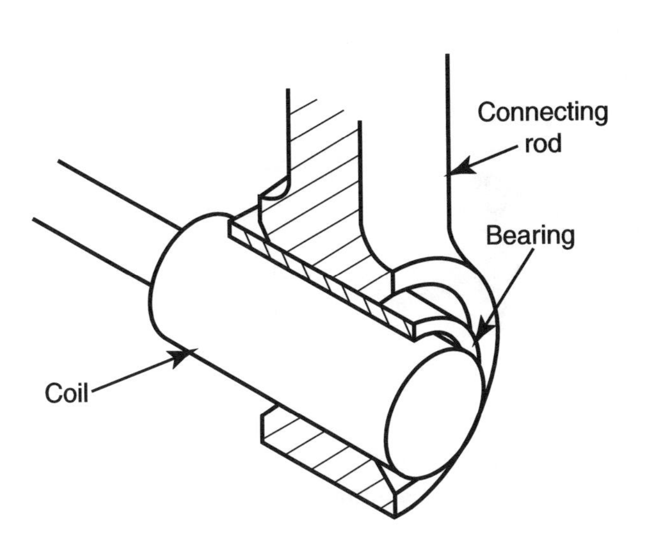

Coil and Bearing Inserted

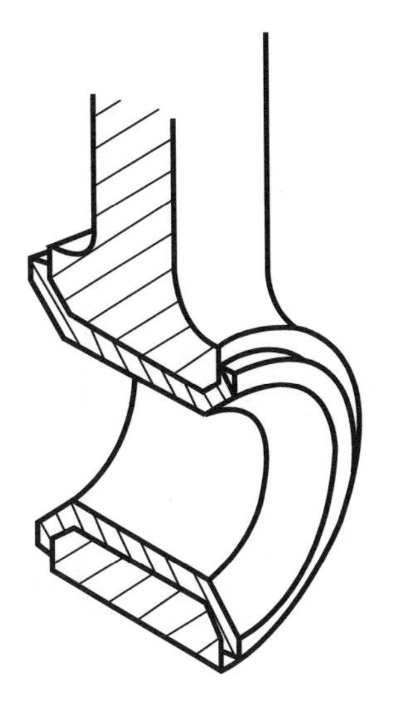

Bearing Expanded into Place

Impact Machining

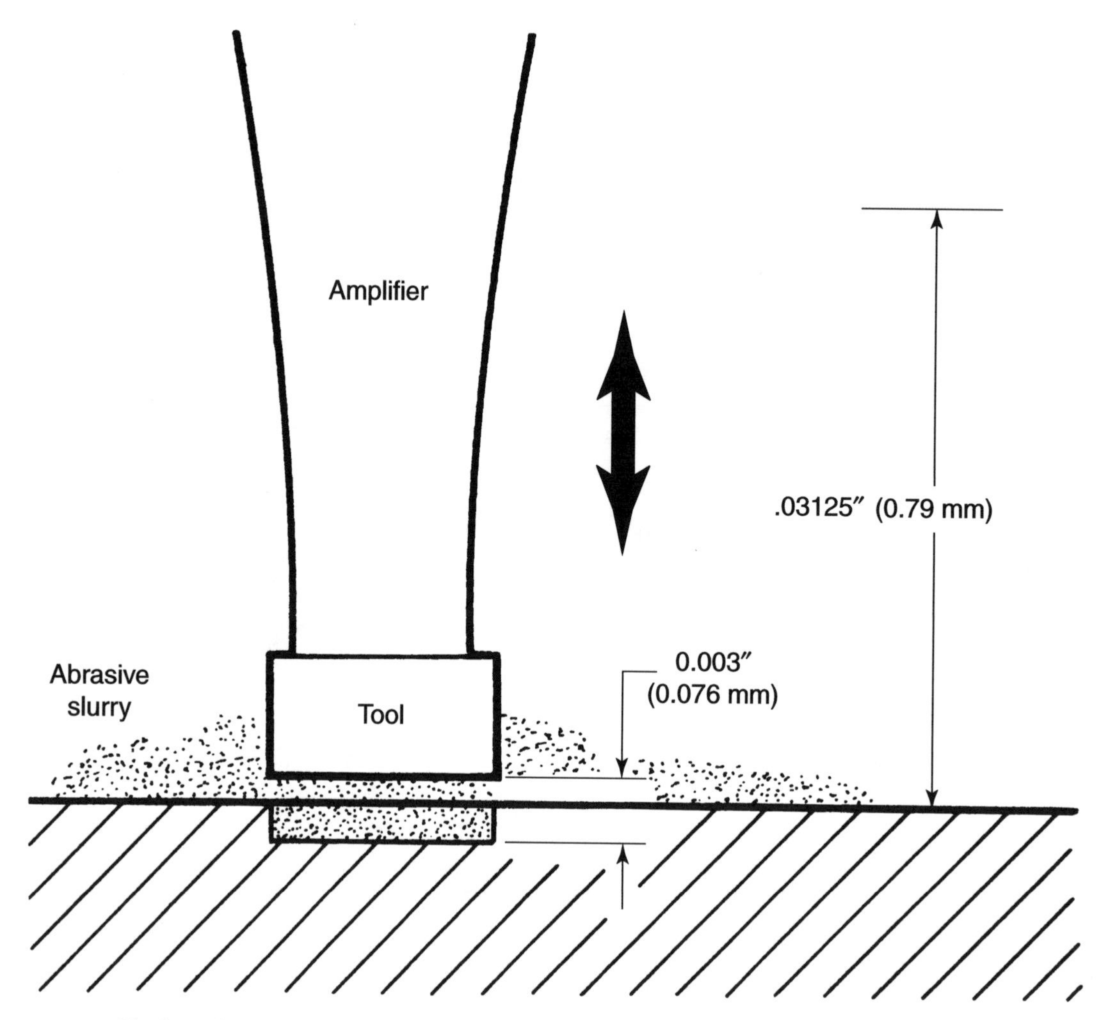

Tool motion in ultrasonic (impact) machining is slight, only 0.003″ (0.076 mm) the .03125″ (0.79 mm) measurement is used to indicate scale.

Chapter 35 Quiz

Nontraditional Machining Techniques

Name: ______________________ **Date:**______________ **Period:**______________

Match the sentence with the appropriate word or phrase.

(a) Electrochemical machining
(b) Laser beam machining
(c) Ultrasonic machining
(d) High-energy-rate-forming
(e) Electromagnetic forming
(f) Waterjet cutting
(g) Chemical blanking
(h) Electrical discharge machining
(i) Electron beam machining
(j) Chemical machining
(k) Impact machining

_________ 1. Uses a high-speed fluid jet to cut material.

_________ 2. Involves total removal of metal from certain areas by chemical action.

_________ 3. Technique that etches away selected areas of metal to produce an accurately contoured part.

_________ 4. Uses a controlled electric spark to erode metal away.

_________ 5. Removes metal by electrolysis. Might be classified as electroplating in reverse.

_________ 6. Must be done in a vacuum. Any known metal or nonmetal can be cut by this technique.

_________ 7. Cutting is done by a focused, narrow, and intense beam of light.

_________ 8. Based on the science of silent sound. A shaped cutting tool pounds a slurry of fine abrasive particles against the work.

_________ 9. Metal is shaped in microseconds with pressure generated by chemical explosives, electric sparks, or electromagnetic force.

_________10. An intense magnetic field is employed to compress or expand metal to produce the desired contours.

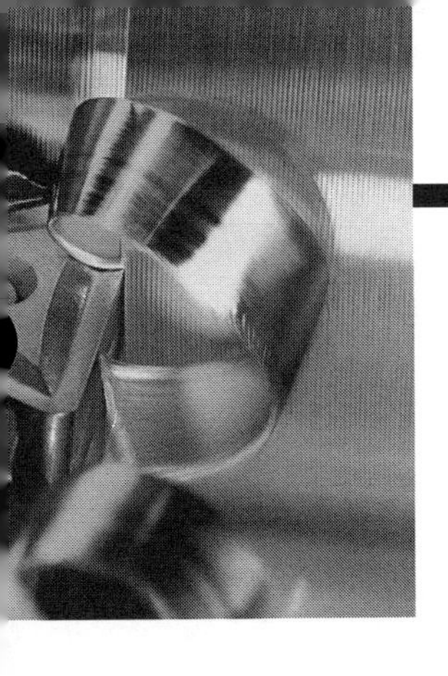

Quality Control

Learning Objectives

After studying this chapter, students should be able to:

- Understand the necessity for quality control.
- Explain the two classifications of quality control.
- List examples of each classification.
- Describe the application of several quality control techniques.

Chapter Resources

Text, pages 563–574
Test Your Knowledge, pages 572–573
Research and Development, page 573
Workbook, pages 155–156
Instructor's Manual
Answer keys for:
Test Your Knowledge Questions
Workbook
Chapter Quiz
Reproducible Master:
36-1 Chapter Quiz

Guide for Lesson Planning

Quality control is one of the most important segments of industry. It plays a vital role in improving and maintaining the competitive position of a manufacturer. Quality control is not used so imperfect parts or products will be detected and discarded. Its main emphasis is to prevent imperfect or defective parts from ever being manufactured.

Have students read and study all or part of the chapter. Review the assignment with them. They should be able to understand and explain:

- The two basic types of quality control.
- Why destructive testing must be used in the development of a new product.
- Types of nondestructive testing.
- Types of quality control measuring devices.
- Why different types of testing techniques are necessary.

Select, or have them volunteer, a team of students to develop a quality control system for a selected class project.

Assignments

1. Assign the Test Your Knowledge questions at the end of the chapter.
2. Assign Chapter 36 in the *Modern Metalworking Workbook.*
3. Assign the chapter quiz. Copy and distribute Reproducible Master 36-1.
4. Permit students to volunteer for one or more of the Research and Development activities at the end of the chapter.

Test Your Knowledge

1. Destructive and nondestructive. Evaluate explanations individually.
2. Any three of the following: measuring, radiographic inspection, magnetic particle inspection (magnafluxing), fluorescent penetrant inspection, Spotcheck, ultrasonic inspection, laser.
3. (d) All of the above.
4. (c) X-rays
5. Evaluate individually. Refer to text Section 36.2.3.
6. Evaluate individually. Refer to text Section 36.2.4.
7. Evaluate individually. Refer to text Section 36.2.5.
8. (b) high frequency sound waves
9. (d) All of the above.
10. profilometer, surface roughness gage

Workbook

1. To seek out and prevent potential product defects in the manufacturing process before they can cause injuries or damage and substandard products.
2. In order: destructive, nondestructive
3. Evaluate individually. Refer to Section 36.1.
4. Evaluate individually. Refer to Section 36.2.
5. A gaging system for the inspection and precise measurement of small parts and sections of larger parts. An enlarged image of the part being inspected is projected on a screen where it is superimposed upon an accurate drawing overlay of the part.
6. Involves passing gamma rays (X-rays) through a part and onto sensitive film to detect flaws in the metal. The developed film has an image of the internal structure of the part or assembly.
7. Evaluate individually. Refer to text Section 36.2.3.
8. A penetrant solution is applied to the part's surface by dipping, brushing, or spraying. Capillary action pulls the solution into the defect. The surface is rinsed clean, and a developer applied. When the part is inspected under ultraviolet light, the defects will glow with fluorescent brilliance.
9. Uses a red liquid dye which soaks into the surface cracks and flaws of a part. The liquid is washed off and the part dried. A developer is dusted or sprayed on the part. Flaws and cracks show up red against the white background of the developer.
10. Uses ultrasonic sound waves to detect cracks and flaws in almost any material that can conduct sound. Sound waves can also be used to measure the thickness from one side of the material.

Chapter Quiz

1. (a) seek out potential product defects
2. Evaluate student responses individually.
3. (g) Radiographic inspection
4. (a) Ultrasonic inspection
5. (h) Fluorescent penetrant inspection
6. (c) Spotcheck
7. (d) Magnaflux
8. (b) Profilometer
9. (e) Optical comparator
10. (f) Destructive testing

Chapter 36 Quiz

Quality Control

Name: ______________________ **Date:**____________ **Period:**____________

1. The primary purpose of quality control is to _____. 1.______________
   (a) seek out potential product defects
   (b) be used when necessary to improve product design
   (c) check how employees are doing their job
   (d) All of the above.
   (e) None of the above.

2. Quality control is divided into destructive and nondestructive testing. Give an example of each.
   (a) Destructive testing: ______________________
   (b) Nondestructive testing: ______________________

Match the sentences with the appropriate word or phrase.

_______ 3. Quality control technique that makes use of X-rays and gamma radiation projected through the object being inspected.

_______ 4. Uses sound waves above the audible range to detect cracks and flaws in material.

_______ 5. Based on capillary action that pulls a dye solution into the surface defect. After being rinsed clean and dried, the part is inspected under black light. The defect area will glow marking its location.

_______ 6. Surface of part is coated with a red dye which soaks into surface flaws and defects. The dye is washed and a developer is sprayed on. Defects show up red against the developer's white background.

_______ 7. Can only be used on metal parts that can be magnetized.

_______ 8. Used to inspect the quality of machined surfaces.

_______ 9. Enlarged image of part is projected on a screen where it is superimposed upon an accurate drawing overlay of the part.

_______10. Controlled crashing of automobiles.

(a) Ultrasonic inspection
(b) Profilometer
(c) Spotcheck
(d) Magnaflux
(e) Optical comparator
(f) Destructive testing
(g) Radiographic inspection
(h) Fluorescent penetrant inspection

Numerical Control and Automation

Learning Objectives

After studying this chapter, students should be able to:

- Explain different methods of controlling machine tools.
- Describe numerical and computer numerical control and how they are used.
- Define automation and its application in manufacturing.
- Discuss the use of robotics in manufacturing.

Chapter Resources

Text, pages 575–592
  Test Your Knowledge, page 591
  Research and Development, page 591
Workbook, pages 157–164
Instructor's Manual
  Answer keys for:
    Test Your Knowledge Questions
    Workbook
    Chapter Quiz
  Reproducible Masters:
    37-1 The Cartesian Coordinate System
    37-2 Tool Positioning—Incremental and Absolute
    37-3 Axes of Machine Movements
    37-4 Contour or Continuous Path Machining
    37-5 Flexible Machining Cell
    37-6 Robot Configuration
    37-7 Chapter Quiz

Guide for Lesson Planning

A week or two before introducing this chapter, request students to research material (review magazines, newspapers, talking with parents or guardians, relatives, or friends who work in industry) on NC machining, automation and mass production. During chapter presentation, invite a speaker from a local industry that uses automated equipment to speak to the class on their company's operation.

Class Discussion and Demonstration

Have students read and study all or part of the chapter. Use the appropriate Reproducible Masters and review the assignment and, using their research material, discuss the following:

- Definition of numerical control (NC).
- The Cartesian Coordinate System.
- NC tool positioning.
- Basic NC tool movement systems including *point-to-point, straight cut,* and *contour* or *continuous path.*
- The importance of computers to control machines.
- The student's definition of mass production and automation.
- Industry definition of mass production and automation.
- The history of mass production. In the United States Eli Whitney in 1798 devised machinery and manufacturing techniques to mass-produce muskets for the United States government. This was the first time parts were made to be interchangeable. What was the importance of this development?
- Development of automation. In 1784, an automated flour mill was constructed in the United States. The Jacquard loom was designed and put into operation in 1801. It used punched paper cards to control the design on cloth. Looms using this technique are in use today.
- How automation is used by industry.
- Robots and their uses.

If time permits, have students design and

mass-produce a project. However. before going into full production, manufacture one or two samples to determine whether the product will perform as designed.

Assignments

1. Assign the Test Your Knowledge questions at the end of the chapter.
2. Assign Chapter 37 in the *Modern Metalworking Workbook.*
3. Assign the chapter quiz. Copy and distribute Reproducible Master 37-7.
4. Permit students to volunteer for one or more of the Research and Development activities at the end of the chapter.

Test Your Knowledge

1. Any order: manual, automatic fixed-cycle, numerical control. Evaluate explanations individually.
2. A method of control that uses coded alphanumeric instructions to automatically direct the operation of a machine tool.
3. Distributed numerical control is a system in which CNC machines are connected to a large mainframe computer to form a two-way communication network. A distributed system allows program changes to be made at both the main computer and at individual machines.
4. A direct system uses a main computer to control one or more machines. Programming changes can only be made at the main computer.
5. open-loop
6. closed-loop
7. Evaluate individually. See Figure 37-6.
8. Evaluate individually. See Figures 37-9 and 37-10.
9. Any order: point-to-point, straight-cut, contour (continuous path). Evaluate descriptions individually.
10. Evaluate individually. See Section 37.4.2
11. A manufacturing procedure that has evolved over many years.
12. Making, inspecting, assembling, testing, packaging.
13. A robot is a programmable, multifunctional manipulator designed to move material, parts, tools, or specialized devices through variable programmed motions for the performance of a variety of tasks.
14. Any order: controller, power supply, manipulator or articulated arm, end-of-arm tooling.
15. Evaluate individually. Refer to Section 37.8.1

Workbook

1. Evaluate individually.
2. Alphanumeric
3. The basis of numerical control programming. It is a system that defines the direction and distance the work and/or tool must move. The position of any point can be defined with reference to a set of axes (X, Y, and Z) that are at right angles to each other.
4. reference to the last tool position
5. a fixed point of origin (zero point)
6. (a) Evaluate individually.
   (b) Evaluate individually.
7. (c) normally used for simple machining operations (drilling, spot welding, punching)
8. (d) All of the above.
9. When machining complex or two- and three-dimensional parts.
10. The drawing is computer-generated and can be used to generate the tool path and related machining commands that will enable the CNC machine to produce the part.
11. (e) None of the above.
12. Flexible manufacturing system (FMS)
13. readily reconfigured to produce a variety of shapes and sizes within a given part family
14. eliminates the need for large inventories of materials and parts. They are scheduled for arrival at the time needed and not before.
15. Any order: a controller, a power supply, a manipulator or articulated arm, end-of-arm tooling.

Chapter Quiz

1. (j) Absolute system
2. (b) Part program
3. (a) Automation
4. (i) Machining program
5. (k) End of block (EOB)
6. (l) Point-to-point system
7. (g) CAD/CAM
8. (h) Robots
9. (d) Continuous path system
10. (m) Flexible manufacturing system (FMS)

The Cartesian Coordinate System

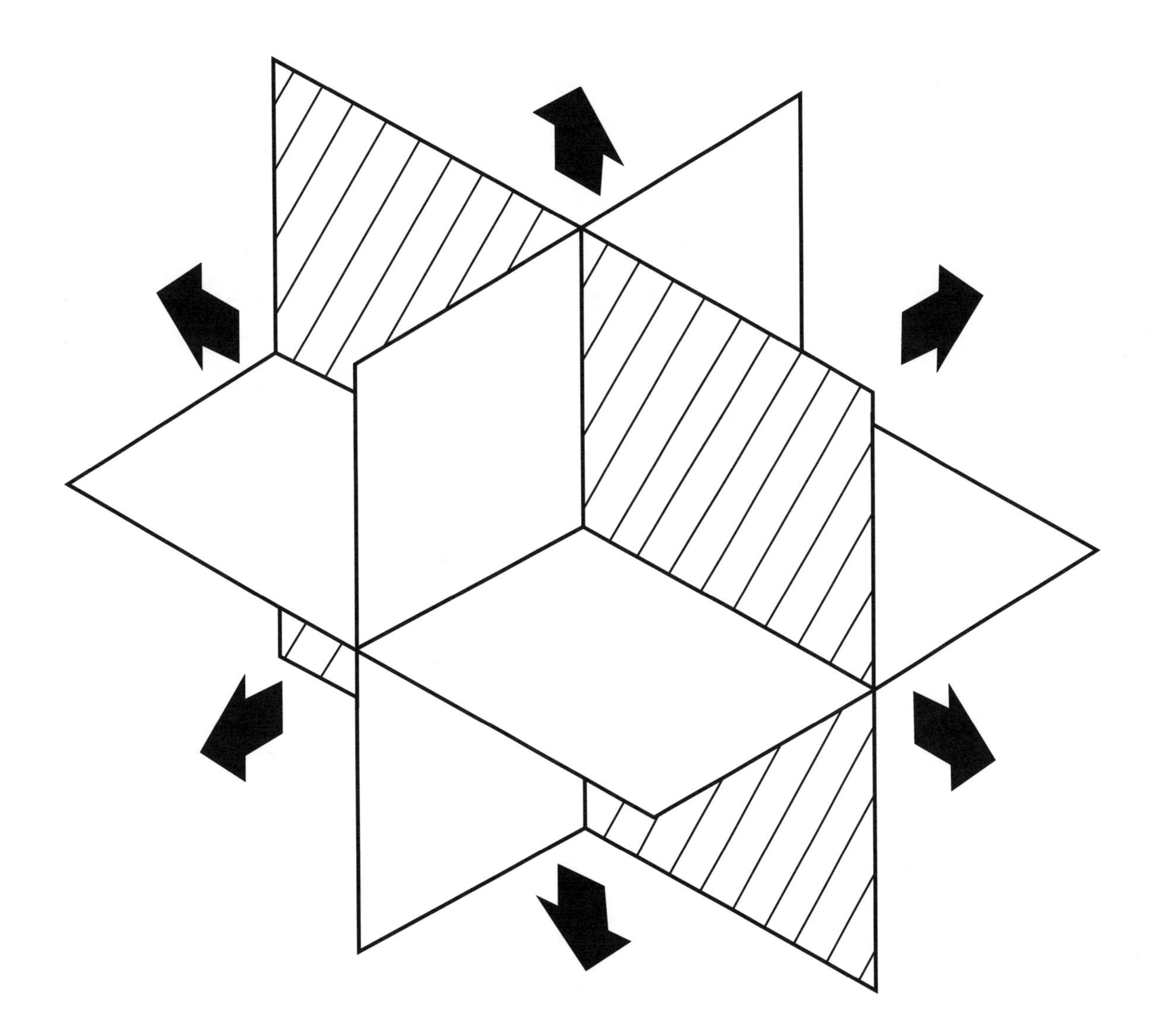

Tool Positioning

37-2

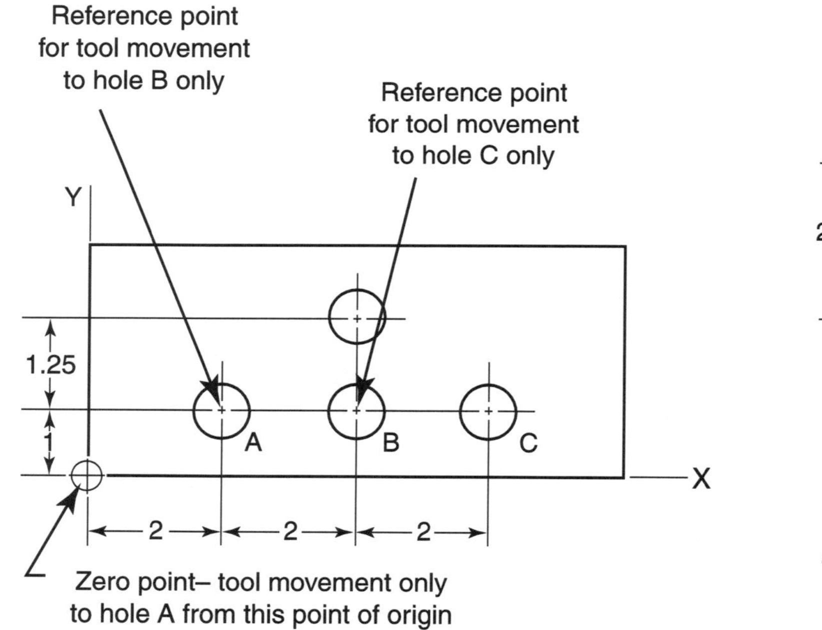

Incremental Positioning
Each set of coordinates has its point of origin from last point established.

Y

X

2.25

1

2

4

6

All tool movement taken from this point

Absolute Positioning
In this system, all coordinates are measured from fixed point (zero point) of origin.

Axes of Machine Movements

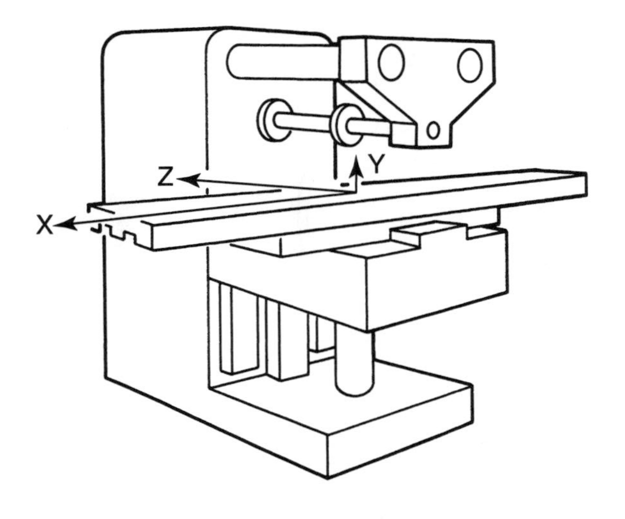

Horizontal Milling Machine

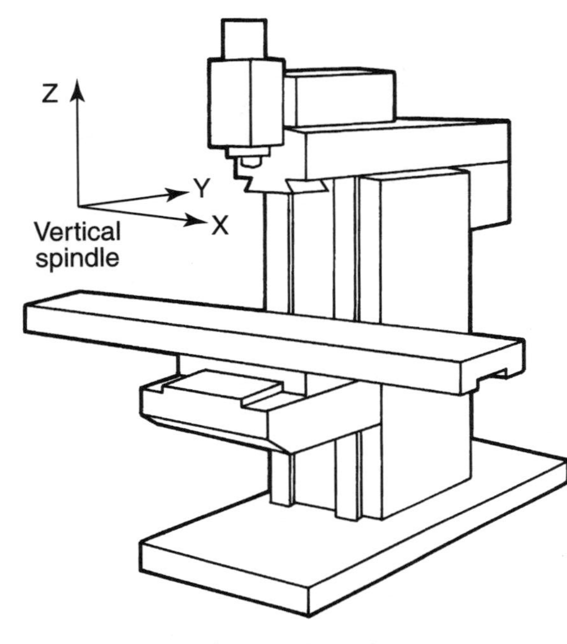

Vertical Milling Machine

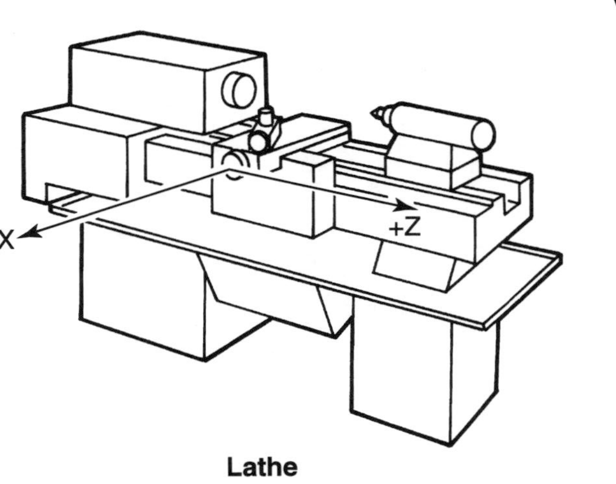

Lathe

Note: Spindle motion is assigned Z axis

37-3

Contour or Continuous Path Machining

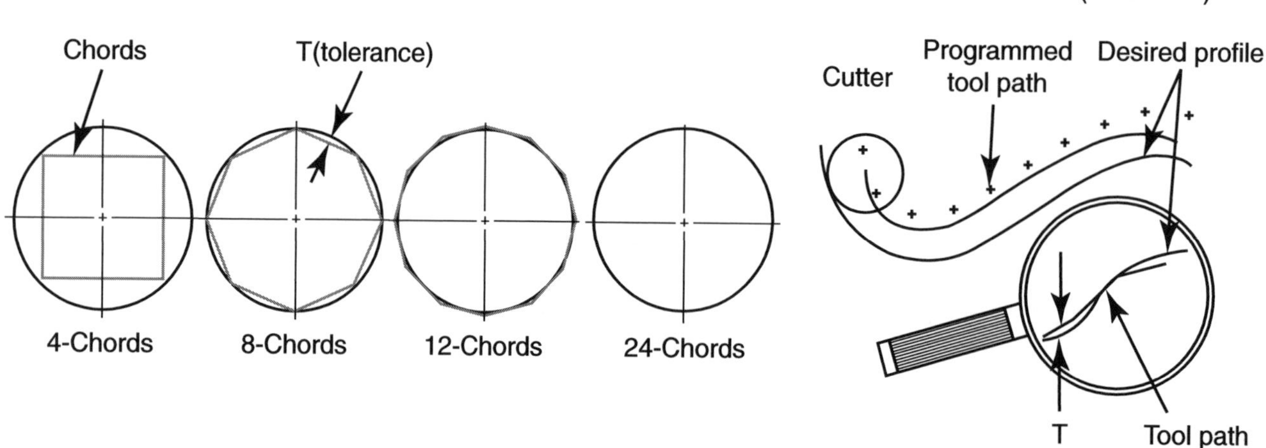

Contours obtained from contour or continuous path machining are result of a series of straight-line movements. The degree to which a contour corresponds with specified curve epends upon how many movements or chords are used. Note how, as number of chords increase, the closer the contour is to a perfect circle. The actual number of lines or points needed is determined by the tolerance allowed between design of the curved surface and one actually machined.

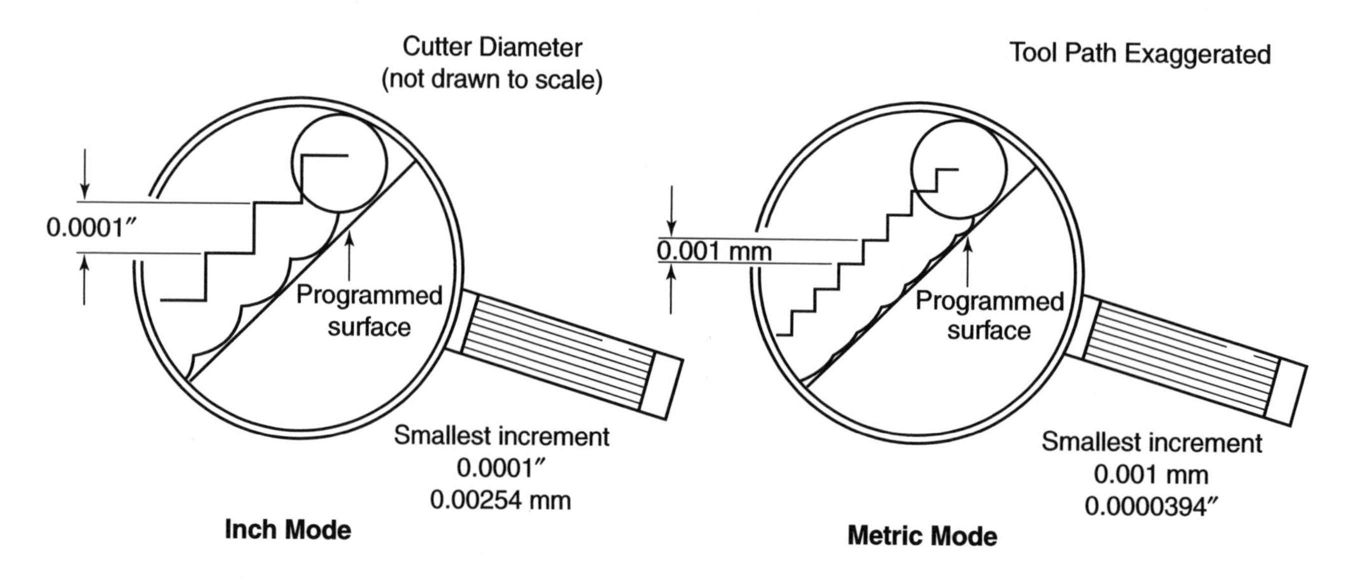

This exaggerated illustration shows why metric machine movement increments are often preferred when contour machining. The benefit has to do with the least input increment allowed in the metric mode. In the inch mode, the least input increment is 0.0001″, which means you can input program coordinates and tool offsets down to 0.0001″. In the metric mode, the least input increment is 0.001 mm, which is less than one-half the least input increment when using the inch mode. The coordinates going into the program will then be much closer to what is desired for accurately machined parts.

Flexible Machining Cell

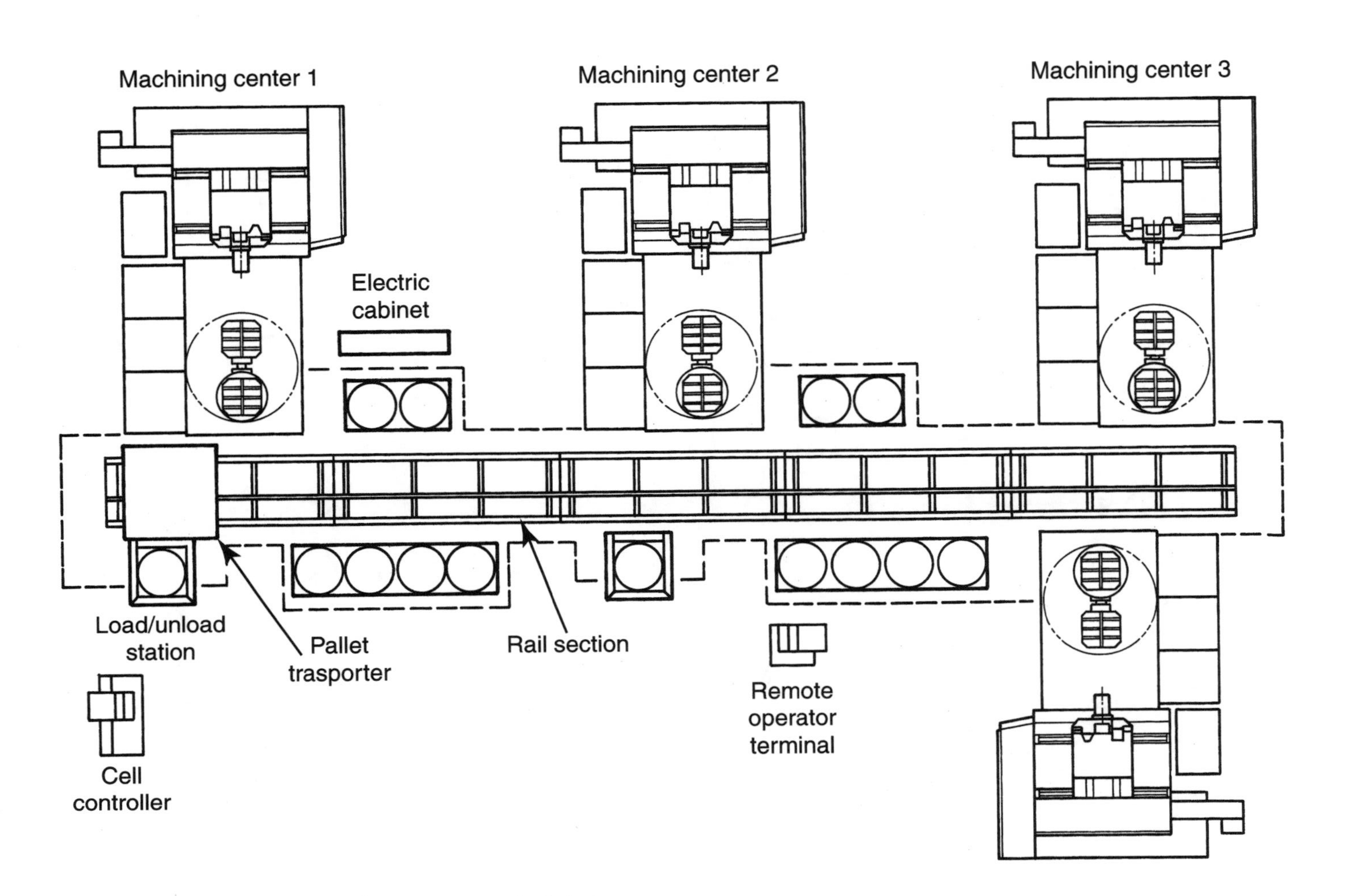

Flexible manufacturing cell that uses a pallet transporter to link the machines. A cell controller automatically queues work for immediate delivery to the next machine available.

37-5

Robot Configuration

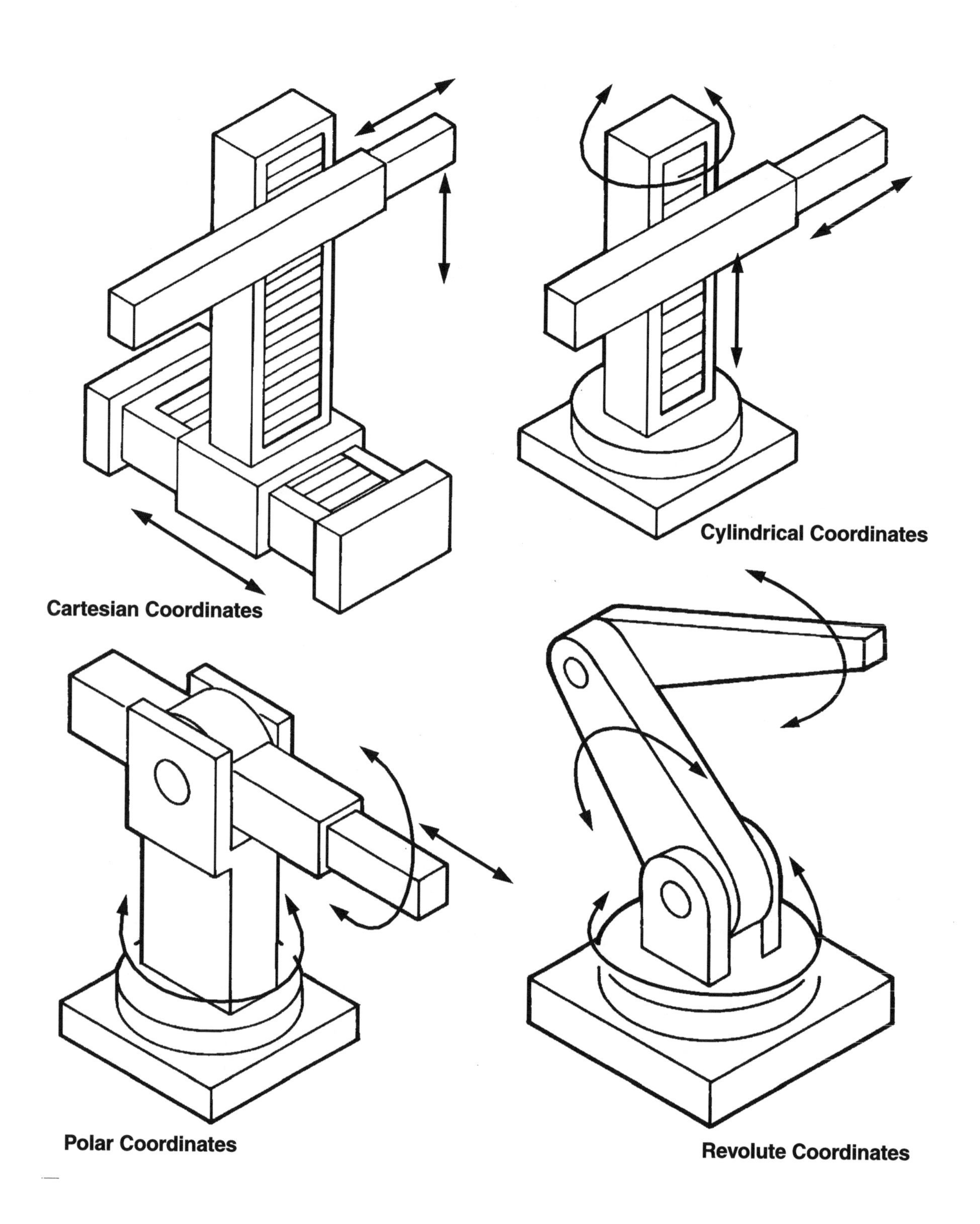

Chapter 37 Quiz

Numerical Control and Automation

Name: ______________________ **Date:**____________ **Period:**____________

Match the sentences with the appropriate word or phrase.

(a) Automation
(b) Part program
(c) Alphanumeric data
(d) Continuous path system
(e) Numerical control
(f) Machine control unit
(g) CAD/CAM
(h) Robots
(i) Machining program
(j) Absolute system
(k) End of block (EOB)
(l) Point-to-point system
(m) Flexible manufacturing system (FMS)

_______ 1. System in which all coordinates are measured from a fixed point of origin.

_______ 2. Consists of letters, punctuation marks, numbers, and special characters. Each identifies a different machine function.

_______ 3. An industrial technique whereby mechanical labor and mechanical control are substituted for human labor and human control.

_______ 4. A sequence of instructions that "tells" a machine what operations to perform, and where on the material they are to be done.

_______ 5. Code that indicates the completion of a program.

_______ 6. System where there is no concern what path is taken when the tool moves from Point A to Point B.

_______ 7. System of design and manufacturing that is computer based.

_______ 8. Many jobs they perform are considered hazardous for human operators.

_______ 9. Movement of the cutting tool is under continuous control of a computer.

_______ 10. System brings together workstations, automated handling and transfer systems, and computer control in an integrated manner.

Product Suggestions

Project 1

Patio Candle/Glass Holder .336

Project 2

Quick Projects .337

Project 3

Cast Nameplate .338

Project 4

Two-Piece Vise .339

Project 5

Photo Frame .341

Project 6

Two-Stroke Cycle Internal Combustion Engine .342

Project 7

Four-Stroke Cycle Internal Combustion Engine 347

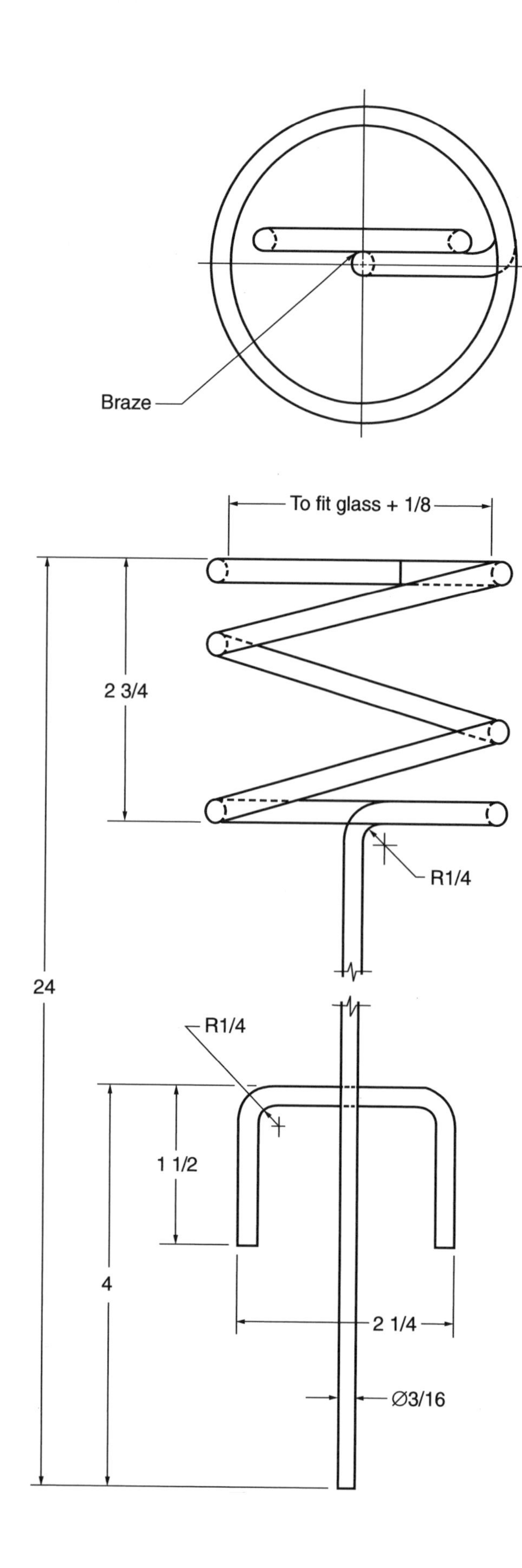

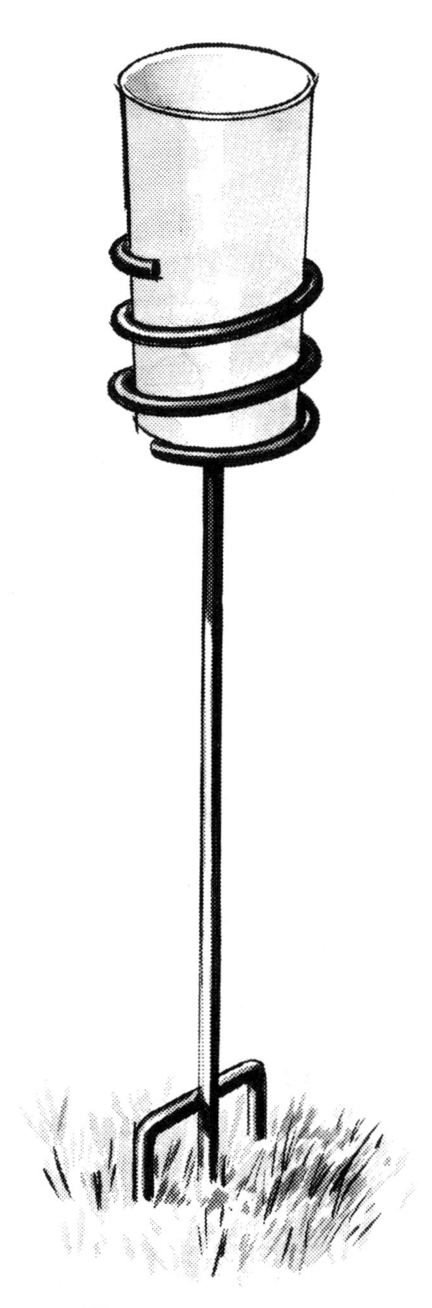

Patio Candle/ Glass Holder

1. Material: Ø3/16 rod.
2. Finish: Paint bright color or cover rod with patterned plastic tubing.
3. Remove all sharp edges.
4. Advanced metals classes can develop a manufacturing technique to produce a uniform spiral.
5. The glass holder may be spun and attached to the rod by threading the end and using two hex nuts or may be brazed to the rod.

Quick Projects

These projects can be made in a few class periods.

METRIC

All dimensions are in mm.

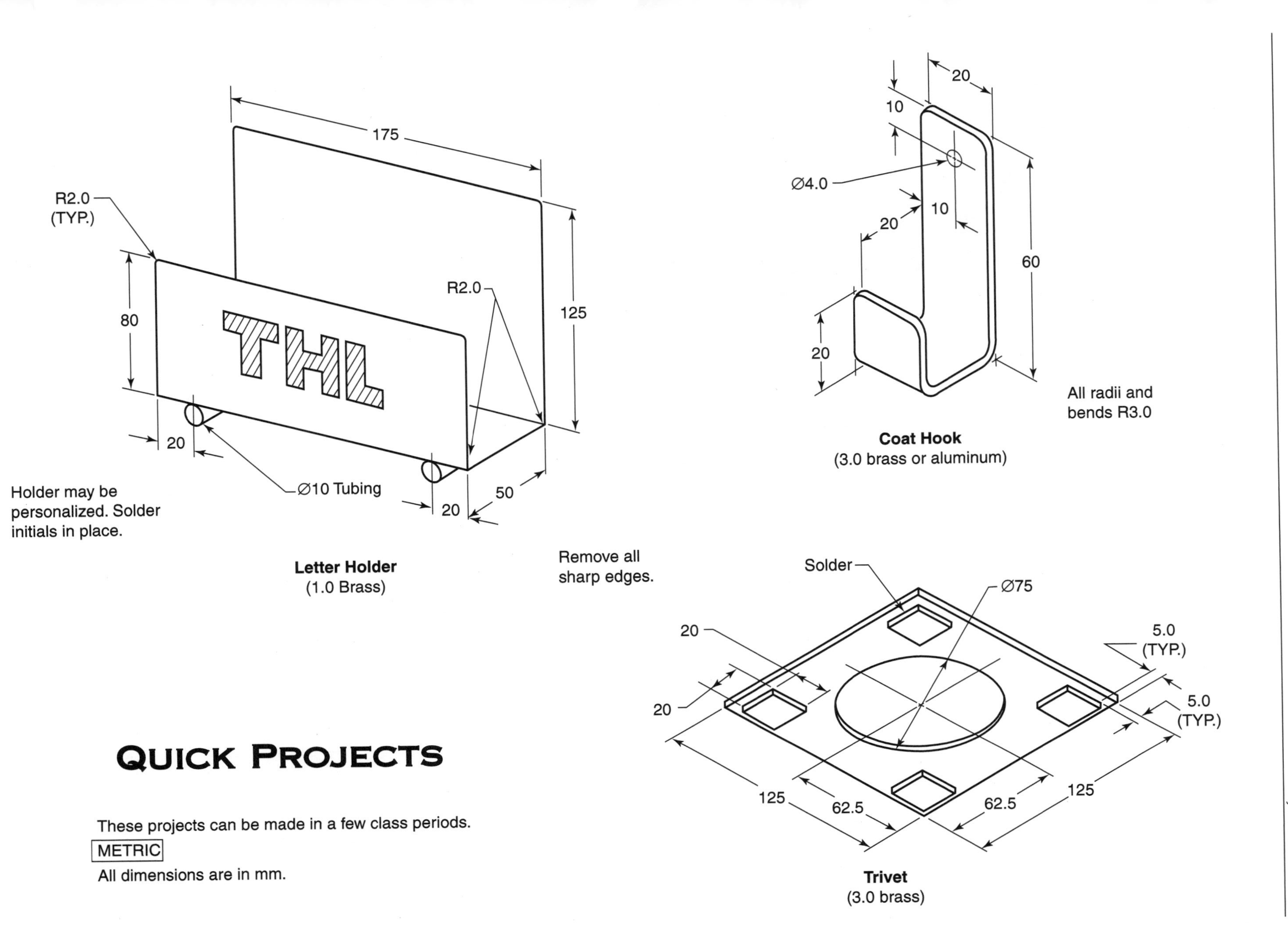

Letter Holder
(1.0 Brass)

Coat Hook
(3.0 brass or aluminum)

Trivet
(3.0 brass)

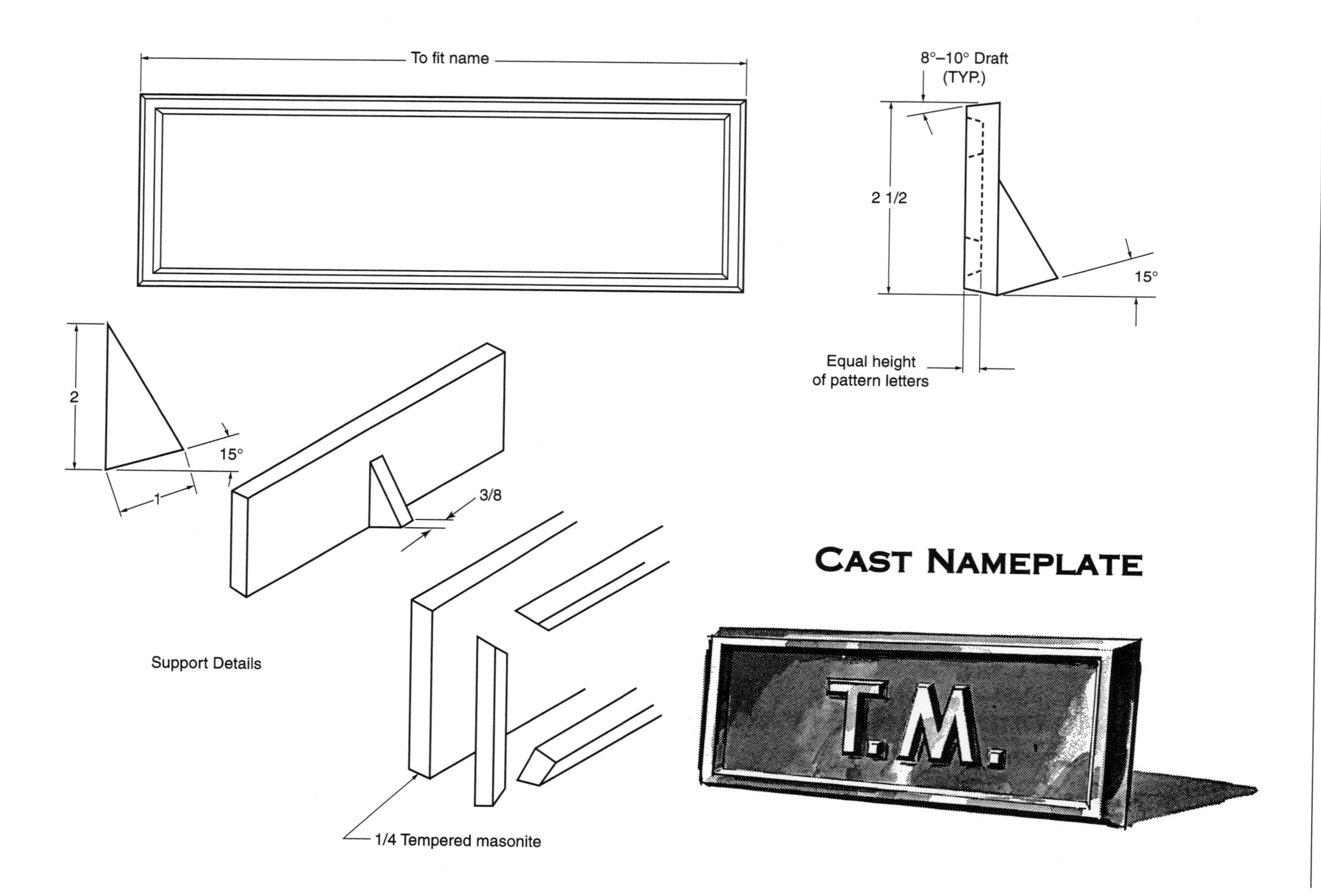

To fit name
8°–10° Draft (TYP.)
2 1/2
15°
Equal height of pattern letters
2
15°
1
3/8
Support Details
1/4 Tempered masonite
CAST NAMEPLATE
T.M.

Two-Piece Vise

The two-piece vise greatly increases the versatility of the machine to which it is attached. The machine will have greater capacity since the opening of the vise is limited to the length of the worktable. The keys in the base ensure accuracy and quick set up.

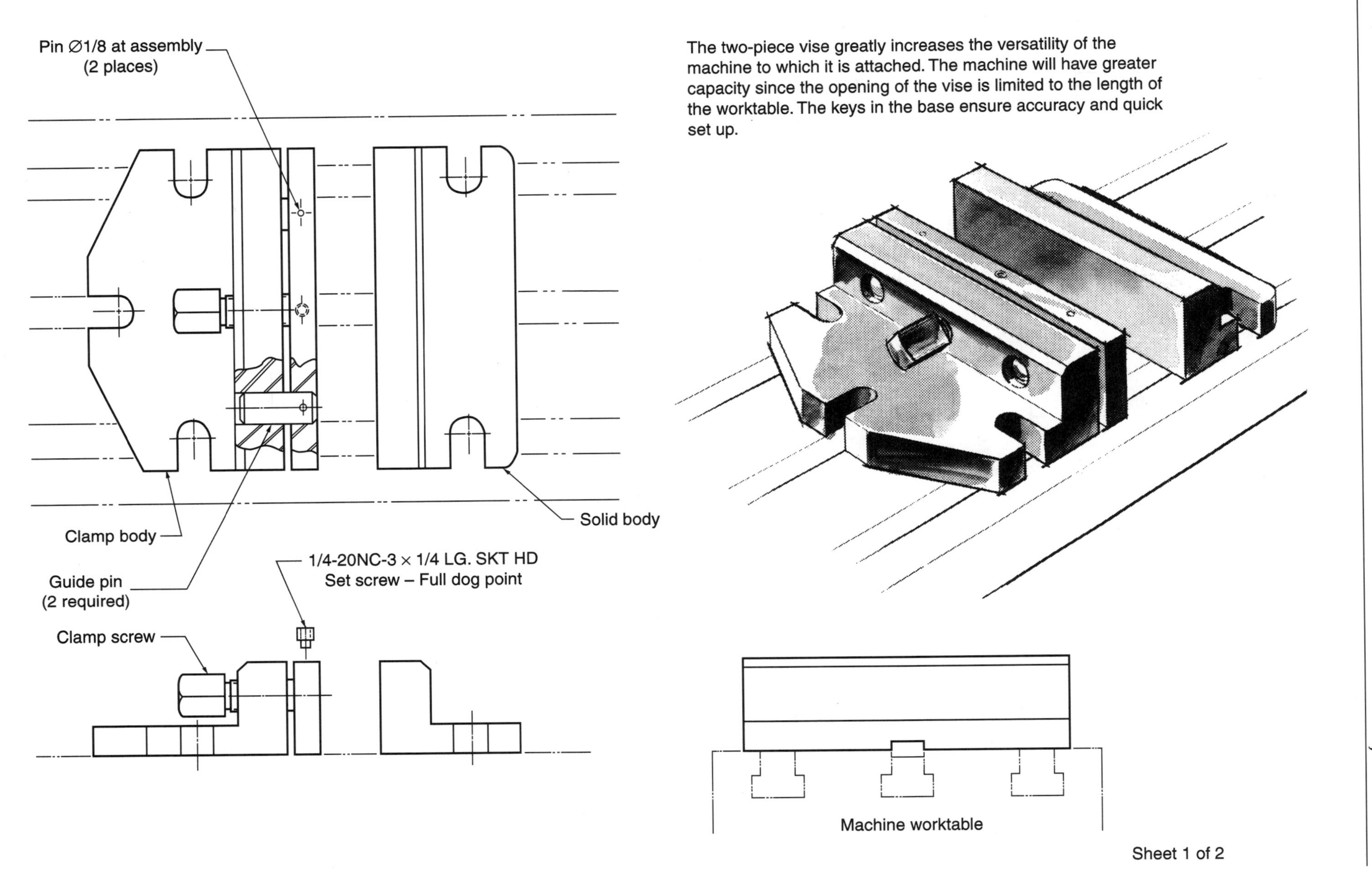

Sheet 1 of 2

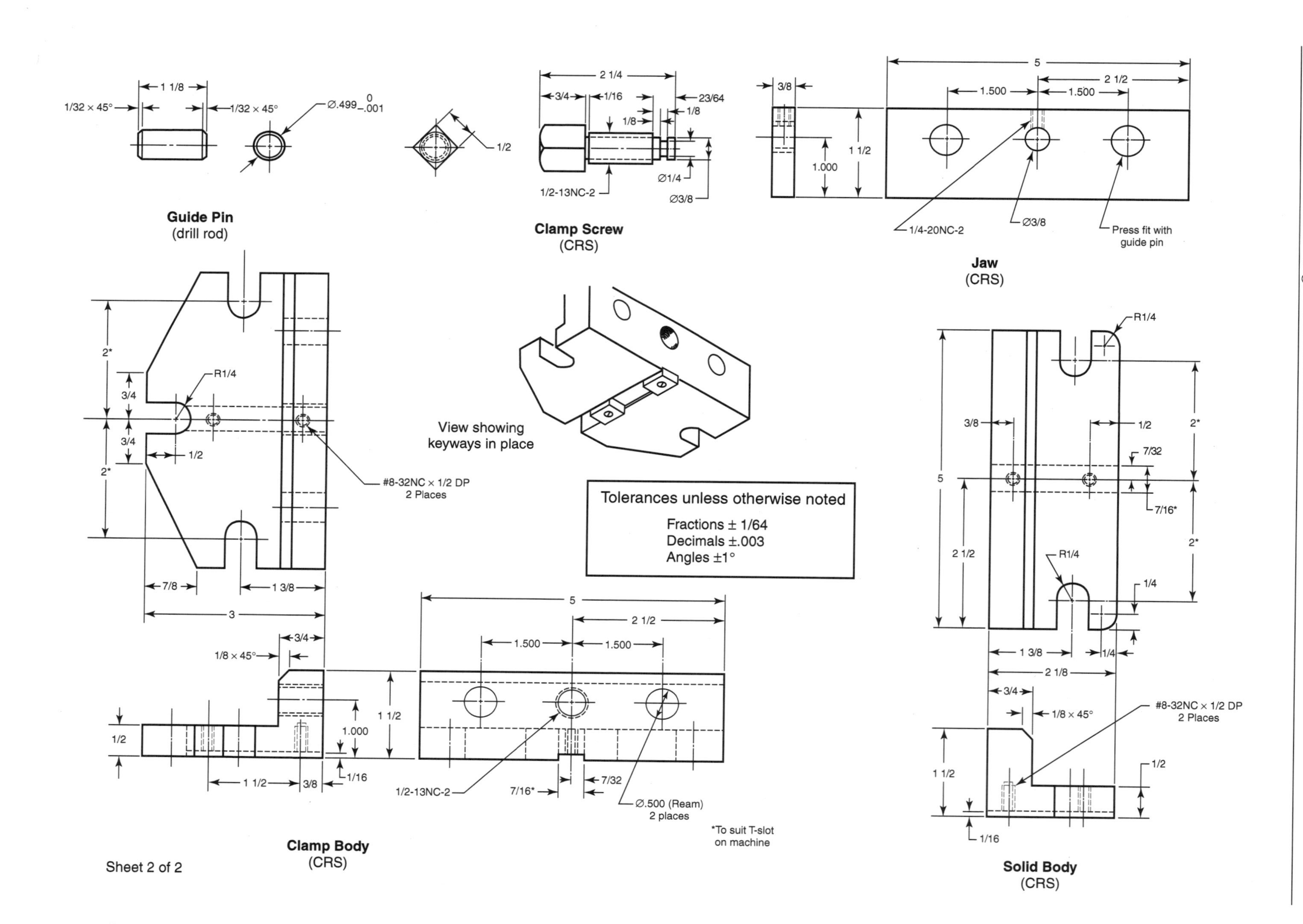

1 1/8
1/32 × 45°
1/32 × 45°
Ø.499 0 -.001
1/2
Guide Pin
(drill rod)
2 1/4
3/4
1/16
23/64
1/8
1/8
Ø1/4
Ø3/8
1/2-13NC-2
Clamp Screw
(CRS)
3/8
1.000
1 1/2
5
2 1/2
1.500
1.500
1/4-20NC-2
Ø3/8
Press fit with guide pin
Jaw
(CRS)
2*
2*
3/4
3/4
R1/4
1/2
#8-32NC × 1/2 DP
2 Places
7/8
1 3/8
3
View showing keyways in place
Tolerances unless otherwise noted
Fractions ± 1/64
Decimals ±.003
Angles ±1°
R1/4
3/8
1/2
2*
7/32
7/16*
2*
5
2 1/2
R1/4
1/4
1 3/8
1/4
2 1/8
3/4
1/8 × 45°
#8-32NC × 1/2 DP
2 Places
1 1/2
1/2
1/16
Solid Body
(CRS)
3/4
1/8 × 45°
1/2
1.000
1 1/2
1 1/2
3/8
1/16
5
2 1/2
1.500
1.500
1/2-13NC-2
7/16*
7/32
Ø.500 (Ream)
2 places
*To suit T-slot on machine
Clamp Body
(CRS)
Sheet 2 of 2

PHOTO FRAME

Material: 14–16 GA. brass
Finish: Polish by buffing

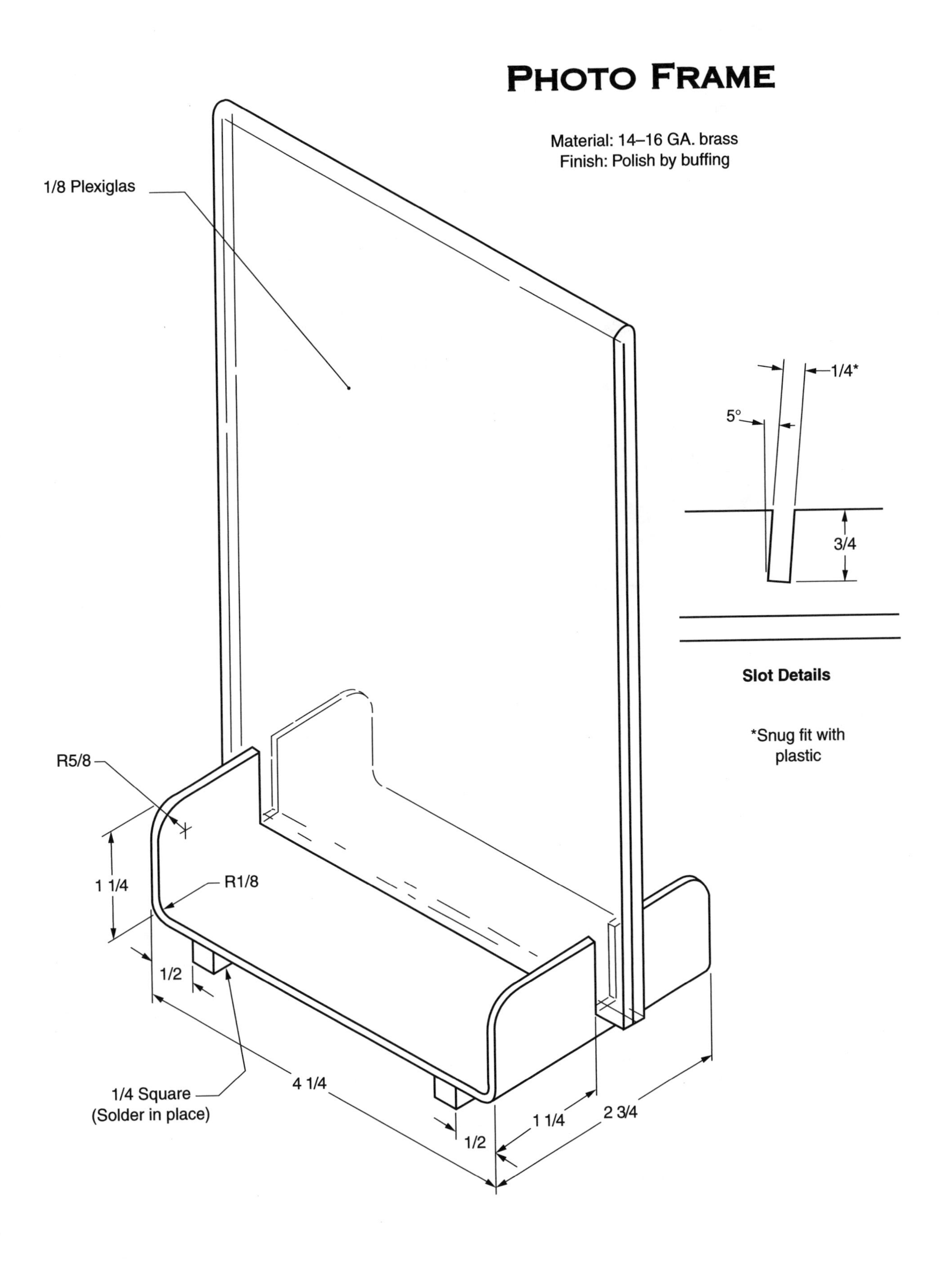

Two-Stroke Cycle Internal Combustion Engine

The two-stroke cycle engine shown in **Figure 1** was originally designed and developed as a mass production problem. Twenty engines plus the jigs and fixtures needed in the manufacturing were constructed at that time.

This particular engine is offered as a basic design. It is possible to increase power and reduce weight through modifications until the engine can be used to power a model airplane with a 5′ wing span. Weight reduction can be accomplished by thinning cylinder walls and rounding off square corners.

Figure 1. Two-stroke cycle engine.

Construction

First make the crankcase, then the cylinder. The crankcase front can be cast or machined and fabricated from solid stock. If cast, a pattern must be made. If you decide to machine the part from solid stock, you have your choice of the three possibilities shown in **Figure 2.** No dimensions are given so *you* can solve the problem.

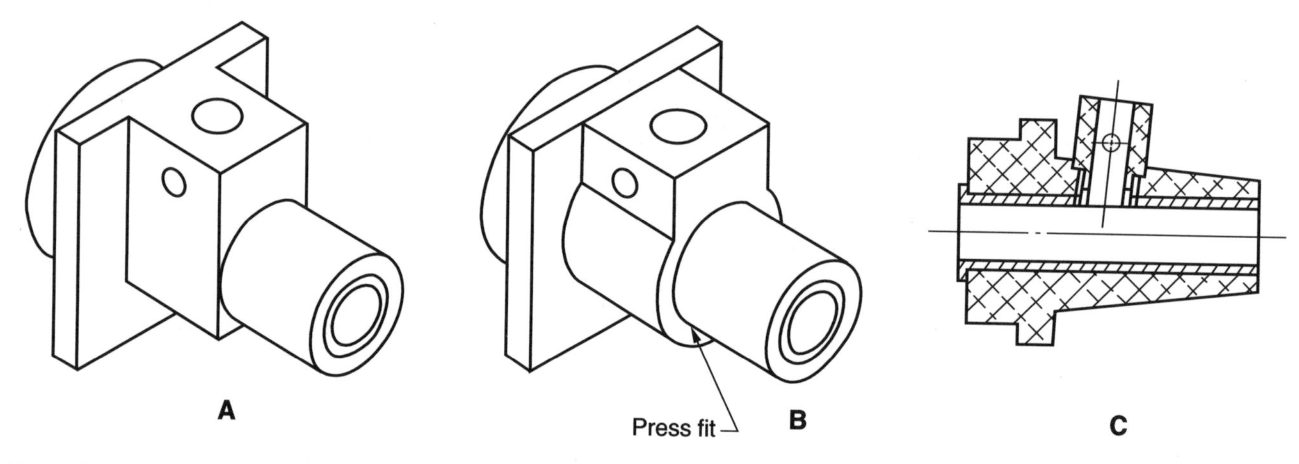

Figure 2. Three methods of machining the two-cycle engine crankcase.

The cylinder may be made for seamless steel tubing, **Figure 3,** or machined from bar stock. Ream or machine the bore to size, being careful to use an ample supply of cutting fluid to ensure a smooth finish.

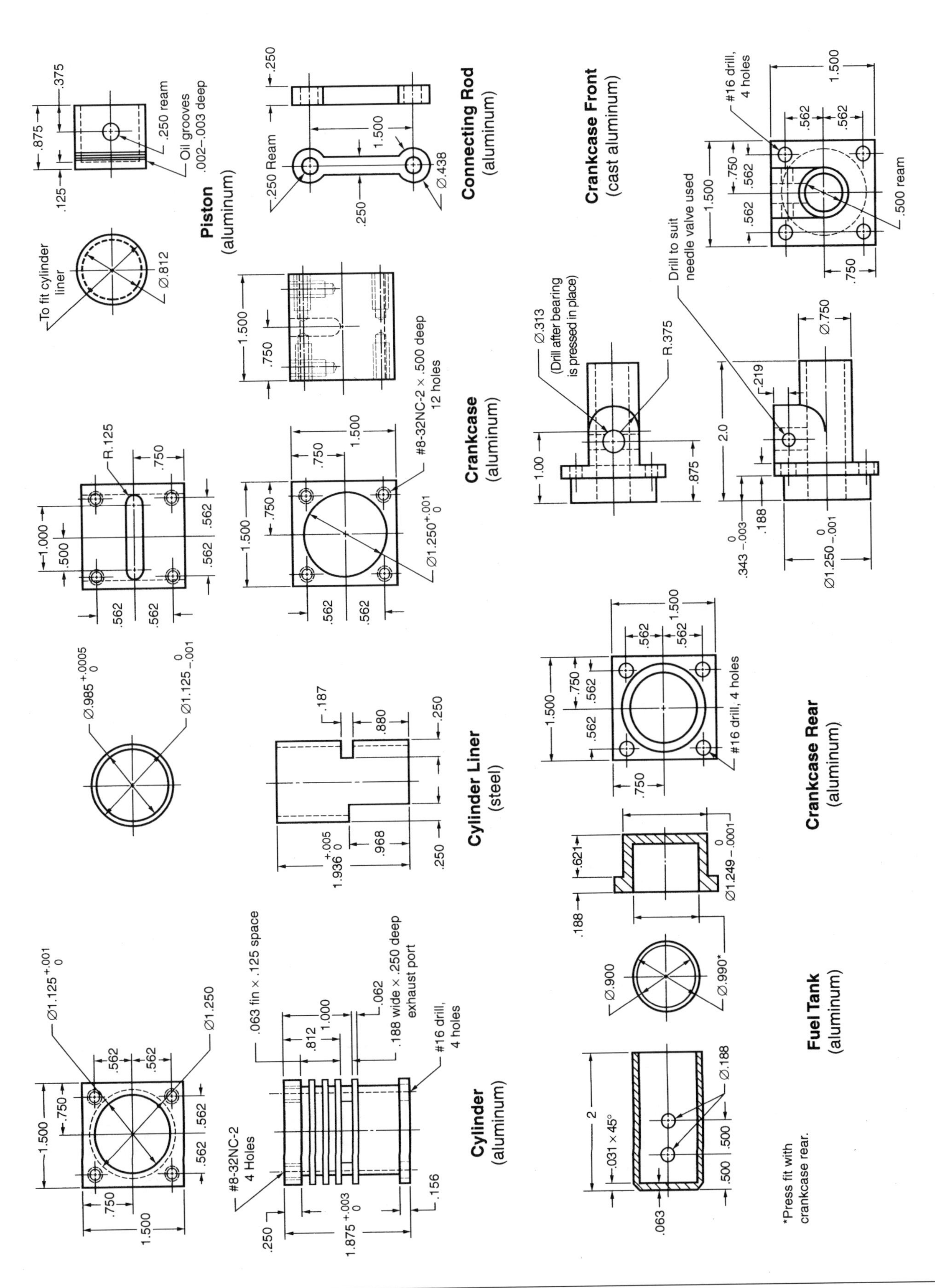

Figure 3. Drawings for two-cycle engine parts: crankcase, piston, cylinder, cylinder liner, and fuel tank.

You can obtain a lapped finish on the bore by making an aluminum or brass lapping bar 0.0005″ smaller than the bore. Machine several shallow grooves about 1/2″ apart on the lapping bar. Use a very fine lapping compound or tooth powder and plenty of cutting oil. Move the charged lapping bar back and forth in the bore.

Protect the machine from surplus lapping compound. From time to time, remove the compound with an approved cleaner and measure the bore. This is done to be sure the bore is uniform in diameter along its length. A very slight "belling" is permitted at the bottom of the cylinder liner.

Make the piston next. It should be a snug fit in the cylinder bore. Use an ample supply of lube oil on the bore and piston when making the fit. The piston will "run in" after several minutes of operation. A good seal between cylinder bore and piston will occur because the piston (aluminum) will expand slightly more than the cylinder liner (steel) as the engine heats up during operation.

There is no set sequence that must be followed to manufacture the remaining parts, **Figures 3** and **4.** However, use care in tapping the small holes. It is almost impossible to remove broken taps from the aluminum engine parts without the use of an electronic tap disintegrator. Securing the 1/4-32NS tap needed to cut threads in the head may be a problem. Most production machine shops have a tap of this size. A local shop owner may loan it to you.

Carefully remove all burrs and chips from the machined parts. Clean them in a good solvent. Place a coating of engine oil on the parts as they are assembled. Generally, no gaskets are needed.

The glow plug, plastic fuel line, and needle valve assembly can be purchased from any hobby shop handling model airplane or boat supplies.

When completely assembled, except for the glow plug, the engine should turn over freely. If not, check for binding parts and correct the problem.

Break-In

Place the engine in a lathe with the chuck gripping the flywheel. Remove the glow plug and place several drops of engine oil in the cylinder. With the plug removed and the lathe adjusted to operate at a slow speed, run the lathe in *reverse*. Grip the engine lightly so that it will easily slip from your hand should the engine "freeze" (seize).

Break in the engine by running the lathe for 10 to 15 minutes, increasing lathe speed during the last few minutes. Then, clean the entire engine in a solvent and apply new engine oil. Fill the fuel tank with a good grade of glow engine fuel that contains a large percentage of castor oil. Fit and tighten the glow plug into the head.

Operation

Mount the engine in a vise, using wood blocks to protect the metal surfaces. Open the needle valve about three full turns. Have someone place a finger over the intake while you rapidly rotate the flywheel in a counterclockwise direction.

1. If fuel is drawn through the fuel line, you are ready to attach the battery terminal to the plug and start the engine.
2. If no fuel is drawn into the line, check the following possible causes and make needed corrections.
   (a) Exhaust port does not "open" before intake port. Correct as shown in **Figure 5.**
   (b) Intake port does not open. Correct as shown in **Figure 6.**
   (c) Intake hole in crankshaft is incorrectly "timed." Hole should start to open as piston starts up in cylinder. It should remain open until piston reaches TDC (top dead center). A minor correction can be made by filing. Otherwise a new crankshaft will have to be made.
   (d) Intake hole in crankshaft does not align with intake hole in crankcase. Correct by redrilling.

With fuel in the fuel line, prime the cylinder (through the exhaust port) with a few drops of fuel. Snap the engine over smartly in a counterclockwise direction. The flywheel should rebound slightly.

Attach the battery terminal to the glow plug and ground the other terminal to the engine. The plug should glow a bright orange (almost a yellow).

Reprime through the exhaust port and snap the flywheel in a counterclockwise direction. A heavy cord wrapped around the flywheel will aid in starting. If available, an electric starter specifically designed to start model airplane engines is best of all. It never fails to start an engine that is properly set up.

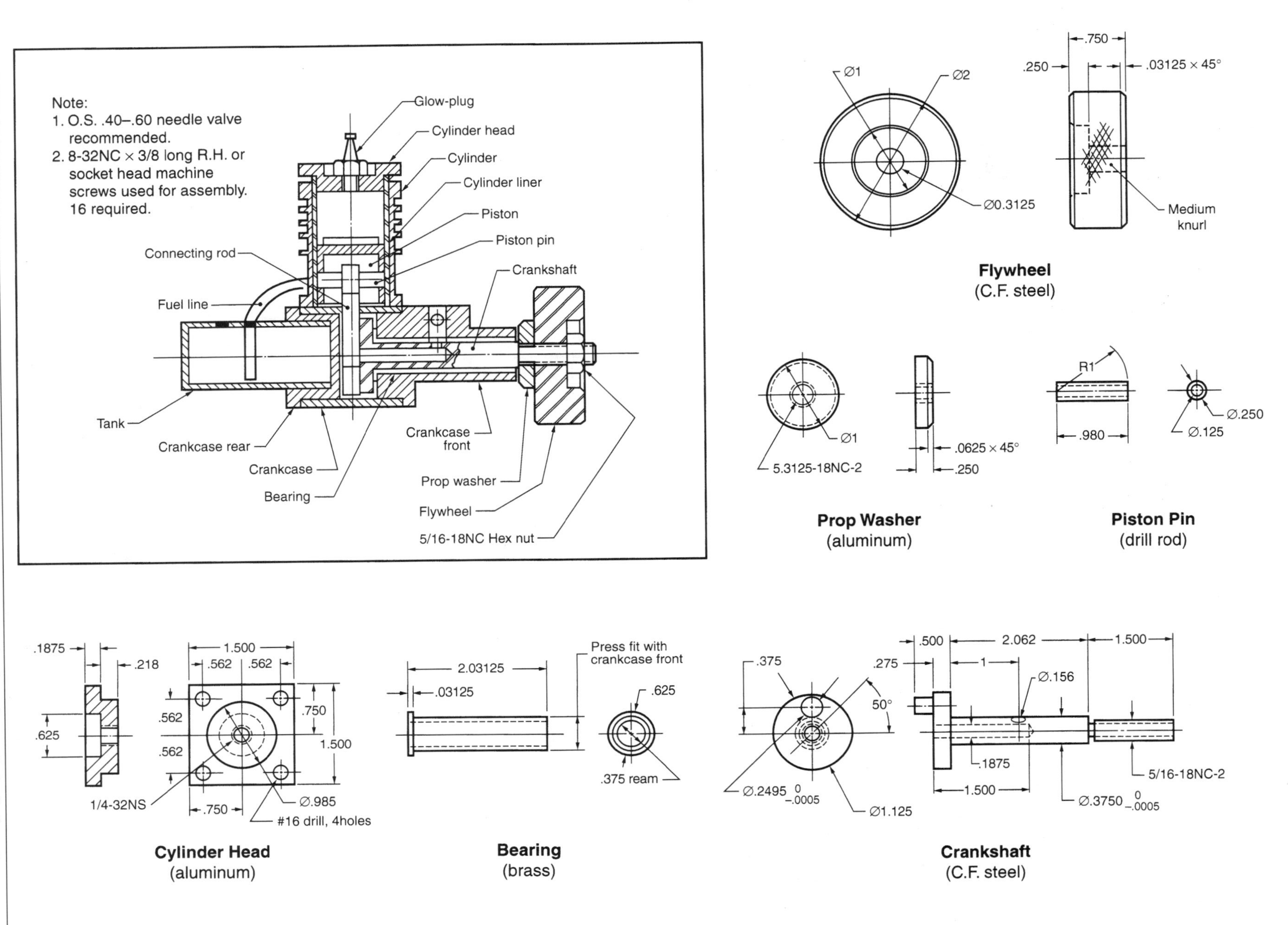

Figure 4. Cross section of assembled two-cycle engine and drawings of cylinder head, crankshaft, bearing, flywheel, piston pin, and prop washer.

1. If the engine will not fire, check for the following possible causes:
   (a) Battery short circuit.
   (b) Glow plug partially or fully burned out.
   (c) Battery run down.
   (d) Little or no compression.
   (e) Old fuel.
2. If the engine "fires" but does not start, it may be flooded.
   (a) Close needle valve completely and turn engine over rapidly. Engine should start and run out of excess fuel.
   (b) If engine does not start, examine spray bar of needle valve assembly. Most have one hole which should be centered in the intake (use small washers as spacers) and face down. If spray bar has two holes, position bar so that holes are centered, and parallel to crankshaft.
   (c) Open needle valve 2 to 2 1/2 turns and try to start engine. Adjust needle valve until engine runs.
3. If engine runs, then "dies" after prime is consumed, open needle valve another 1/4 turn and try again.
4. Check to see if engine is drawing fuel by noting movement of small bubbles in fuel line. If no fuel is being drawn, check for the following problems and make necessary corrections.
   (a) Dirt in needle valve.
   (b) Fuel hole pressed against bottom of fuel tank.
   (c) Dirt in fuel line.
   (d) Joints leak and engine may require gaskets.
   (e) Glow plug loose in head.
   (f) No compression.
   (g) Old fuel.

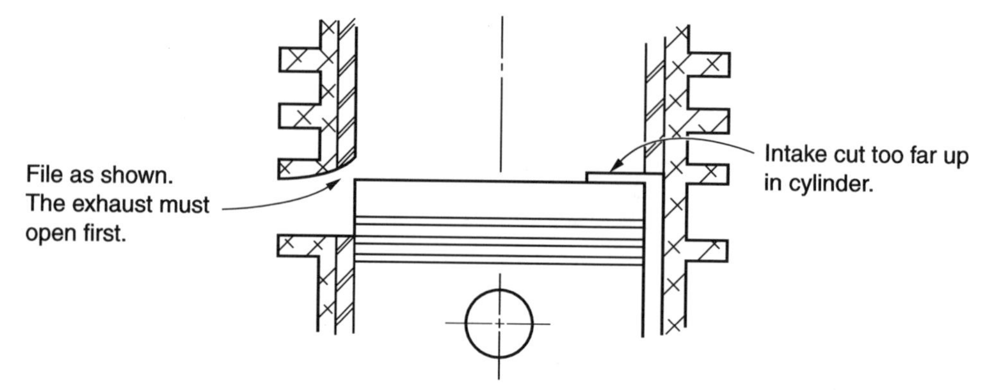

Problem: Intake opens before exhaust. The intake or exhaust is improperly located.

Figure 5.

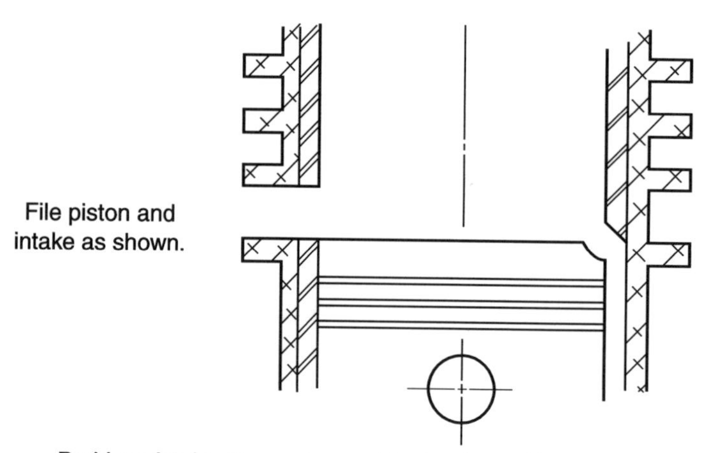

Problem: Intake does not open. The intake and/or exhaust is improperly located.

Figure 6.

FOUR-STROKE CYCLE INTERNAL COMBUSTION ENGINE

The four-stroke cycle engine shown in **Figure 1** is a project intended for more experienced machine shop students. This particular engine is not a complex piece of machinery. The machining operations are simple and straightforward. Most parts can be made on the lathe and drill press.

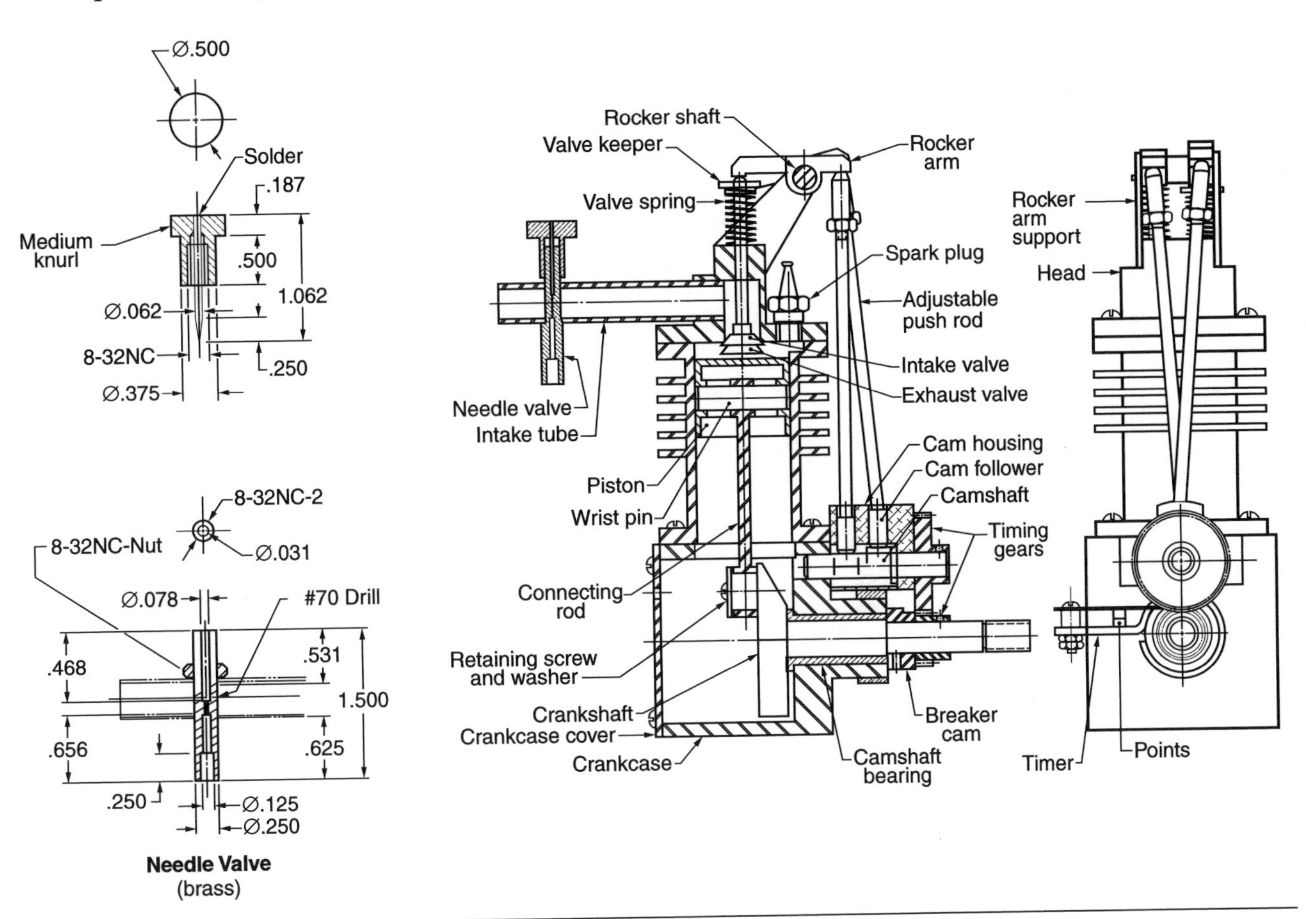

Figure 1. A simple four-stroke cycle gasoline engine.

Start by making the crankcase, **Figure 2,** from aluminum or steel. Use care in tapping the holes for mounting the cylinder, camshaft housing, and rear cover plate.

Next, make the cylinder, **Figure 3.** Note on the assembly drawing that the spark plug hole in the head overlaps a portion of the cylinder top. A small section of the cylinder top must be cut away to permit the spark plug to ignite the fuel charge. Be sure that the cutaway section is properly located.

The piston is cast iron, which permits you to lap the piston into the cylinder for a good fit. Turn the piston on the lathe until it fits closely in the cylinder. Then, make the connecting rod and assemble it to the piston.

Next, prepare a lapping compound composed of tooth powder and light machine oil. Carefully grip the cylinder in a lathe chuck and apply lapping compound to the bore. Fit the piston into the cylinder, using twisting motion. Then, move the piston back and forth in the bore while the cylinder is turning at a slow speed.

Thoroughly clean *all* traces of lapping compound from the cylinder and piston assembly.

The crankshaft can be fabricated from several pieces of steel. See **Figure 4.** However, machining from solid stock is recommended if time permits.

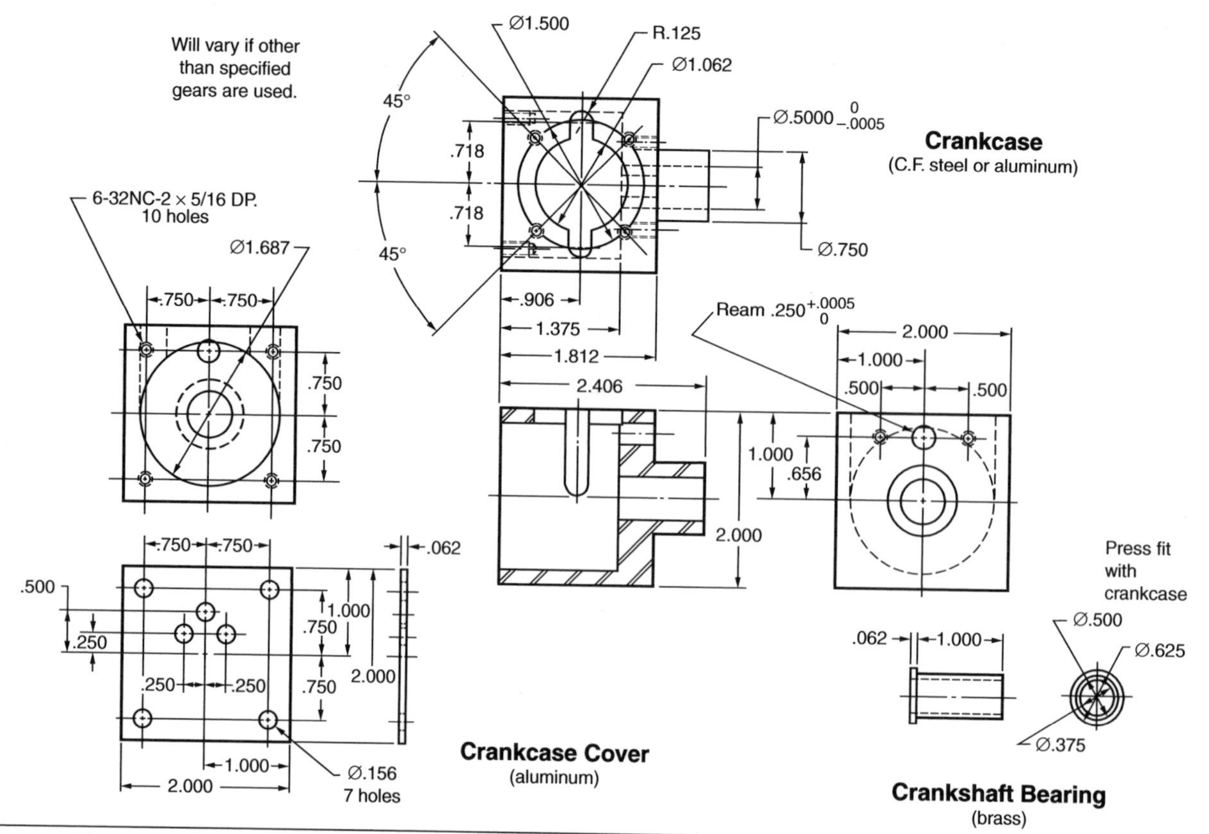

Figure 2. Plans for the crankcase.

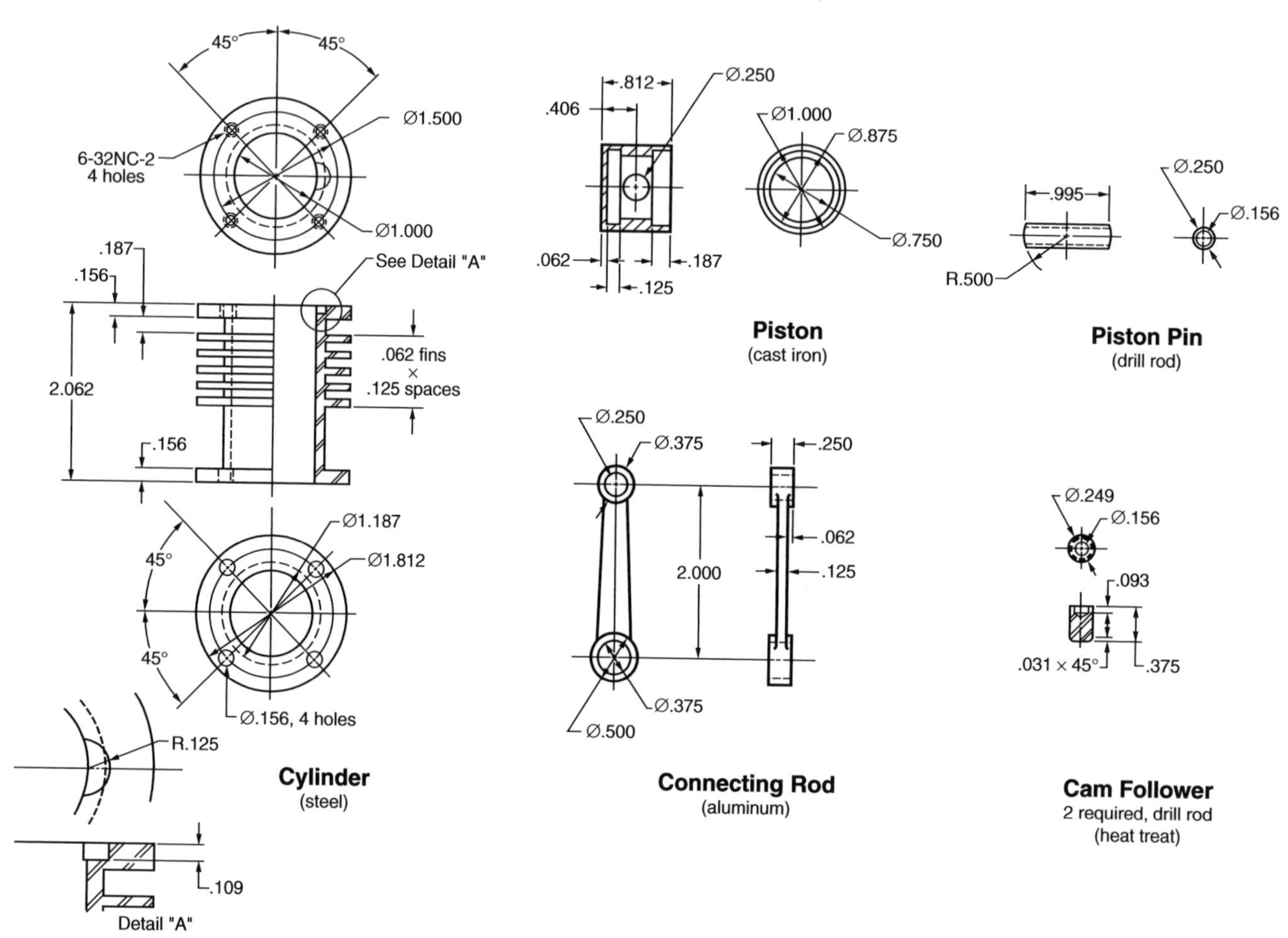

Figure 3. Plans for the engine cylinder.

An aluminum head will dissipate heat better than a steel head, **Figure 5.** However, an aluminum head requires the installation of steel valve seats. Use care in the manufacture of the valves. They are fabricated from two pieces of steel and MUST be machined true for an effective seal with the valve seats. See **Figure 4.**

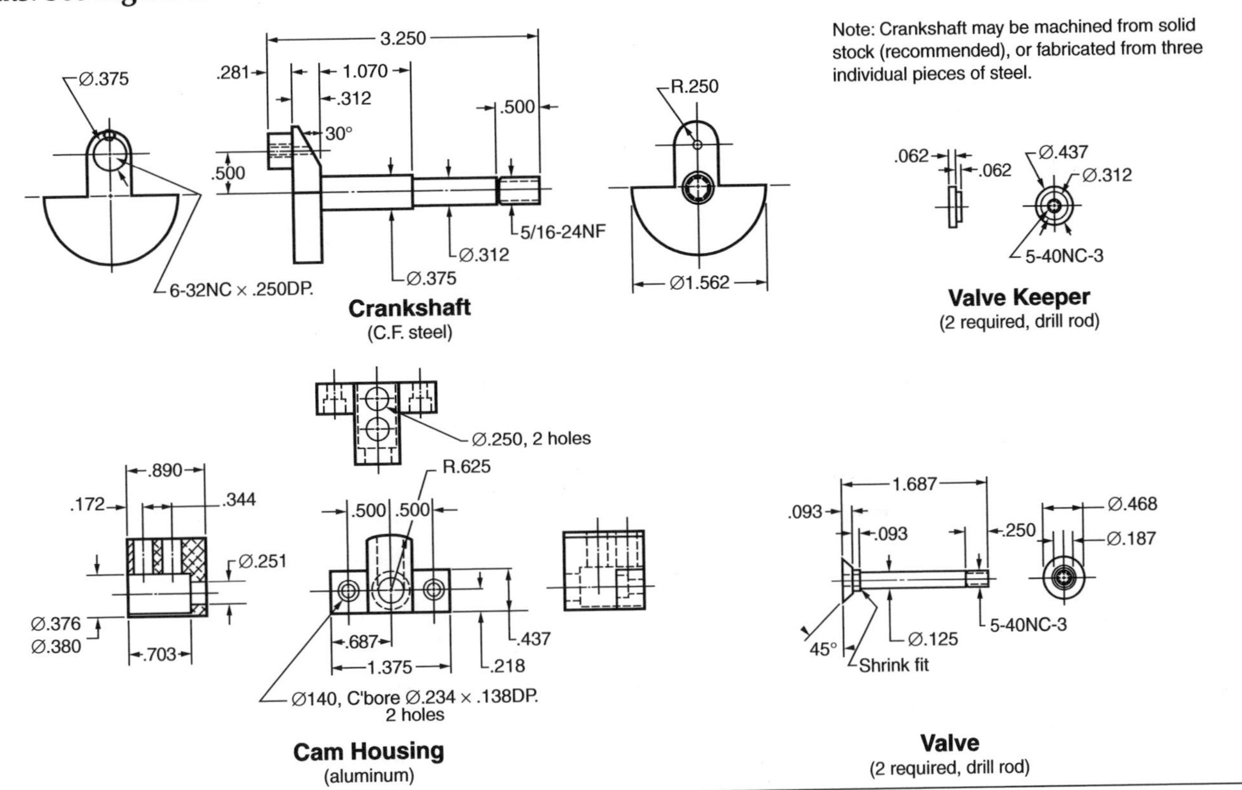

Figure 4. Plans for the crankshaft.

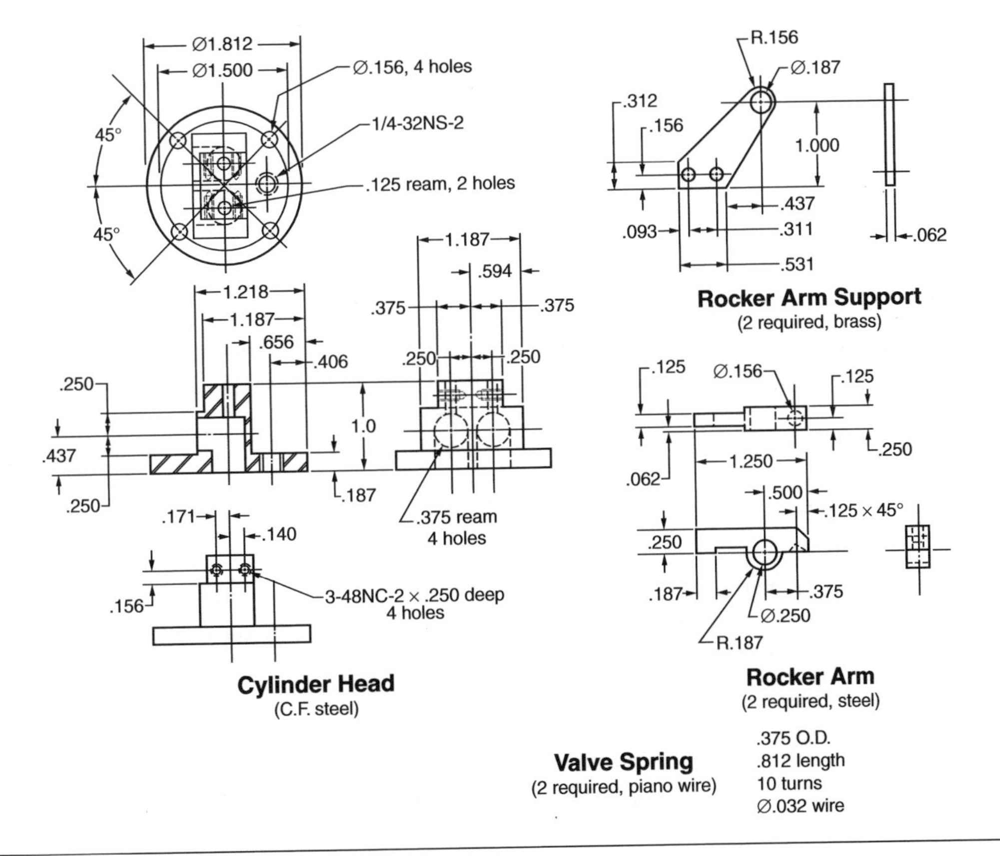

Figure 5. Plans for the four-cycle engine head.

The camshaft, **Figure 6,** is partially machined and partially filed from drill rod. The lobes should be positioned to permit the exhaust valve to open at 15° to 20° before BDC (bottom dead center) and close at TDC (top dead center). The intake valve opens at TDC and closes at 5° to 10° after BDC. The camshaft must be heat-treated.

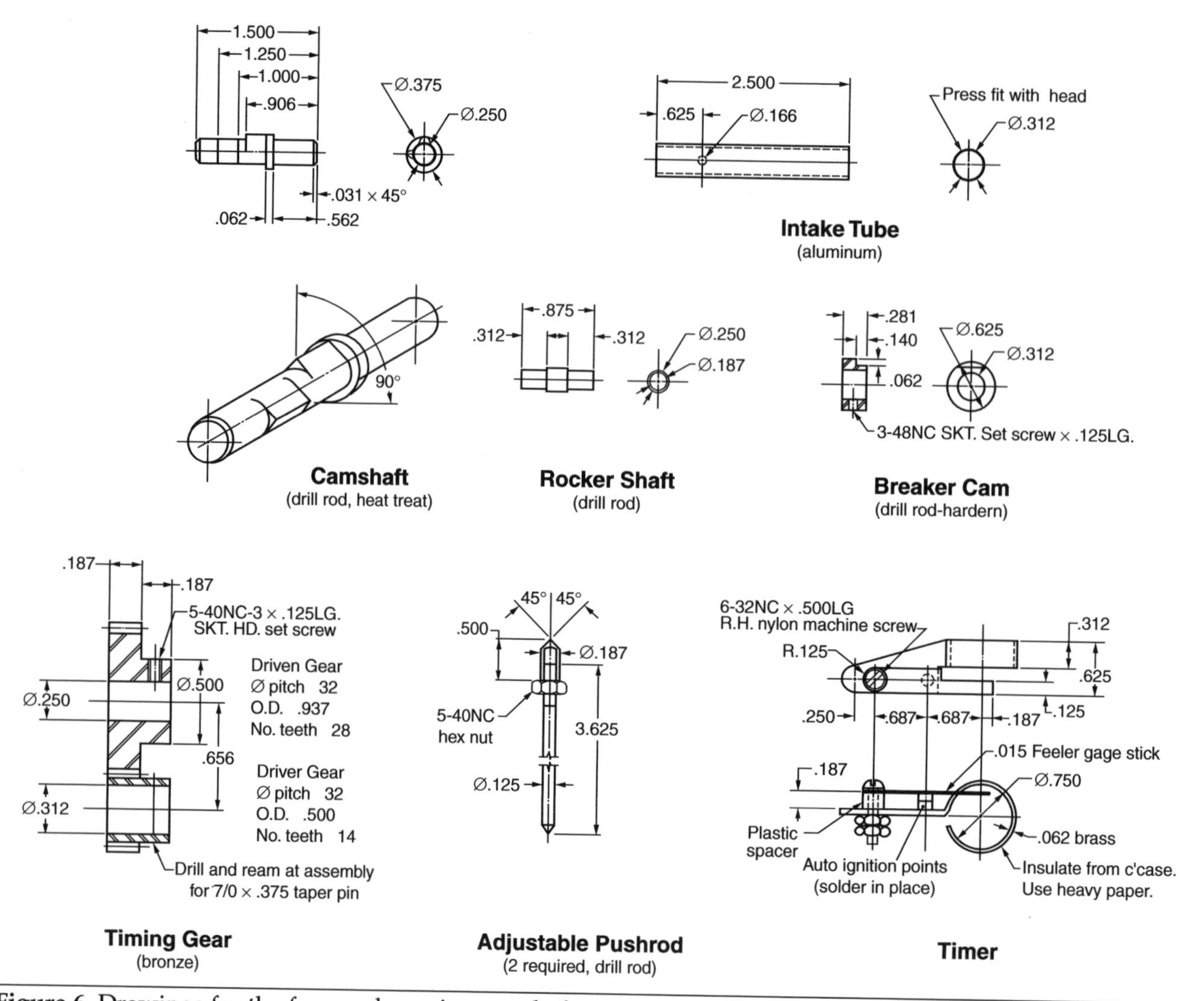

Figure 6. Drawings for the four-cycle engine camshaft.

Gears may be made in the shop or purchased from one of the gear manufacturers. Look in the yellow pages of the telephone book or in the advertising index of home craftsman magazines for sources of suitable gears. Gear sizes may vary slightly from those specified but they must be 2:1 ratio. Location of the camshaft housing will change if other than specified gears are used.

The carburetor must be made exactly as shown in **Figure 1.** The jet hole is made with a No. 70 drill and *must* be toward the engine when the spray bar is installed.

Ignition timing is a bit unique on this engine because the spark plug fires every revolution. The timer is made as specified in **Figure 6.** It must be insulated from the crankcase by a piece of index card stock. Timer points can be taken from an automobile timing unit. Solder the points in place.

Before final adjustments can be made, the valves must be seated. Use a very fine lapping compound to work them into the seats. Valves must seat tightly if the engine is to operate.

Completely assemble the engine, **Figure 1.** Apply a thin coating of engine lubricating oil to all mating parts before fitting them together. Check for binding parts. Make corrections if necessary.

In operation, the engine rotates counterclockwise. Position the gears so that the cam will start to open the exhaust valve (lobe in front of housing) at about 15° before BDC.

Ignition is timed by adjusting the timer cam.

The original engine ran on a mixture of four parts unleaded gasoline and one part No. 70 lubricating oil. Since No. 70 oil is difficult to find, No. 50 oil has been substituted with no apparent damage to the engine.

Place some of the mixture in the crankcase for initial lubrication in that area of the engine.

The spark plug (not a glow plug, which will not work), and ignition coil were taken from an old ignition type model airplane engine. They can be purchased from advertisers in one of the model airplane magazines.

To start the engine, open the needle valve about two turns. Rotate the flywheel two times while choking the intake tube with your finger. Place a pull cord around the flywheel and rapidly pull the engine over. If it runs with a short burst of power and then stops, the carburetor is set lean. Open the needle valve another turn.

The carburetor is set rich if the engine starts but seems to have little power. Close down the needle valve until the engine picks up power and runs smoothly.

The author can take no credit for the design of this engine. Plans were made from an engine acquired with donated material. The engine is unique. The designer deserves credit for such a clever piece of machinery.